Immittance Spectroscopy

Scrivener Publishing
100 Cummings Center, Suite 541J
Beverly, MA 01915-6106

Publishers at Scrivener
Martin Scrivener (martin@scrivenerpublishing.com)
Phillip Carmical (pcarmical@scrivenerpublishing.com)

Immittance Spectroscopy

Applications to Material Systems

Mohammad A. Alim
Department of Electrical Engineering and Computer Science
Alabama A & M University
Huntsville, Alabama, U.S.A.

Scrivener Publishing

WILEY

This edition first published 2017 by John Wiley & Sons, Inc., 111 River Street, Hoboken, NJ 07030, USA and Scrivener Publishing LLC, 100 Cummings Center, Suite 541J, Beverly, MA 01915, USA

For more information about Scrivener publications please visit www.scrivenerpublishing.com.

Wiley Global Headquarters
111 River Street, Hoboken, NJ 07030, USA

For details of our global editorial offices, customer services, and more information about Wiley products visit us at www.wiley.com.

Library of Congress Cataloging-in-Publication Data
ISBN 978-1-119-18485-0

Cover image: the author
Cover design by Russell Richardson

Set in size of 13pt and Minion Pro by Exeter Premedia Services Private Ltd., Chennai, India

10 9 8 7 6 5 4 3 2 1

بسم الله الرحمن الرحيم

In the name of Allah, the Beneficent, the Merciful

Dedicated to

- **Father**: **Mohammad Alim Biswas**, M.Sc., Ph.D., D.I.C., (1927 - 2009) son of *Moulavi* Mohammad Bhadu (Bahadur) Biswas (1885 - 1961) and *Mosammat* Saleha Khatun (1894 - 1970), who provided inspiration to comprehend general science beside navigating mathematics, chemistry, physics, history, geography, philosophy, and exposing world religions including exercising english and other global languages;
- **Mother**: *Mosammat* **Jinnatun Khatun** (1934 -), daughter of *Moulavi* Mohammad Boytullah Mondol (1894 - 1976) and *Mosammat* Rowshan Ara Khatun (1904 - 1990), who guided to comprehend Bangla language making literature easy and walking through mathematics;
- **Wife**: **Sanjida Khanam**, M.Sc., M.S., M.S.M.E., (1957 -), daughter of Muhammad Aminuddin (1919 - 2001) and *Mosammat* Anosha Khatun (1930 - 1989), who helped in achieivng academic and professional goals; and
- **Four Children**: Nasiha Muna, Sabiha Runa, Maliha Lubna, and Mohammad E. Alim who helped in many ways to materialize this work.

Contents

Background of this Book

While the complicated multi-component polycrystalline system was hunting the answer for the underlying operative mechanisms (UOMs), a few questions emerged a few decades ago despite speculated theories bouncing all over reputed journals. At that time *impedance spectroscopy* began walking. What is the meaning of the term 'spectroscopy' for the measured *impedance* or the *admittance* or the *phasor data*? Is it just data acquisition as a function of applied frequency or something else? Is it only the *impedance plot* that causes the term *impedance spectroscopy*? Why is there no presence of an *admittance plot* when the same data are acquired and used though the name *admittance spectroscopy* is used? In order to answer these questions the materialization of this book emerged to help the understanding of the investigators and the new learners for the generations to come. As of today, the three existing books on the subject demonstrated successfully and efficiently *single complex plane* (SCP) analysis for the measured data along with elegant theories, relevant or non-relevant, but therein it remained difficult to comprehend the meaning of *spectroscopy* when it involves *impedance* or *admittance* or *phasor* form of the measured data.

In the most recent decades, a separate assertive parasitic name emerged as the *electrochemical impedance spectroscopy* (EIS) among the users of the immittance data. What is EIS? In general, *impedance spectroscopy* is in use for a variety of solid material systems that include electronic or ionic type single crystals and polycrystals, hybrid semiconductors, electrolytes, composites, etc., beside the liquid and biological systems emphasizing reactions in many electrochemical materials. Each material system is not limited to the use of *impedance spectroscopy*. A widespread frequency range strengthens the horizon of the investigation. Therefore, using an emphasized "word" in front of this term does not increase the horizon or density or intensity or importance of the in-depth investigation.

The generalized form of *impedance spectroscopy* is already adequate or self-contained for every material system as the fathom depends only on the acquisition of the *ac small-signal electrical data*. Does one need *hybrid impedance spectroscopy* as a hybrid system investigator, or *quality impedance spectroscopy* as a quality control professional? The entire theory and basis of *impedance spectroscopy* (IS) are the same regardless of the material system. The acronym "IS" is ready to alternate between *impedance spectroscopy* and *immittance spectroscopy* though it is used in the first book for *impedance spectroscopy* in 1987.

The terms *impedance spectroscopy* and *admittance spectroscopy* (AS) are explained vividly using relevant situations and cases. The real-world situation is more complicated with respect to the nature of the material systems. For the polycrystalline systems, obtaining two consecutive semicircles in the impedance plane does not mean that grains and grain boundaries are explored. What about the role of the defects and traps having discontinuities in grain-to-grain contacts though ohmic contacts for the electrodes are ignored? Are these defects relaxed or non-relaxed? Are these traps distributed over a range of energy levels? What is the distribution of time constant?

It is amazing that the conceptual term for impedance or admittance of a circuit came to the scientists and engineers shortly after Michael Faraday offered *modern electricity* to mankind. Nearly a hundred years after Michael Faraday, *immittance* emerged as a hybrid word originating from *impedance* and *admittance*. Though the common sense wisdom serves better, it is known that the *impedance plot* did not float before the classic Cole-Cole plot in 1941. This breakthrough indicated that the material systems, liquids or solids, require a broader window for investigation to understand the science or engineering aspects of learning. How does the *impedance plot* become the sole single complex plane (SCP) analytical tool for the entire material systems? How does *admittance spectroscopy* never use an admittance plot but instead uses some other plot?

Despite the existence of the three separate books on *impedance spectroscopy* covering both solid and liquid systems, this book is prepared to serve a broad range professionals handling *concurrent relaxations in multiple complex planes* for the same set of measured *ac small-signal electrical data*. Through this book the use of *multiple complex plane analysis* in a simultaneous fashion for the same set of measured electrical data is strongly encouraged, and exercised with demonstrations to surmount every stack of problems. The pedagogical or tutorial contents available today on this subject do not address adequately the use of *concurrent multiple complex planes* (CMCP) apart from the advantages or disadvantages of apprehending complicated underlying operative mechanisms (UOMs) within the device under test (DUT). The extraction of the meaningful and logical equivalent circuit model possessing interpretation follows from the *concurrent multiple complex plane analysis* (CMCPA) of the data.

This book is certainly designed to serve the advanced knowledge seekers dealing with benefits of multiple complex plane representations of the data to unravel the UOMs for the DUT. This book encourages to wide-open the window of investigation. It is also designed to serve as a broad range reference on the *immittance spectroscopy* where complexity of the material system is severe. Thus, this book is directed toward the advanced researchers as well as the learners comprising of graduate and undergraduate students in several disciplines at any global corner. The undergraduate students of all disciplines of engineering, physics, chemistry, and applied sciences can use this book as a reference for the fundamentals of electrical circuits, basics of semiconductor physical electronics, electrical properties of material systems, and characterization of engineering materials besides undertaking senior design and analysis projects in this area of technology.

This book is organized in such a way that there is no duplication or repetition or redundancy as the concept of *Concurrent Multiple Complex Plane Formalisms* (CMCPFs) is established serving applications and purpose of the DUT. A few problems are incorporated at the end of each chapter to serve practice sessions. It is expected that the learners and the investigators are used to with the computers and spreadsheets beside the access to the available software for data acquisition and necessary analyses. Overall the content provides a worthy source of information for the investigators working in multidisciplinary areas. Background principles, basics of measurements, knowledge on instrumentation, and detailed analysis or analytical procedures are not included in this book because similar content already exists.

The *immittance spectroscopy* is such a subject that one has to get absorbed or immersed in it with heart and mind possessing strong inspection capability, and subsequent assessment for any set of frequency window possessing number of data points. There is no scope of guessing or speculation for extrapolation of the observed behavior of the DUT. It is a fact that with the level of experience it is possible to visualize the duality of the circuits or duality of the complex planes just seeing or inspecting the feature of one complex plane. Bode plots often mask such tangential behavior. This achievement requires heavy practice and understanding the *walking nature* of the data points on the semicircle for any DUT. This book has attempted to provide that kind of concept to the interested readers, learners, and investigators.

The CMCPFs apply to all four complex planes in a simultaneous fashion for the same set of measured data. Representation of the measured data in *one selected choice* of a complex plane via *impedance* or *admittance* or *capacitance* or *modulus* plot is referred to as the SCP plot besides using a number of favorable and non-favorable Bode plots. This approach somehow paves the initial road for understanding *lumped* contribution for the DUT. When all four complex planes are used concurrently the road becomes wider. Thus, the effect of using *lumped parameter/complex plane analysis* (LP/CPA) and *Bode plane analysis* (BPA) is emphasized together with *spectroscopic analysis* (SA).

The meaning of the duality of the equivalent circuit models (ECMs) obtained from one or more complex planes in a simultaneous fashion is explained. The complexities of these equivalent circuits originated from a single set of measured data are not averted but dealt with by way of published examples. Converting one form of circuit to another form clearly demonstrates how LP/CPA can influence each complex plane. Nevertheless, two simultaneous circuits obtained from two complex planes allow a more visionary window than a single circuit obtained from a single complex plane via better exposure of the underlying operative phenomena. Among the highlighted topics, concurrent multi-plane analysis and their usefulness of developing equivalent circuit is an important issue, particularly, when no usual semicircular relaxation appears in the template impedance plot or classic Cole-Cole plot.

The *first book* entitled "Impedance Spectroscopy: Emphasizing Solid Materials and Systems" edited by J. Ross Macdonald was published by John Wiley & Sons, Inc. in 1987 comprises a collection of articles using impedance data and findings for the solid material systems. This book used IS as the acronym for the subject, making complete sense for the users. It attracted the attention of various deep root researchers in this area as the users had readily access to the presented theories supported by the experimental data. As a first generation text, most of the demonstrations emphasized SCP preferably in the impedance plane. As the beginning text in this subject certain basic principles along with the data acquisition and analysis were useful. The second edition of the same book was published in 2005 with Evgenij Barsoukov and J. Ross Macdonald as the editors. Both the editions addressed multiple relaxations in the same preferred SCP, and proposed *concurrent complicated equivalent circuit models* beside the simplistic viewpoints. A few additional articles were included in the later edition to enhance the content, and dimension of the research areas that reflected the same template demonstration with SCP. In reality, often the measured data exhibit relaxation in more than one complex plane. Handling such a situation is not addressed as the preferred SCP analysis skillfully adopted satisfying feasible format by keeping completely away the relaxation observed in another complex plane. Intermixing of the complex planes for a single set of data or handling the same set of data for the CMCPFs is not addressed or emphasized in these two editions though conversions within the complex planes are mentioned regardless of Debye or non-Debye parameters.

The *second book* entitled "Electrochemical Impedance Spectroscopy" authored by Mark E. Orazem and Bernard Tribollet appeared in 2008, published by John Wiley & Sons, Inc. This book is more than likely to be designed as a potential textbook serving the course content per outline. It appears from the content that this book has created a limited window emphasizing electrochemical systems rather than serving the broader range material systems. As the background-serving content, this book is excellent for the beginners in the classroom setting. It addressed the electrochemical aspect of the impedance or admittance or phasor data representing sole *impedance plane plotting* mostly on the liquid systems. It is observed that the liquid system, electrochemical or not, often exhibits relaxation in more than one complex plane in a simultaneous fashion. Information about the solid electrochemical species or electrolytes is difficult to squeeze from this book. This book completely eliminates or ignores observed relaxation in another complex plane for the same set of data while template SCP format via impedance plane is all over. Using the prescribed format there is no scope to evaluate a second equivalent circuit model outside the impedance plane for the same DUT. The *second edition* of the same book published in 2017 is not different than the previous edition. Several basic principles are efficiently used along with the instrumentation and necessary background of the data collection. Intermixing of the complex planes for a single set of data or handling the same set of data for the CMCPFs is not addressed in this book.

The *third book* entitled "Impedance Spectroscopy; Applications to Electrochemical and Dielectric Phenomena" authored by Vadim F. Lvovich was published by John Wiley & Sons, Inc. in 2012. The content describes electrochemical and dielectric phenomena of the measured data representing primarily SCP analysis comprising of both solid and liquid systems. Many basic principles are addressed efficiently concerning the demonstration of the data. Relaxations are considered in the SCP mode via impedance plot for the DUT as if no relaxation exists in other complex planes. This book does not address intermixing of the complex planes for a single set of data or handling the same set of data for the CMCPFs.

Since the existing three books do not provide concurrent treatment of the single set of data in the CMCPFs, there is a room to fill the gap with another book. Considering this situation, this book is adding a new dimension focusing CMCP approach, and presents subsequent meaning of the corresponding equivalent circuit model for each complex plane. The relevant inter-conversion of the equivalent circuit elements allows ultimate selection of the equivalent circuit. The limitation and validity of this conversion is based on Debye or non-Debye response of the DUT. This aspect of the analytical method for a single set of data is adding a meaningful avenue in addition to the merits of the foregoing three books. Overall the eventual equivalent circuit model derived from the CMCPFs reflecting contributed elements originated from the UOMs of the DUT.

Nevertheless, these three books are serving a wide range of audience for handling the data via SCP analysis using some sort of backing of the conventional Bode plot. Basically these three books attempted to educate the audience with the SCP analysis that facilitated the propagation of the subject. The demonstration of the SCP analysis specifically addresses the issues as if UOMs are only related to the template impedance plane apart from the Bode plot. Knowing a broad range of background of the data acquisition procedure, the impedance plane alone does not always provide a complete answer for the underlying operative phenomena.

The limitation and scope of the impedance plane are not noted. Thus, the investigators propagate in the public domain *two semicircles in the impedance plane means grain and grain-boundary behavior* for the heterogeneous multi-phase polycrystalline systems. These investigators cannot think outside this concept, having tons of defects and traps that always vitiate the Mott-Schottky plot for these materials. This is regardless of the nature of the solid or the type of liquid in conjunction with the associated transport processes whether attributing to the electrochemical species or not. The data representation in the template impedance plane form is an extremely simplistic lumped picture, though mathematically satisfied multiple circuits can satisfy the relaxation therein. This is because a simple dielectric behavior is always more than likely to be shunted by a conductance which normally receives support from the impedance plane.

When the dielectric portion is predictable by the *concurrent multiple phenomena* as F. A. Grant demonstrated in 1958, and each of them is distinguishable in the frequency domain, then the capacitive component emerges as the *lumped* component with its resistive counterpart in the impedance plane. This drawback cannot be delineated or understood when only impedance plane alone is used. Concurrent multiple dielectric phenomena may not be present in many simplistic liquid electrochemical systems. Thus, the measured data often can be presented in the impedance plane. All aspects of the measured data remained deficient for the deep-level inquisitive researchers and learners as satisfactory explanation often encounters limited room for the complicated material systems.

In order to fulfill the shortage of the real-world problems, several aspects of the origin of the multiple equivalent circuits have been addressed in this book using CMCPFs. Utilizing CMCPFs can lead to an effective equivalent circuit model by exploiting LP/CPA attributing to the complete set of underlying operative mechanisms for the DUT. The novel tools/techniques using LP/CPA satisfying CMCPFs are emphasized for various DUTs. The meaning of the CMCPFs is explained and elucidated for the extracted elements of the ECMs. Subsequent interpretation of the analytical approach is presented in a meaningful way for the simple single-crystals and the heterogeneous multi-component polycrystalline material systems. The UOMs are explained in terms of the window of visibility and situation or condition. The LP/CPA allows a *spectroscopic* analytical window via CMCPFs in conjunction with several types of Bode plots. This approach is invariably different than just measuring or acquiring the data as a function of *applied frequency* or *measurement frequency* in the form of real and imaginary parts of a *complex electrical quantity*. Such a definition of *impedance spectroscopy* in the *Annual Review of Materials Science* has confused the investigators and new learners since 1986.

A thorough guideline on the development of the software for the CMCPFs is outlined beside the adoption of the ultimate ECMs satisfying UOMs for the DUT. In this way the ECMs are not restricted to the SCP formalism and/or in conjunction with the limited window of the Bode plots. The user of this book will eventually come to realize the purpose and meaning of the CMCPFs. Due to the lack of demonstration of CMPCFs in the existing books, the investigators often do not realize the importance of having CMCPFs for the measured data. Through the demonstration of CMCPFs the user will be accustomed with the semicircles of the frequency window reflecting the data-handling criteria. The correlation of the *relaxation time* (or *time-constant*) for each complex-plane semicircle is unique and establishes equivalent circuit in an effective manner covering broad meaning of the eventual model portraying UOMs. Thus, the lumped parameter obtained in a SCP can be further subdivided into the contributing elements forming the ultimate equivalent circuit model. Furthermore, this book explains and distinguishes the shortcomings of using *idealized Debye* concept for the observed *non-Debye relaxations*.

This book addresses the following contents for the as-measured data but not limited to the:

1. Understanding *immittance spectroscopy* via precise definition, theory, applications, etc.;
2. Clarification of the complexities of the devices and material systems;
3. Attentive simplistic forms of the complicated devices and material systems;
4. Use of LP/CPA for the CMCPFs in conjunction with the Bode plots and spectroscopic approach;
5. Derived ECMs without mixing Debye and non-Debye elements;
6. Distinction between *impedance spectroscopy* and *admittance spectroscopy*;
7. Concept of the ECMs for the UOMs;
8. Possible interpretation of the UOMs;
9. Identification of the flaws, error, and inaccuracy;
10. Justification, effect, and impact of using geometrical parameters to the measured quantities;
11. Explanation of the spectroscopic nature of the extracted parameters;
12. Ideal (Debye) and non-ideal (non-Debye) relaxations via depression parameter;
13. Guidance on the software development along with the impact of the methods of the analysis; and
14. Data-Handling Criteria (DHC) using Complex Nonlinear Least Squares (CNLS) fitting procedures.

This book has used generic plots or graphs all throughout. Thus, traditional or non-traditional behavior of the DUT is described or illustrated. Unlimited references have been digested and consulted but only a handful is listed at the end of each chapter. The erroneous or ambiguous content of the piled-up papers are not considered as they never felt worthy since the contents therein are heavily diluted or distorted or deformed. Frequently many published documents are found to be flawed on a regular basis. It is expected that this book can help the investigators detecting the flaws, and filter those efficiently so that scholarly waste is eliminated from the society.

The content of this book is attractive in satisfying a broad range graduate students, post-doctoral scientists and engineers, established and beginner-level researchers in the disciplines of solid state physics, chemistry, electrochemistry, metallurgy, corrosion, biology, biological sciences, bioengineering, biotechnology, materials science and engineering, electrical engineering, mechanical engineering, civil engineering, chemical engineering, metallurgy, reliability engineering, corrosion science, quality control, etc. covering semiconductors, solid state junction devices, solid and liquid electrolytes, electrochemical power sources, biomaterials, etc. Based on the specific content and pedagogical approaches contained herein this book stands by itself. Overall *Immittance Spectroscopy* has firmly established a sophisticated novel tool/technique to unravel complicated material systems toward understanding. This book is a useful source of information for established researchers, including the learners working in the aforementioned areas. From the manufacturing standpoint, the quality control aspect of the end-product is also addressed.

Acknowledgments

The concept of this book was originally discussed with Dr. Martin A. Seitz sometime around early 1985, shortly after his 1984 paper was published, while I was in the process of leaving his laboratory. In this paper the concept of an equivalent circuit was noted for each polarization mechanism which was the essence of the course that Dr. Seitz used to offer every year. By that time he led to develop two separate projects for complex nonlinear least squares (CNLS) fitting procedures. He executed mainframe-based *FORTRAN language* in 1979 for the first one. It utilized the Marquardt-Levenberg algorithm and he demonstrated the results of the biological samples in his 1980 paper. I was involved in the next project in early 1980s which was based on an HP86 computer employing the *BASIC language*, having the capability of fitting two consecutive semicircles that employed an iteration-selection method. Thus, analytical software was refined in two stages before achieving its final form in early 1984. While progressing this work, simultaneously a dual circle fitting paper of Tsai and Whitmore appeared in 1982, which encouraged developing the iteration-selection method. My early papers were based on the usage of HP86.

My involvement with the development of the analytical software was again in 1992, using the 486 computer while serving as the external thesis committee member from the industrial corner for the group of Dr. Sheikh A. Akbar at The Ohio State University. This software remained identical with the previous one developed at Marquette University. Several papers had been published together in 1990s using this software.

Dr. Seitz indicated that commercialization of a book on this subject requires plenty of labor and effort to materialize as it is addressing the research audience. At that time there was no book on this subject in the academic arena. As a graduate student, I felt an acute shortage of a basic text to comprehend the fundamentals of complex plane analysis. Dr. Seitz taught a senior-level course carrying graduate credit where he emphasized complex plane analysis for about four to five weeks and covered applications of the dielectrics. To the best of my knowledge, I recollect that no partial course content existed at that time at any university. Dr. Seitz referenced from his own publications beside a few other papers. Though there was an official text for the course, practically no conceptual explanation existed there about the complex plane analysis other than emphasis on polarization mechanisms stating a variety of electronic materials.

The idea of a book originated from the use of concurrent multiple complex plane analysis (CMCPA) of a large number of ZnO-based varistor materials that provided substantial meaning of trapping behavior, and resonance phenomenon for the measurement frequencies ranging $10^{-2}\ Hz \leq f \leq 10^{9}\ Hz$. Trapping behavior was rationalized using the classic work of F. A. Grant and J. E. Bauerle amalgamated with the classic paper of K. S. Cole and R. H. Cole beside frequency-dependence of the series components of J. Maserjian. At that time it was very rarely noticed that investigators used concurrent multiple plane analysis, despite relaxation behavior was obtained in more than one complex plane. Surely Grant could not win the popularity of the investigators with the demonstrated concurrent multiple plane analysis. In general, investigators followed the easiest route emphasizing template format for the single complex plane analysis, preferably the impedance plane, and completely ignored meaningful response observed in another complex plane. This kind of attitude of using single complex plane analysis did not help explain the underlying operative mechanisms for the multi-phase polycrystalline ZnO varistors. Interested readers can examine varistor papers as how the single semicircular relaxation of the impedance plane got lumped for the multiple operative mechanisms in the complex capacitance plane. In this way, concurrent multiple plane analysis is proven to be very important for the complicated multi-component material system but this never attracted Dr. Seitz to generate a text.

Not too long after our discussion, an edited book was published by Dr. J. R. Macdonald. Upon digesting the content, shortcomings and deficiencies were discussed soon thereafter with Dr. Seitz. Ultimately my intention received no further attention except citing the massive labor and unworthy effort. As a fresh graduate, I felt it inappropriate at that time proceeding alone though several papers were in preparation or in the review process or in the press. I did not give up my desire as I continued discussing with many other professionals at a later time while producing subsequent papers. Meanwhile I noticed flawed papers were surfacing in the journals. Sometimes professional societies violated the quality requirements, possibly without knowing, by incorporating erroneous manuscripts. Seeing this, at one point I rejected offers from the Associate Editors to review the manuscripts as I had cited several erroneous papers to them. I was asked to write comments but I did not agree to do so. This is because I knew that such massive volumes of publications cannot be filtered with corrections by just taking a single person's viewpoint unless the Editorial desk asserts necessary demand.

Around the early 1980s, I noticed a top-rated European journal published a semicircle with three experimental points contributed from one of the populous countries. These points were fitted using a protractor. Later I met the contributor almost two decades later. This contributor admitted that somehow the group needed a paper in a reputed journal using the impedance data so that exposure takes place in a new arena. On another occasion, I came across a few contributors in the meetings who demonstrated extraction of the circuit parameters from any shape of the curve in the complex plane as if the CNLS fitting procedure doesn't exist. Thus, an easy route for the investigators led to confusion and created chaotic evidence for the audience. In the 1990s, I noticed in several meetings that the polycrystalline materials investigated via impedance plane had grain and grain-boundary relaxations. These investigators were so single-sided in attitude that they could never accept the same valuable data in another complex plane to find some added information about the traps. Later on I realized that these contributors specialize in defect chemistry, but trapping in the polycrystalline materials is very delicate for them. Such contributors began educating the audience, meeting after meeting, describing the novelty of the two semicircular relaxations in the impedance plane.

I wrote a comment paper in 1991, and thereafter I noticed that similar flawed manuscripts were getting published globally almost frequently. I had no intention to develop professional difference by producing comment papers. However, users or investigators continued establishing the concept of the Nyquist plot, impedance plane stamped as the Cole-Cole plot, electrochemical impedance spectroscopy, constant phase element for the depression angle of the impedance plane, inappropriate circuit parameters, etc. These concepts propagated rapidly during the past three decades or so. Despite the flawed papers, in my opinion, I continued enjoying reading the content to stay updated but never kept listing of those contents as reference. One chapter is based on the corrective effort of the flaws as those non-listed contributors *en masse* helped in generating them.

I must acknowledge profusely Dr. Martin A. Seitz for his initiation in teaching CMCPA for the same data as he supported with his 1980s papers. These mini texts provided guidelines for the interested readers. After leaving his laboratory the unfinished definition of *impedance spectroscopy* appeared in 1986. Dr. Seitz always explained the meaning of spectroscopy in the light of physics and chemistry as I hunted for it. The difference between the two perspectives helped assessing the horizon of *impedance spectroscopy*. Exploration of ZnO based varistors never got influenced with the economical definition. Dr. Seitz always handled polycrystalline materials but never asserted on grain and grain boundary behavior only. Thus, he extended the edge by inducting me exploring thermally stimulated current (TSC) in 1980 and deep-level transient spectroscopy (DLTS) in 1984 to support the study of the defects and traps. I had a scope of learning the novel tools/techniques in his research feature. Lumped behavior and subsequent verification of the extracted parameters beside quantification was the exposure to the users of the CMCPA. My freedom to learn CMCPA paved the avenue for initiating this book.

When I decided to write this book only a few years ago my family members became supportive of the idea. My wife, Sanjida Khanam, is familiar with the subject matter as she was exposed under the guidance of Dr. Seitz for her M.S.M.E. thesis where she used impedance data related plots. She encouraged me to undertake this task along with my four children. Due to their encouragement, I decided to start this massive work. Because of the change in my professional career I had to wait for the available time to put several papers together as a concise book. My children helped in many ways gathering various latest references beside little typing of the manuscript at the initial stage.

I sketched all the generic illustrations as I did not feel using the actual graphs in many cases. I advanced steadily with patience utilizing available time having single-handed effort. Among the four children, my son Mohammad E. Alim took serious interest in reading the manuscript. Occasionally I received help from my colleague Dr. Sudip Bhattacharjee while working together. A significant contribution giving detailed description on the geometrical and CNLS fitting methods for the semicircular relaxation curves came from him. We have co-authored an article on this subject in the *Journal of Materials Science*: *Materials in Electronics*, and generated a few papers in other journals using the immittance data. The section on the geometrical and CNLS fitting methods is updated using that paper, and demonstrated beneficial aspects of these two methods. Dr. Bhattacharjee is gratefully acknowledged for the segment that I used in this book. This is my third data-handling analytical software development for the complex plane analysis.

In this book I avoided the fundamentals or basics and the tutorial aspect of the subject which are nicely provided in the previous three books. Thus, I trust there is no redundancy. I expect that the investigators and the graduate students will benefit from the content. Senior undergraduate students can extract essence for their design and analysis projects. Eventually I forwarded the manuscript to Mr. Martin Scrivener of Scrivener Publishing for the publication of this book following the review process with John Wiley. Mr. Scrivener took serious interest in this work and was kind enough to allow me time to complete the task.

Mohammad A. Alim
Alabama A & M University
Huntsville, Alabama 35762
U.S.A

1

Introduction to Immittance Spectroscopy

1.1 Basic Definition and Background

The definition and background of immittance spectroscopy has always been very important. Using the term "spectroscopy" emphasizes a broad meaning of handling measured immittance data. The commonly used term impedance spectroscopy or admittance spectroscopy actually belongs to the immittance spectroscopy where the hybrid word "immittance" presumably coined by H. W. Bode [1] is in use for at least five decades applying to either impedance or admittance. Commercially this hybrid word was in propagation with the General Radio products. In atomic spectroscopy, elements of the Periodic Table are diagnosed by their characteristic frequencies or corresponding wavelengths [2, 3]. Each frequency or wavelength refers to the specific transition of the electrons outside the nucleus. The specific transition of the electrons from one shell to another shell gives rise to frequency and wavelength. This atomic level investigation or understanding provides firm identification of the elements that refer to the spectroscopic approach for the same element. Due to this concept of "spectroscopic approach" a trace element in a matrix can be determined during the chemical analysis. The detection limit of a trace element can be ascertained by the capability of the diagnostic tools and methods utilized.

Therefore, a term or nomenclature bearing "spectroscopy" indicates an in-depth investigation or evaluation process. It may be noted that for light, the product of frequency and its corresponding wavelength is equal to its speed. The atomic spectroscopic transition from one level to another level of energy is noted in this case to occur at the speed of the light.

Emphasizing the material system or subject matter or specific mechanism in front of the term as found in electrochemical impedance spectroscopy (EIS) brings a new nomenclature for the readers. This nomenclature does not import new meaning or interpretation when immittance (impedance or admittance) spectroscopy is viewed in a broad sense. In reality, the term EIS does not reflect any change in the nature of the data or the type of analysis other than the elaboration of the confusing redundant term. Overall, using the term electrochemical impedance spectroscopy for electrochemical systems neither enhances nor broadens the underlying meaning of the entire investigation other than spreading redundancy or confusion among the learners and the users. Therefore in this text, only the first three aforementioned terms will be used as these are synonyms for the purpose regardless of the material system or subject matter or specific mechanism.

Importing new nomenclatures for the same term like capacitance (complex capacitance) spectroscopy, modulus spectroscopy, dielectric spectroscopy, etc. are new in recent years. This sort of nomenclature is a misnomer emphasizing redundancy or ambiguity or confusion for both learners and users. None of these terms serve the underlying purpose of achieving scientific or technical goals concerning interpretation of the measured data. The term capacitance spectroscopy emphasizes the quantity expressed as complex capacitance. Both capacitance spectroscopy and modulus spectroscopy are straightforward derivatives of the first three terms obtained via conversion of the data. Using "geometrical factors" to obtain the state of normalization leads to the so-called dielectric spectroscopy for the measured data obtained in the phasor form or in the form of impedance or admittance. Essentially dielectric spectroscopy does not alter the subject matter by introducing the phrase dielectric in the term. By the same token there is no term like resistivity (complex resistivity) spectroscopy and conductivity (complex conductivity) spectroscopy. Again, these terms can be obtained from impedance or admittance or phasor data through conversion. In each case the type of data analysis is not affected but the data acquisition may have specific conditions due to the type and scope of instrumentation than many other material systems.

By the foregoing token it is not worthwhile to propagate device impedance spectroscopy or solid state impedance spectroscopy or sensor impedance spectroscopy or biological impedance spectroscopy or any similar derived term emphasizing devices or solid state materials or sensors or biological materials. Discovering such a term does not bear meaningful sense in the real world applications. Specific material systems should not influence adventurous nomenclatures or terms without adding anything to the knowledge or altering the basics of the fundamental concepts. The measured data usually reflect known non-equilibrium variables such as temperature, voltage, pressure, ambient gas, processing method or fabrication route, composition or recipe, etc.

In the existing literature there are very few texts that explain detailed meaning of the immittance spectroscopy. Often explained is "... ... ... the term impedance spectroscopy is used to describe any technique in which data, equivalent to the real and imaginary parts of a complex electrical quantity, are measured as a function of frequency." This type of definition [4] of impedance spectroscopy is a misnomer though it got propagated for years within the community of the impedance data users. These users accepted such a definition as if a correct message is conveyed for the subject. Certainly the acquisition of the ac small-signal electrical data as a function of applied frequency expressed in the form of real and imaginary parts of a complex electrical quantity is not the impedance spectroscopy. This faulty definition has been frequently used in the published research papers keeping totally devoid of the spectroscopic approaches of the measured impedance (or admittance) data. This type of definition is far from satisfying in-depth analytical perspective of the data except for displaying single complex plane (SCP) representation via "impedance plane" plots beside occasional Bode plots and tentative equivalent circuit modeling. In many cases impedance plane plot has become preferred mode of displaying the data for which the term impedance spectroscopy is used. Again this is absolutely misleading as such a display never contains in-depth analysis or interpretation except for matching the traditional equivalent circuit model. Sometimes Bode plots are used as a supportive avenue for the equivalent circuit model.

The measurement frequency is applied to the system while acquiring the electrical data. The unit of frequency is Hz corresponding to the cycle/second whereas the unit of angular frequency ω (= $2\pi f$) is radian/second. Most commercial machines are capable of acquiring the data in three forms: (1) impedance or series mode components or in-phase and out-of-phase components, (2) admittance or parallel mode components or in-phase and out-of-phase components, and (3) phasor. Any of these measurements provide impedance or admittance as each measured form is convertible to the other form including complex capacitance and modulus. These conversions are found in any standard text on linear circuits [5]. The dielectric representation of the data can be obtained when the geometrical factors are incorporated to any of the three measured forms of the data [6–8].

The term impedance spectroscopy has become very familiar in the last five decades or so. It is often used in ambiguous style where the word "spectroscopy" for the measured immittance data remains unclear or unexplained or unaddressed. The realistic scenario of the immittance spectroscopy contains far more analysis than the use of the SCP analysis including limited Bode plane analysis (BPA) of the measured data. This is because immittance or phasor form of the ac small-signal data acquisition has direct connection with the extracted or derived parameters from the observed relaxation behavior. In general, the extracted parameters from the SCP analysis are frequency-independent in nature, and noted as the:

a) relaxation time or time constant;
b) peak-frequency corresponding to relaxation time or time constant;
c) left-intercept and right-intercept of the semicircular relaxation;
d) location of the center of the semicircular relaxation;
e) associated depression parameter measured by the depression angle (if any); and
f) equivalent circuit elements.

Obtaining the role of the aforementioned parameters beside ambient variables and other non-equilibrium conditions provide complete impedance spectroscopy of the measured data. Invariably information from the BPA is no exception from the SCP analysis. Apart from single complex plane analysis, it is important to include rest of the complex plane analysis in a simultaneous fashion to obtain complete scenario of the investigated material system. Interpretation of such broad range information will be conducive for the investigator to establish structure-property-processing interrelationships. Only then spectroscopic analysis [9–17] of the immittance data can be successful. immittance spectroscopy for a device under test (DUT) certainly refers to the spectroscopic approach of the immittance data bearing similarity with that of the atomistic level in-depth profile. This approach [9–17] includes handling and dealing the measured quantities under various conditions besides the extraction of the parameters or derived elements and their dependences, characteristics, subsequent interpretations of the responses under certain non-equilibrium conditions.

1.2 Scope and Limitation

The ac small-signal electrical data refers to the as-measured electrical parameters or electrical quantities as a function of measurement frequency (f). These data can be transformed to the immittance function having real and imaginary parts attributing to the in-phase and out-of-phase components. The immittance function represents complex plane formalism by:

$$I = I' \pm jI'', \tag{1.1}$$

where $j = \sqrt{-1}$, and I′ and I″ are the electrical quantities ascribing real and imaginary parts of I, respectively. Each of the electrical quantities contains as-measured electrical parameters where the form depends on the type of the immittance function using proper ± sign for the imaginary quantity. The complex plane representation of the data employs a plot of I′ on the x-axis (real) and I″ on the y-axis (imaginary). This approach refers to the lumped-parameter/complex-plane analysis (LP/CPA) that allows extraction of the parameters. These extracted parameters certainly derive from the lumped nature of the contributing elements within itself.

The BPA involves a number of immittance functions derived from the real (I′) and imaginary (I″) parts of I. These parameters may include $|I|, \frac{I'}{I''}, \frac{I''}{I'}$, etc. as a function of measurement frequency where |I| = I.I* with I* as the complex conjugate of I. In the same way the spectroscopic analysis (SA) involves similar investigation of these parameters as a sequential function of possible experimental variables. Using LP/CPA, BPA, and SA together it is possible to extract total information for an unknown DUT. Additional tools/techniques will provide further supporting information to delineate the complexity of the DUT. Thus, these analytical approaches provide vast scientific/technical information to understand a DUT thoroughly. Sometimes these analytical approaches become limited when the relaxation processes are unclear and become uncertain despite using all resources of the complex plane analysis. Often BPA or SA alone cannot unravel the complexity of the DUT if the analytical patterns do not follow usual template features. In such a case, additional investigative tools/techniques will be necessary to understand the total feature of the DUT.

1.3 Applications of the Immittance Studies to Various Material Systems

Immittance studies can be conducted for a number of material systems that are not limited to the items listed in Figure 1.1. The health condition of the material systems can be ascertained by this study as it is non-destructive. The presence of hidden cracks or fractures may not be delineated but the existence will provide altered data violating normal behavior. Thus, immittance measurements can be a quality control tool assuring product standard and customer satisfaction. Because of certainty in the properties determined via immittance measurements reverse engineering of a DUT is possible.

The usefulness of immittance spectroscopy emphasizes the ability to distinguish dielectric behavior from the electrical properties of individual contributions of the DUT under investigation. The modeling procedure utilizes electrical circuits built from components such as resistors and capacitors to represent the total behavior of the DUT. Changes in the immittance values as a function of applied frequency for the DUT indicate ultimate behavior and performance characteristics. Overall, immittance spectroscopy is a non-destructive tool/technique. Thus, the time-dependent information about the properties can be obtained in addition to the ongoing processes such as corrosion, discharge of batteries, degradation of the semiconductors, time-dependent behavior of the biological materials, electrochemical reactions in fuel cells or batteries, etc. Overall the advantages or usefulness of the immittance measurements include:

1. high resistance materials such as semiconductors, paints, coatings, etc.;
2. accessibility or availability of the time-dependent data;
3. totally non-destructive tool due to very low signal-voltage (μV – mV range);
4. quality control/assurance tool;
5. reverse engineering tool;
6. diagnostic tool;
7. possibility of quantitative analysis;
8. non-equilibrium environment;

Systems	Effects to be diagnosed
Semiconductors and devices	• Dopant effects in the bulk • Additive effects in the bulk • Conduction mechanism • Transport properties • Single junction behavior • Multi-junction behavior • Defect characterization • Time-dependent processes • Damage detection • Degradation and thermal runaway • Aging behavior • Performance characteristics
Sensors	• Response assessment • Material sensitivity • Porosity and surface area effect • Reversibility or hysteresis
Composites: (a) Inorganic-inorganic (b) Inorganic-organic (c) Organic-organic	• Percolation threshold analysis • Porosity and surface area effect • Damage detection • Defect characterization • Performance characteristics
Powdered material processing	• Flow characteristics • Additive effects • Trace and contamination effects
Corrosion and inhibitor	• Rate determination • Reaction kinetics and degradation • Galvanized surfaces • Inhibitor performance
Batteries	• Materials selection • Electrode design • Electrolytic dissociation • Performance characteristics
Electrolytes (a) Solid electrolytes (b) Liquid electrolytes	• Charge storage • Internal reactions • Materials selection
Electrodeposition and coating	• Formulation • Deposit characterization • Size and/or thickness effect
Biological systems	• Cells and tissues • Properties and structure • Physiological events • Biological transitions • Functions of living organisms • Morphological and internal changes • Decomposition and degradation
Fuel cells	• Material selection • Functional behavior
Electroceramics	• Processing • Characterization
Materials testing and evaluation (non-destructive characterization)	• Quality control/assurance • Reverse engineering
Crystallization	• Growth and deposition rate
Nanoparticles	• Flow rate • Size effect • Surface area

Figure 1.1 List of material systems as DUT and effects to be diagnosed by the immittance measurements.

9. thermodynamic equilibrium;
10. small perturbation considered as linear;
11. different processes with variety time constants;
12. instruments available covering large frequency range (mHz – GHz);
13. availability of various competitive models for data analysis; and
14. data analysis accessible with or without presumed model.

A few minor disadvantages or drawbacks of the immittance measurements include:

1. instruments often considered as expensive though in the long run not really so;
2. very often measurements at low frequencies not available, particularly below mHz;
3. often data analysis for quantification considered too complicated;
4. require familiarity and understanding of the data for quality control/assurance; and
5. need well-acquainted personality to operate the machine beside obtaining/analyzing the data.

Since immittance spectroscopy employs wide range of frequencies the information content is certainly much higher than dc ($f \rightarrow 0$ Hz) measurement or just single-frequency measurement. The immittance spectroscopy is a novel powerful tool/technique that can:

1. test or evaluate assembled systems,
2. identify diffusion limited reactions, that is diffusion through passive film,
3. distinguish between two or more distinct electrochemical reactions,
4. delineate complicated systems,
5. use analytical expressions,
6. validate data,
7. allow applicability in the time domain data,
8. provide information about the electron transfer rate of the reaction,
9. provide information on the capacitive and resistive behavior or response of the system, and
10. develop equivalent circuit model for the system.

1.4 Concept of the Linear Circuit Elements: Resistance, Capacitance, and Inductance

The fundamental concept of resistance comes from the Ohm's law: temperature and pressure remaining constant, applied voltage across a conductor is proportional to the current flowing through the same conductor [6]. Thus,

$$V \propto I \quad \text{gives} \quad V = R \cdot I, \tag{1.2}$$

where R is the proportionality constant defined as the resistance opposing the flow of current through the conductor. The unit for R is ohm (or Ω), for V is Volt (or V), and for I is Ampere (or Amp or A). Equation (1.2) explains the dc condition whereas for the ac condition both voltage and current are in-phase for a pure resistance. The terms resistor and resistance are often synonyms and regarded as the passive circuit element. Similar terms are capacitor or capacitance, and inductor or inductance as the passive circuit elements. The circuit symbol for a resistor, a capacitor, and an inductor are shown in Figure 1.2.

The capacitance is the ability of a device or element or material system to store electrical charges. The simplistic form of capacitance of a capacitor can be considered between the two opposite electrodes having uniform geometrical shape or structure of parallel plates separating by a constant distance. Let the capacitance of this parallel plate configuration under vacuum be C_O. When the same two electrodes are placed across a material system then the capacitance turns to C. This means that the material system is causing the change in the measurement of the capacitance. Two situations are illustrated in Figure 1.3. The ratio of these two capacitances gives rise to the relative permittivity or static relative permittivity or permittivity or dielectric constant or relative dielectric constant, ε_r, of the material system between the two parallel plates:

$$\varepsilon_r = \frac{C}{C_o} \quad \text{or} \quad C = \varepsilon_r C_o . \tag{1.3}$$

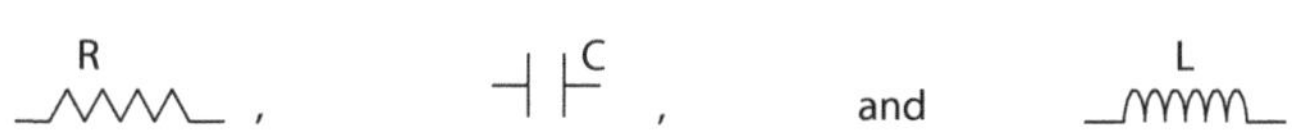

Figure 1.2 Circuit symbol for the passive elements: resistor (R), capacitor (C), and inductor (L).

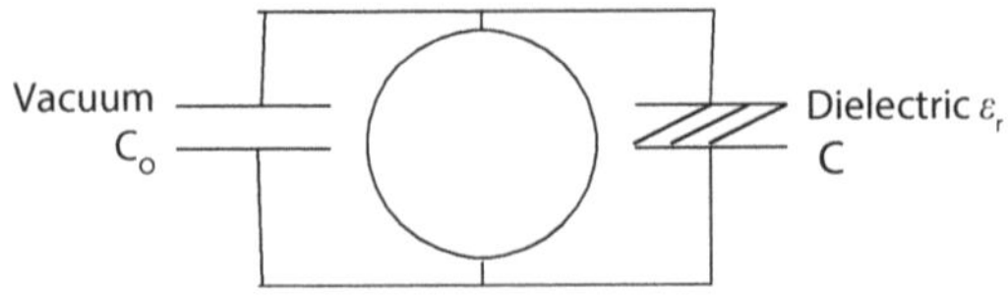

Figure 1.3 The dc power source provides the same voltage to the parallel plate capacitor in vacuum on left, and to the parallel-plate capacitor with dielectric ε_r on right.

The capacitor is a linear circuit element. Equation (1.3) is valid for the single-phase system such as single crystal where dielectric constant refers only to such an ideal or perfect or near perfect system. For an amorphous and polycrystalline system the same relation may not be valid. In the case of multi-phases or multi-junctions or polycrystal systems the term dielectric constant becomes a misnomer as explained later.

Equation (1.3) can be visualized by setting two separate experimental arrangements. In the first arrangement, a dc power source can be placed between two identical sets of parallel plate capacitors: one in vacuum and the other with a material system having dielectric constant ε_r as shown in Figure 1.2. In this case the dielectric material is causing accumulation of more charges than the other parallel plates in vacuum [18].

In Figure 1.4 the same two capacitors are arranged separately where the capacitance is the same using two separate voltage sources. The capacitor in vacuum shows capacitance C_O at voltage V_O whereas the capacitor with the dielectric constant ε_r shows the same capacitance C_O at voltage V. The capacitor in vacuum requires high voltage whereas the capacitor with dielectric requires low voltage to achieve the same level of charges. In this case two voltage ratios provide the dielectric constant ε_r given below:

$$V_o = \varepsilon_r V \quad \text{or} \quad V = \frac{V_o}{\varepsilon_r} \quad \text{or} \quad \varepsilon_r = \frac{V_o}{V}. \tag{1.4}$$

Thus, the dielectric constant can be defined as the ratio of the applied voltage or electric field strength in vacuum to that in the material for the same distribution of charges:

$$\varepsilon_r = \frac{C}{C_o} = \frac{V_o}{V} \tag{1.5}$$

yielding

$$CV = C_o V_o. \tag{1.6}$$

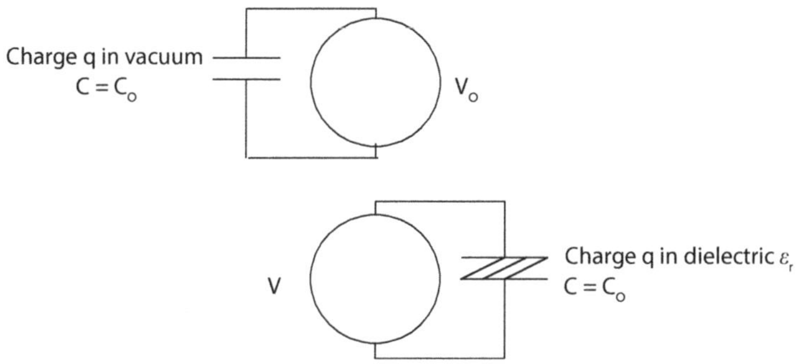

Figure 1.4 The capacitor in vacuum is at the top and the capacitor with dielectric is at the bottom. Both capacitors possess the same charges. The capacitor in vacuum requires high voltage whereas the capacitor with dielectric requires low voltage to achieve the same charges.

Studying the phasor relations in resistor-inductor R-L and resistor-capacitor R-C circuits excited with the ac source the voltage-current relations for the three basic elements [6] are the following:

$$V(t) = R\,I(t) \qquad \text{for resistance,} \tag{1.7}$$

$$V(t) = (1/C) \int I(t)\,dt \qquad \text{for capacitance, and} \tag{1.8}$$

$$V(t) = L\,[dI(t)/dt] \qquad \text{for inductance.} \tag{1.9}$$

In equations (1.7) through (1.9) V(t) and I(t) denote the voltage and current at time t, respectively, with R as resistance, L as inductance, and C as capacitance. When the current I(t) is given by:

$$I(t) = I_m \cos(\omega t), \tag{1.10}$$

the voltage across each of the three elements becomes [6]

$$V(t) = R\,I_m \cos(\omega t) \qquad \text{for resistance,} \tag{1.11}$$

$$V(t) = (1/\omega C)\,I_m \cos(\omega t - 90°) \qquad \text{for capacitance, and} \tag{1.12}$$

$$V(t) = (\omega L)\,I_m \cos(\omega t + 90°) \qquad \text{for inductance,} \tag{1.13}$$

where $\omega = 2\pi f$ with f as applied frequency. Thus, the voltage and current are in-phase for the resistance implying no phase shift between voltage and current. The quantity $(1/\omega C)$ is the capacitive reactance and the quantity (ωL) is the inductive reactance. Each reactance causes 90° phase shift between voltage and current. The voltage lags the current by 90° or current leads the voltage by 90° for the capacitance, and the voltage leads the current by 90° or current lags the voltage by 90° for the inductance. These passive elements are illustrated in Figure 1.5. The voltage-current relations can be written in phasor form as:

$$V = R\,I_m = Z_R\,I_m \qquad \text{for resistance,} \tag{1.14}$$

$$V = [1/(j\omega C)]\,I_m = Z_C\,I_m \qquad \text{for capacitance, and} \tag{1.15}$$

$$V = j(\omega L)\,I_m = Z_L\,I_m \qquad \text{for inductance,} \tag{1.16}$$

where Z_R is the impedance of the resistance R, Z_C is the impedance of the capacitance C, and Z_L is the impedance of the inductance L. The impedance of these circuit elements used in basic circuit theory are simply low-frequency asymptotes of the overall frequency responses. These circuit elements are listed in Figure 1.6. In equation (1.15) j is "– 90°" for the capacitance implying voltage lags the current by 90°, and in equation (1.16) j is "+ 90°" for the inductance implying voltage leads the current by 90°. In this way switching of capacitive behavior to inductive behavior implies – 90° to + 90°. Similarly switching of inductive behavior

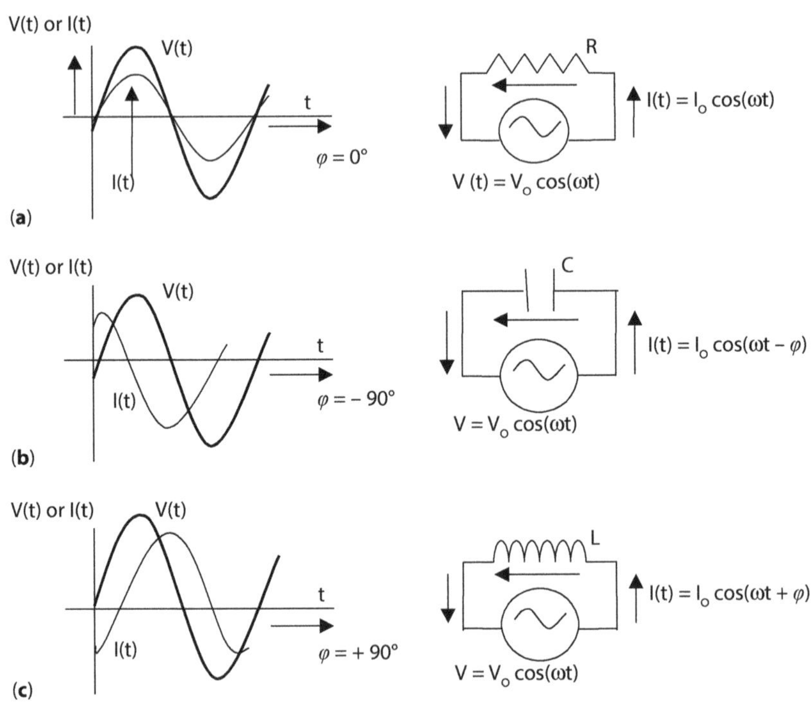

Figure 1.5 Voltage – current [V(t) – I(t)] response: (a) for a resistor R showing φ = 0°, (b) for a capacitor C showing φ = - 90°, and (c) for an inductor L showing φ = +90°.

Components	Current-voltage relation	Impedance Response	Function
Resistor	$V(t) = R\ I(t)$ $I(t) = (1/R)\ V(t)$ $R = V(t)/\ I(t)$	$Z = Z_R = R$ $Z_R = R < 0°$ $V = R\ I < 0°$	Dissipation of energy per Ohm's law
Capacitor	$V(t) = (1/C) \int I(t)\ dt$ $I(t) = C\ (dV(t)/dt)$ $C(t) = dQ/dV = I/(dV/dt)$	$Z = Z_C = 1/(j\ \omega\ C)$ $Z_C = -1/(\omega\ C) < 90°$ $I = j\ \omega\ C\ V < 90°$	Storage of charge per Coulomb's law
Inductor	$V(t) = L\ [dI(t)/dt]$ $I(t) = (1/L)\int(V(t)dt)$ $L = d\Phi/dI = V/(dI/dt)$	$Z = Z_L = j\ \omega\ L$ $Z_L = \omega\ L < 90°$ $V = j\ \omega\ L\ I < 90°$	Storage of magnetic energy per Faraday's law

Figure 1.6 Usual passive circuit elements with current-voltage relation and their impedance response.

to capacitive behavior implies + 90° to – 90°. An ideal resistor has zero reactance whereas an ideal inductor consists of pure reactance having zero resistance. By the same token, an ideal capacitor consists of pure reactance having infinite resistance. An ideal resistor follows Ohm's law at all currents and voltages, ac voltage and current for a resistor are in-phase with each other, and the value of a resistor is independent of frequency.

At resonance, when capacitance and inductance in the R-L-C circuit do not contribute to the impedance implies the effect of pure resistance then the reactance of the capacitance $1/(\omega C)$ and the reactance of the inductance ωL become equal at certain frequency in the opposite direction possessing mathematical sign. Thus, both the reactance cancel each other as exploitation of the situation yields $j = \sqrt{-1}$.

To describe the response of a DUT on the alternating current three passive circuit elements (R, C, and L) are introduced. An alternating current can be phase shifted with respect to the voltage. The phase shift depends on the type of DUT that allows current to pass. The current through a resistor R is not phase shifted with the applied voltage and thereby is not frequency-dependent.

1.5 Concept of Impedance, Admittance, Complex Capacitance, and Modulus

Several electrical parameters were developed by Oliver Heaviside using complex numbers/quantities that included impedance in 1886 and admittance in 1887 along with the constituting components. In order to explain each of these electrical parameters it is necessary to comprehend each term explicitly. Impedance is a general expression for electrical resistance used for alternating currents. From Ohm's law, the impedance Z(t) is given by the ratio of voltage V(t) and current I(t). The magnitude of Z(t) is represented in a two-dimensional space spanned by real and imaginary vectors. In addition, the number of degrees that current leads or lags voltage in an ac circuit is called the phase angle. The presence of phase angle [6] refers to phase change or phase shift between the current and voltage. Since the ac signal is periodic sinusoidal, it is convenient to express the phase shift as an angle in degrees.

In the immittance spectroscopy a DUT treated as linear system is perturbed with an alternating current by the application of a voltage signal of small amplitude (magnitude) to observe the steady state response. A linear system possesses properties of Ohm's law satisfying the principle of superposition. The impedance Z(t) of the DUT is given by:

$$Z(t)=\frac{V(t)}{I(t)}. \tag{1.17}$$

When V(t) or I(t) is applied as sinusoidal function to the DUT both V(t) and I(t) can be represented by a sum of the sinusoidal functions with phase shift shown in Figure 1.7. The response signal I(t) is shifted in phase by the amount φ and has a different amplitude $I_O(t)$. Thus, current for the DUT is given by:

$$I(t)=I_o\cos(\omega t-\varphi) \tag{1.18}$$

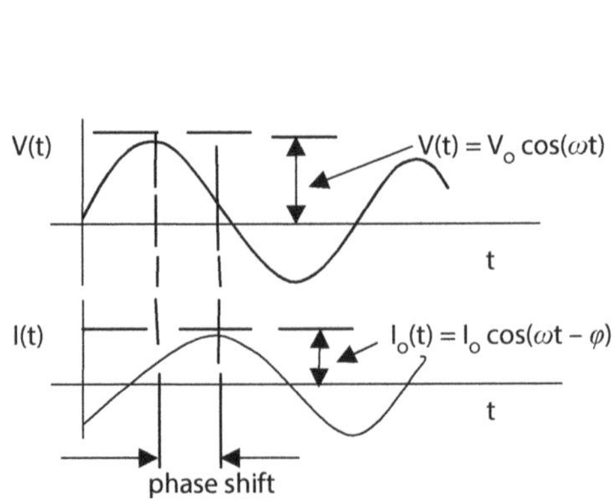

Figure 1.7 Phase shift between voltage and current for a DUT containing passive elements.

where ω (= 2πf) is the angular frequency with f is the applied frequency of the sinusoidal signal. Using equation (1.17) impedance Z(t) of the DUT displaying the phase shift [5] is given by:

$$Z(t)=\frac{V(t)}{I(t)}=\frac{V_o\cos(\omega t)}{I_o\cos(\omega t-\varphi)}=Z_o\left[\frac{\cos(\omega t)}{\cos(\omega t-\varphi)}\right]. \tag{1.19}$$

At this conjecture introducing complex function for voltage and current for the DUT it is obtained:

$$V(t) = V_o \exp(j\omega t) \tag{1.20}$$

and

$$I(t) = I_o \exp[j(\omega t - \varphi)] = I_o \exp(j\omega t - j\varphi)]. \tag{1.21}$$

Then the impedance is given by:

$$Z = Z(t)=\frac{V_o\exp(j\omega t)}{I_o\exp(j\omega t-j\varphi)}=Z_o\exp(j\varphi). \tag{1.22}$$

Introducing Euler's formula for $\exp(j\varphi) = \cos\varphi + j\sin\varphi$ impedance becomes:

$$Z(t)=Z_o\exp(j\varphi)=Z_o\cos\varphi+jZ_o\sin\varphi=Z_o'+Z_o''. \tag{1.23}$$

The magnitude of Z(t) is given by:

$$|Z|=\sqrt{Z_o'^2+Z_o''^2}. \tag{1.24}$$

The real part of the impedance is resistance (R) where voltage and current are in-phase, and for the imaginary part of the impedance is reactance (X) where voltage and current are out-of-phase. Both resistance and reactance are the constituting components of impedance, Z. Thus,

$$Z = R - jX, \tag{1.25}$$

where Z is the impedance, R is the resistance, and X is the reactance ($X = 1/\omega C$ as capacitive reactance or $X = -\omega L$ as inductive reactance) with $j=\sqrt{-1}$. The unit of each of these three parameters is ohm (Ω). In phasor analysis, reactance is used to compute amplitude and phase change of the sinusoidal alternating current passing through the circuit element.

The constituting components of admittance (Y) are conductance (G) introduced in 1885 and susceptance [$B = \omega C$ for capacitive susceptance or $B = 1/(\omega L)$ for inductive susceptance] introduced in 1887 having unit for each of these three parameters as Siemens (S). The conductance is the reciprocal of resistance whereas susceptance is the reciprocal of the reactance. Thus,

$$Y = G + jB, \tag{1.26}$$

where ωC is the capacitive susceptance and $(1/\omega L)$ is the inductive susceptance. The unit of these three parameters is Siemens. The early nomenclature of susceptance was permittance. The term inductance was introduced in the literature in 1886.

The relationship between Z and Y are reciprocal to each other given by:

$$Z = 1/Y \text{ or } Y = 1/Z. \tag{1.27}$$

Complex capacitance is derived from the equation (1.26) by defining:

$$C = Y/(j\omega). \tag{1.28}$$

Thus,

$$(G + jB)/(j\omega) = -j(G/\omega) + B/\omega = B/\omega - j(G/\omega). \tag{1.29}$$

Complex capacitance can be expressed as:

$$C = B/\omega - j(G/\omega). \tag{1.30}$$

Modulus can be derived from equation (1.25) or equation (1.30) by defining:

$$M = (j\omega)Z \text{ or } M = 1/C. \tag{1.31}$$

Thus,

$$M = (j\omega)Z = (j\omega)\,(R - jX) = j\omega R + \omega X. \tag{1.32}$$

Modulus can be expressed as:

$$M = \omega X + j\omega R. \tag{1.33}$$

Equations (1.25), (1.26), (1.30), and (1.33) are the four complex plane quantities. Each of these parameters defined so far as Z, Y, C, and M represent as-measured complex quantity. No asterisk is used in these parameters to signify complex quantity. However, often each of these complex quantities is denoted with asterisk as in Z^*, Y^*, C^*, and M^*. Using asterisk along with single prime for the real part and double prime for the imaginary part these four complex planes appear as:

$$Z^* = R - jX = Z' - jZ'', \tag{1.34}$$

$$Y^* = G + jB = Y' + jY'', \tag{1.35}$$

$$C^* = B/\omega - j\,(G/\omega) = C' - jC'', \tag{1.36}$$

and

$$M^* = \omega\, X + j\omega R = M' + jM''. \tag{1.37}$$

When equations (1.34) through (1.37) are normalized using physical geometric parameters (electrode area and thickness) of the DUT then these equations become using appropriate symbol in respective order:

$$\rho^* = \rho' - j\rho'', \tag{1.38}$$

$$\sigma^* = \sigma' + j\sigma'', \tag{1.39}$$

$$\varepsilon^* = \varepsilon' - j\varepsilon'', \tag{1.40}$$

and

$$m^* = m' + jm''. \tag{1.41}$$

where ρ^*, σ^*, ε^*, and m^* are resistivity, conductivity, permittivity, and electric modulus. The term complex is not preferred for each of these quantities unless each needs distinction for some specific reasons. The term complex capacitance is meaningful as there is no redundancy or duplication in it. Similarly the term modulus or electric modulus can be used as measured quantities and normalized quantities, respectively. Invariably impedance and admittance parameters are inherently complex quantities. Therefore, using the term complex before impedance or admittance is simply redundant or unnecessary emphasis.

Each complex plane is plotted with real part on the x-axis and imaginary part on the y-axis. The negative sign for the imaginary part of impedance and complex capacitance is usually ignored as the magnitude is plotted in the first quadrant using the positive y-axis. In that case there is no basic difference in the interpretation of the results whether plotted in the first quadrant or in the fourth quadrant. In reality this is now the regular practice of using all the four complex plane analyses.

Usually there are three preferred measured forms to obtain these four complex plane parameters. In Figure 1.8 each complex plane parameter is illustrated using the measured form: series, parallel, and phasor [6–8]. Each complex plane equation is listed there, and associated conversions are shown by the arrow direction using the multiplying factor. The following observations can be summarized for a circuit:

A. Preferred measured form:

(a) Series:

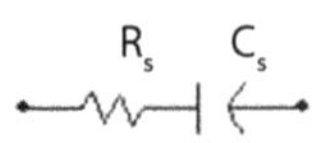

(b) Parallel:

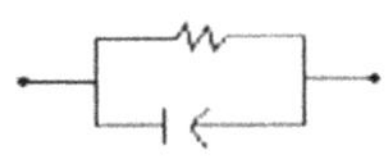

(c) Phasor: $|Z^*|$ or $|Y^*|$ and loss tangent $\tan\delta = \dfrac{G_p}{\omega C_p} = \omega R_s C_s$

B. Four complex plane equations with angular frequency $\omega = 2\pi f$, f = applied frequency, and $j = (\sqrt{-1})$:

1. Impedance: $Z^* = (1/Y^*) = [M^*/(j\,\omega)] = R_s - j\,1/(\omega\,C_s)$

2. Admittance: $Y^* = (1/Z^*) = j\,\omega\,C^* = G_p + j\,(\omega C_p)$

3. Capacitance: $C^* = (1/M^*) = [Y^*/(j\,\omega)] = C_p - j\,(G_p/\omega)$

4. Modulus: $M^* = [1/C^*] = j\,\omega\,Z^* = 1/C_s + j\,(\omega R_s)$

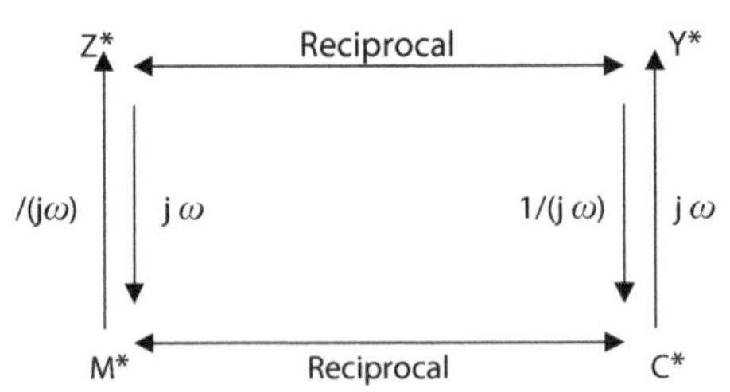

Fundamental equations in four complex planes representing the measured form of the as-measured data and its transformation related to electrical characteristics consisting of Z*, Y*, C*and M*. Vertical arrow direction implies multiplying factor for conversion between

Figure 1.8 Choice of measured forms of the ac small-signal electrical data acquisition and four complex plane formalisms along with their inter-convertible relationships.

1. To describe the response of a DUT on the alternating current (ac) three passive circuit elements (R, C, and L) are introduced;
2. An alternating current can be phase shifted with respect to the voltage depending on the elements;
3. The phase shift depends on the type of DUT used for passing the current;
4. The current and voltage through a resistor R is not phase shifted;
5. The current and voltage through a capacitor C is phase shifted;
6. The current and voltage through an inductor L is phase shifted;
7. A resistor only contributes to the real part of the impedance which is not dependent on frequency;
8. A capacitor contributes to the imaginary part of the impedance which is dependent on frequency; and
9. An inductor contributes to the imaginary part of the impedance which is dependent on frequency.

Summary of the four basic immittance functions and their conversions of the as-measured parameters without geometrical factors are listed in Figure 1.9(a). The normalized parameters of the same immittance

(a)

Parameters ↓ →	$Z^* =$	$Y^* =$	$C^* =$	$M^* =$
Z^*	Z^*	$1/Z^*$	$1/(j\omega Z^*)$	$j\omega Z^*$
Y^*	$1/Y^*$	Y^*	$Y^*/(j\omega)$	$(j\omega)/Y^*$
C^*	$1/(j\omega C^*)$	$j\omega C^*$	C^*	$1/C^*$
M^*	$M^*/(j\omega)$	$(j\omega)/M^*$	$1/M^*$	M^*

(b)

Parameters ↓ →	$\rho^* =$	$\sigma^* =$	$\varepsilon^* =$	$m^* =$
ρ^*	ρ^*	$1/\rho^*$	$1/(j\omega\rho^*)$	$j\omega\rho^*$
σ^*	$1/\sigma^*$	σ^*	$\sigma^*/(j\omega)$	$(j\omega)/\sigma^*$
ε^*	$1/(j\omega\varepsilon^*)$	$j\omega\varepsilon^*$	ε^*	$1/\varepsilon^*$
m^*	$m^*/(j\omega)$	$(j\omega)/m^*$	$1/m^*$	m^*

Figure 1.9 Summary of the four basic immittance functions along with conversions without normalization in (a), and with normalization in (b).

functions are listed in Figure 1.9(b). At this conjecture it is worthwhile noting complex conjugate of each of these complex quantities that may be used in determining the absolute value of each of these four immittance ($|Z^*|$ or $|Y^*|$ or $|C^*|$ or $|M^*|$) functions by taking the square root of the resulting quantity.

A few simple circuits along with corresponding frequently used parameters [19, 20] are listed in Figure 1.10. These listed parameters are not limited in numbers if manipulation continues for making combinations if those are convenient to use. Sometimes misnomer term is used frequently without any hesitation besides advocating the use of the as-measured modulus M^* and normalized electric modulus m^*. In this case M^* and m^* are not distinguished when conversion is used from ε^*. Precisely the quantities M^* and m^* are related to specific terms by:

$$M^* = \frac{1}{C^*} \qquad \text{and} \qquad m^* = \frac{1}{\varepsilon^*} \tag{1.42}$$

Circuits	$Z^*(\omega) = Z' - jZ''$	$\lvert Z(\omega)\rvert = \sqrt{Z'^2 + Z''^2}$	$Y^*(\omega) = Y' + jY''$	$\lvert Y(\omega)\rvert = \sqrt{Y'^2 + Y''^2}$	$\tan\delta = \cot\phi = \frac{Z'}{Z''} = D$	$\tan\phi = \cot\delta = Q = \frac{1}{D}$
R_s C_s 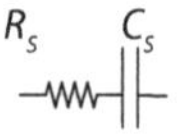 *dc* blocking	$R_s - j(\frac{1}{\omega C_s})$	$\sqrt{R_s^2 + \frac{1}{\omega^2 C_s}}$	$Y' = \frac{1}{R_s(1+Q^2)}$ $Y' = \frac{Q}{R_s(1+Q^2)}$	$\frac{1}{R_s\sqrt{1+Q^2}}$	$\omega C_s R_s$	$\frac{1}{\omega C_s R_s}$
$R_p = 1/G_p$ 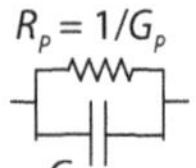 C_p non-blocking	$Z' = R_s = \frac{R_p}{1+Q^2}$ $Z'' = \frac{1}{\omega C_s} = \left(\frac{Q}{1+Q^2}\right)R_p$ $C_s = (1+D^2)\,C_p$	$\frac{R_p}{\sqrt{1+Q^2}}$	$G_p + j(\omega C_p)$ $= \frac{1+jQ}{R_p}$	$\sqrt{G_p^2 + \omega^2 C_p^2}$ $= \frac{\sqrt{1+Q^2}}{R_p}$	$\frac{1}{\omega C_p R_p}$	$\omega C_p R_p$
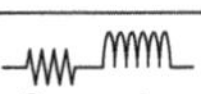 R_s L_s *dc* blocking	$R_s + j(\omega L)$	$\sqrt{R_s^2 + \omega^2 L^2}$	$Y' = \frac{1}{R_s(1+Q^2)}$ $Y'' = -\frac{Q}{R_s(1+Q^2)}$	$\frac{Q}{\sqrt{R_s^2(1+Q^2)}}$	$\frac{R_s}{\omega L_s}$	$\frac{\omega L_s}{R_s}$
$R_p = 1/G_p$ 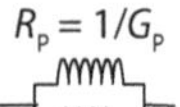 L_p Non-blocking	$Z' = R_s = \frac{R_p}{1+Q^2}$ $Z'' = \frac{1}{\omega C_s} = \left(\frac{Q}{1+Q^2}\right)R_p$	$\frac{R_p}{\sqrt{1+Q^2}}$	$G_p - j(\frac{1}{\omega L_p})$ $= \frac{1+jQ}{R_p}$	$\sqrt{G_p^2 + \frac{1}{\omega^2 L_p^2}}$ $= \frac{\sqrt{1+Q^2}}{R_p}$	$\frac{\omega L_p}{R_p}$	$\frac{R_p}{\omega L_p}$

Figure 1.10 Simple circuits with corresponding frequently used parameters.

Thus, non-normalized as-measured parameter M* is not directly reciprocal of the normalized parameter ε* as demonstrated in the standard legendary texts [19–21]. Another text [22, 23] lacks elaboration of this complex quantity but demonstrates non-scientific unit Ω-cm² for potentially misnomer Z*-plot. However, plenty of peripheral basics are available there.

1.6 Immittance Functions

The immittance function in the series configuration of the measurement process becomes:

$$I = Z = \text{impedance} = R_S - j\left(\frac{1}{\omega C_S}\right) \tag{1.42}$$

with

$$I' = R_S = \text{series resistance, and } I'' = \frac{1}{\omega C_S} = \text{reactance.} \tag{1.43}$$

The capacitive reactance in equation (1.43) may be termed as series reactance. Using the same series measurement configuration the immittance function can be re-arranged as:

$$I = M = j\,\omega\, Z = \text{modulus} = \frac{1}{C_S} + j\,(\omega\, R_S) \tag{1.44}$$

with

$$I' = \frac{1}{C_S} = \text{elastance or inverse series capacitance or real modulus,} \tag{1.45}$$

and

$$I'' = \omega\, R_S = \text{imaginary modulus.} \tag{1.46}$$

The unit of elastance is Daraf (reverse order Farad) or F^{-1}. The immittance function in the parallel configuration of the measurement process becomes:

$$I = Y = \text{admittance} = G_p - j(\omega C_p) \tag{1.47}$$

with

$$I' = G_p = \text{parallel conductance,} \tag{1.48}$$

and

$$I'' = \omega C_p = \text{parallel susceptance.} \tag{1.49}$$

Using the same parallel measurement configuration the immittance function can be re-arranged as:

$$I = C = \frac{Y}{j\omega} = \text{Complex capacitance} = C_P - j\left(\frac{G_P}{\omega}\right) \tag{1.50}$$

with

$$I' = C_P = \text{parallel capacitance or real capacitance,} \tag{1.51}$$

and

$$I'' = \frac{G_P}{\omega} = \text{imaginary capacitance.} \tag{1.52}$$

The terminal parameters R_S and C_S are the as-measured series resistance and series capacitance, respectively, and the terminal parameters G_P and C_P are the as-measured parallel conductance and parallel capacitance, respectively. Thus, these four as-measured electrical parameters are the basic constituents of the immittance function. For inductance the reactance is ωL and susceptance is $1/(\omega L)$. The reciprocal of inductance is reluctance introduced around 1888 which is sometimes denoted with the symbol ξ or ζ having unit H^{-1}.

1.7 Series Resonant Circuit

A series LC circuit has impedance that is the sum of the impedances of the inductor and the capacitor given by:

$$Z = j\omega L + \frac{1}{j\omega C} = j\left(\omega L - \frac{1}{\omega C}\right) \tag{1.53}$$

with

$$X = X_L - X_C = \omega L - \frac{1}{\omega C} \tag{1.54}$$

where X_C is the capacitive reactance, and X_L is the inductive reactance. Although X_C and X_L are both positive by convention, the capacitive reactance makes a negative contribution to the total reactance.

Three situations arise for the LC circuit as:

- $X > 0$, the reactance of the total impedance is inductive;
- $X = 0$, the impedance is purely resistive; and
- $X < 0$, the reactance of the total impedance is capacitive.

For $X = 0$, the circuit experiences a resonance for which essentially $X_L = X_C$. At low frequencies the reactance is dominated by the capacitor and so is large and negative. This monotonically increases towards zero as the magnitude of the capacitive reactance is becoming smaller. The reactance passes through zero at the point where the magnitude of the capacitive reactance and inductive reactance are equal occurring at resonant frequency (f_r), and then continues to monotonically increase as the inductive reactance becomes progressively dominant.

1.8 Parallel Resonant Circuit

A parallel L-C circuit has admittance as the sum of the admittances of the inductor and the capacitor given by:

$$Y = j\omega C + \frac{1}{j\omega L}, \tag{1.55}$$

and the impedance function from equation (1.27) is given by:

$$Z = j\left(\frac{\omega L}{1 - \omega^2 LC}\right). \tag{1.56}$$

At low frequencies the reactance is dominated by the inductor and is small and positive. This monotonically increases towards a pole at the anti-resonant frequency where the susceptance of the inductor and capacitor are equal and opposite, and cancel. Past the pole the reactance is large and negative and increasing towards zero where it is dominated by the capacitance.

1.9 Capacitance and Inductance in Alternating Current

When two running parameters are going through a cycle with respect to another parameter, falling together and rising together then these two parameters are called in-phase. For the case of out-of-phase the lead or lag of one of the parameters is a measure of the amount they are out of the sequence or step. The time it takes the current in an inductor to build up to maximum and to fall to zero is important for another reason. It illustrates a very useful characteristic of inductive circuits for which the current through the inductor always lags the voltage across the inductor. A circuit having pure resistance would have the alternating current through it and the voltage across it rising and falling together. This is invariably becoming in-phase response having current and voltage not the same amplitude.

In the case of a circuit having inductance, the opposing force would be enough to keep the current from remaining in-phase with the applied voltage. A capacitor will appear to conduct in alternating current and the same will not conduct in direct current. In a direct current circuit containing pure inductance the current took time to rise to maximum even though the full applied voltage was immediately at maximum. Any standard text describes the complete behavior [5].

Problems

1. Calculate the capacitance of a Si thin-film sandwich having thickness 100 μm and electrode area 10 μm^2.
2. In a series *R-C-L* circuit, a capacitance of 10 μF shows equal impedance with that of an inductor of 10 μH at a certain frequency having signal-voltage 1 V_{ac}. What is the reason for equal impedance of the capacitor and also inductor at that frequency? What is the impedance of the *R-C-L* series circuit in this situation?
3. In Figure 1.3, a Si wafer is placed with a 12 V battery as the power source. What will be the value of C if C_O is 1 pF?
4. A Si wafer (implies to only single-crystal) is subject to measure capacitance and resistance using one state-of-the-art high-precision *L-C-R* meter. The semiconductor is ion bombarded using dose 10^{12} $ions/cm^2$ at the surface so that by diffusion process ion migration took place to the interior of the semiconductor. The same semiconductor is annealed at

400 °*C* and cooled down to room temperature. Again for the same semiconductor, capacitance and resistance are measured. What would be the difference in the values before and after annealing? Explain the result.

References

1. H. W. Bode, *Network Analysis and Feedback Amplifier Design,* Princeton: Van Nostrand, 1959.
2. A. Beiser, *Perspectives of Modern Physics,* McGraw-Hill Book Company, New York, 1985, and International Edition, 1969.
3. A. Beiser, *Concepts of Modern Physics,* 6th Edition, McGraw-Hill, New York, 2003.
4. I. D. Raistrick, Application of Impedance Spectroscopy to Materials Science, *Annual Review of Materials Science*, 16, 343–370, 1986.
5. J. D. Irwin and R. M. Nelms, *Basic Engineering Circuit Analysis,* 10th Edition, John Wiley & Sons, Inc. New York, 2011.
6. M. A. Seitz, Study of Heterogeneous Composite Materials via Lumped Parameter/Complex Plane Analysis, *International Journal of Hybrid Microelectronics* (*I.S.H.M.*), 3, 1–7, 1980.
7. J. J. Ackmann and M. A. Seitz, 'Methods of Complex Impedance Measurements in Biologic Tissue," *CRC Critical Review of Biomedical Engineering*, 11, 281–311, 1984.
8. M. A. Alim, Electrical Characterization of Engineering Materials, *Active and Passive Electronic Components*, 19, 139–169, 1996.
9. D. L. Losee, Admittance Spectroscopy of Deep Impurity Levels: ZnTe Schottky Barriers, *Applied Physics Letters*, 21, 54–56, 1972.
10. D. L. Losee, Admittance Spectroscopy of Impurity Levels in Schottky Barriers, *Journal of Applied Physics*, 46, 2204–2214, 1975.
11. G. Vincent, D. Bois, and E Pinard, Conductance and Capacitance Studies in GaP Schottky Barriers, *Journal of Applied Physics*, 46, 5173–5178, 1975.
12. G. Vincent, A. Chantre, and D. Bois, Electric Field Effect on the Thermal Emission of Traps in Semiconductor Junctions, *Journal of Applied Physics*, 50, 5484–5487, 1979.
13. C. X. Qiu and I. Shih, Investigation of Electrodeposited $CulnSe_2$ Films by Admittance Spectroscopy, *Journal of Applied Physics*, 64, 758–760, 1988.
14. L. V. Hmurcik and R. A. Serway, Frequency Dispersion in the Admittance of the Polycrystalline CU2S/CdS Solar Cell, *Journal of Applied Physics*, 61, 756–761, 1987.
15. K. Kobayashi, M. Takata, Y. Fujimura, and S. Okamoto, Investigation of Trapping States in a Nb-Doped Rutile by Admittance Spectroscopy, *Journal of Applied Physics*, 6, 4191–4198, 1986.

16. M. D. Giulo, G. Micocci, and A. Tepore, Admittance Spectroscopy of Traps in Au-InSe Schottky Cells, *Solid State Electronics*, 27, 1015–1019, 1984.
17. P. Krispin, Impurity Detection by Admittance-Frequency Dispersion, *Physica Status Solidi (a)*, 4, K81–K84, 1977.
18. R. Resnick and D. Halliday, *Physics I & II*, John Wiley & Sons, New York, 1966.
19. V. F. Lvovich, *Impedance Spectroscopy: An Application to Electrochemical and Dielectric Phenomena*, John Wiley & Sons, Inc., Hoboken, New Jersey, 2012.
20. E. Barsoukov and J. R. MacDonald (Eds), *Impedance Spectroscopy: Theory, Experiment, and Applications*, John Wiley & Sons, Inc., Hoboken, New Jersey, 2005.
21. J. R. MacDonald (Ed.), *Impedance Spectroscopy: Emphasizing Solid materials and Systems*, John Wiley & Sons, Inc., Hoboken, New Jersey, 1987.
22. M. E. Orazem and B. Tribollet, *Electrochemical Impedance Spectroscopy*, John Wiley & Sons, Inc., Hoboken, New Jersey, 2008.
23. M. E. Orazem and B. Tribollet, *Electrochemical Impedance Spectroscopy*, 2nd Edition, John Wiley & Sons, Inc., Hoboken, New Jersey, 2017.

2

Basics of Solid State Devices and Materials

2.1 Overview of the Fundamentals of Physical Electronics

In order to fabricate solid state devices it is necessary to get familiar with the solid materials such as metals, semiconductors, and insulators. These materials are classified in terms of the concept of band gap shown in Figure 2.1, and used in achieving desired performance of the devices. The description of band gap or energy band is found in many texts [1–6]. The parameter Φ in eV is the work function of a material. It is defined as the energy required in isolating an electron from the Fermi-level (E_F) to the vacuum level (E_O). The work function is a critical parameter for forming junctions involving semiconductors.

Electrons in a completely filled band cannot move, since all states occupied per Pauli's exclusion principle. The only way to move would be to "jump" into the next higher band which needs energy to do so. Totally filled or totally empty bands do not allow current flow. This is similar to no motion of liquid in a totally filled bottle. Electrons in a partly-filled band can move since there are free states to move to.

An unoccupied electronic state in the valence band is called a hole. At temperature $T = 0$, there are no electrons in the conduction band, and the semiconductor does not conduct (due to a lack of free charge carriers). At

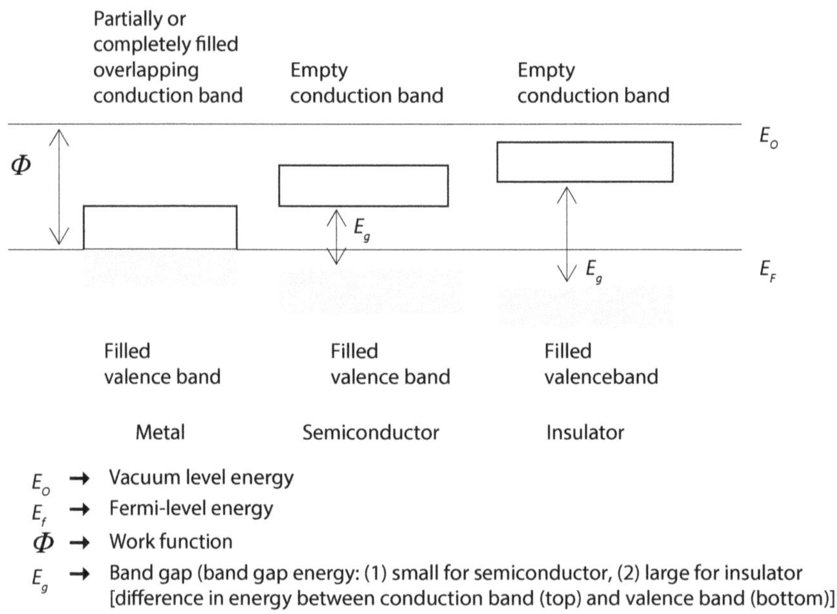

Figure 2.1 Distinguishing metal, semiconductor, and insulator using the *band gap* concept.

$T > 0$, some fractions of electrons have sufficient thermal kinetic energy to overcome the gap and jump to the conduction band whereas the fraction rises with temperature.

Metals have half-filled or partially filled conduction band, and become conductors having overlapped *band gap*. This implies no precise *band gap* but the charges move freely without storing to give finite resistance for the metals. Ohm's Law for such a conductor states that at a given temperature and pressure, the applied voltage across a current carrying conductor is proportional to the current flowing through the same conductor. This implies:

$$V \propto I \qquad \text{giving} \qquad V = R\,.\,I, \tag{2.1}$$

where I is the current flowing through the conductor and V is the applied voltage across the same conductor having R is the proportionality constant called the *resistance* of that current carrying conductor. The resistivity of the same resistor is given by:

$$R = \rho\cdot\left(\frac{l}{A}\right), \tag{2.2}$$

where l is the length and A is the cross-sectional area of the conductor, respectively. Incorporating equation (2.2) in equation (2.1) gives:

$$V = \rho \cdot \left(\frac{l}{A}\right) \cdot I. \tag{2.3}$$

Rearranging the geometrical parameters in equation (2.3), Ohm's Law becomes:

$$\frac{V}{l} = E = \rho \cdot \left(\frac{I}{A}\right) = \rho \cdot J = \frac{1}{\sigma} \cdot J, \tag{2.4}$$

where E is the electric field, J $(= \sigma E)$ is the current density, and σ is the conductivity.

The same equations apply for the semiconductors possessing finite *band gap*. For the insulators the resistivity is extremely high, and normally considered $\rho \approx 10^{17}$ *ohm-cm* or higher. From practical standpoint, the resistivity $\rho \geq 10^{13}$ *ohm-cm* may be treated as an insulator with *band gap* higher than the semiconductors. Large *band gap* does not always assure insulating behavior as found in semiconducting diamond. However, *high resistivity* and large *band gap* are usually observed for the insulators.

The *Fermi-level* plays an important role in the band theory of solids. The term Fermi-level is used to describe the top of the electron energy levels at absolute zero and shown in Figure 2.2 for an intrinsic semiconductor. This concept comes from the Fermi-Dirac statistics [1–6]. Electrons are fermions and, by the Pauli exclusion principle, cannot exist in identical energy states. At absolute zero, electrons pack into the lowest available energy states and build up a *Fermi sea* of electron energy states.

The *Fermi-level* is the surface of that sea at absolute zero where no electrons will have enough energy to rise above the surface. The concept of the Fermi energy is a crucially important concept for the understanding of

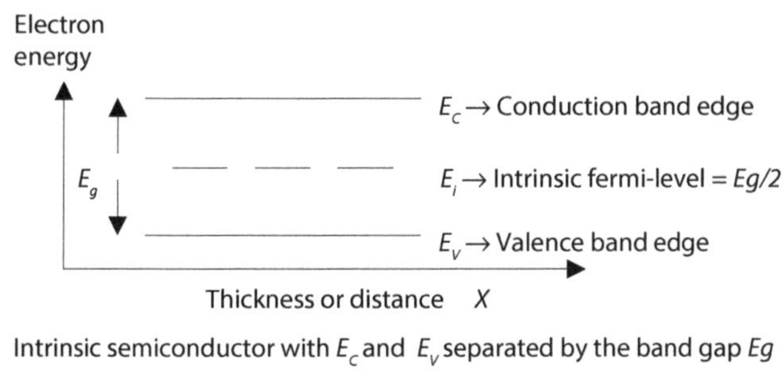

Figure 2.2 The context of Fermi-level of a semiconductor at absolute zero (0 °K) using the *band gap* concept.

the electrical and thermal properties of solids. Both thermal and electrical processes involve energies of a fraction of an electron volt (eV). But the Fermi energies of metals are on the order of electron volts. This implies that the vast majority of the electrons cannot receive energy from those processes because there are no available energy states for them to go to within a fraction of an eV of their present energy. Limited to a tiny depth of energy these interactions are limited to the ripples on the Fermi sea.

In metals, the Fermi energy gives information about the velocities of the electrons which participate in ordinary electrical conduction. The amount of energy which can be given to an electron in such conduction processes is on the order of micro-electron volts (μeV), so only those electrons very close to the Fermi energy can participate. The Fermi velocity (V_f) of these conduction electrons can be calculated from the *Fermi energy* by [1–6]:

$$V_f = \sqrt{\frac{2E_f}{m}}, \tag{2.5}$$

where m is the mass of the charge carrier electron. The Fermi energy also plays an important role in understanding the mystery of why electrons do not contribute significantly to the specific heat of solids at ordinary temperatures, while they are dominant contributors to thermal conductivity and electrical conductivity. Since only a tiny fraction of the electrons in a metal are within the thermal energy kT of the Fermi energy, they are frozen out of the heat capacity by the Pauli's exclusion principle. At very low temperatures, the electron specific heat becomes significant.

The Fermi-level is the highest energy of the occupied states for electrons in a solid at absolute zero temperature. The Fermi-level in conductors or metals lies in the conduction band (E_C); in insulators or non-conductors it lies in the valence band (E_V); and in semiconductors, whether elemental or compound, it lies in the gap between the conduction band (E_C) edge and the valence band (E_V) edge.

The Fermi-level is a measure of the energy of the least tightly held electrons within a solid. It determines the electrical and thermal properties of solids. The value of the Fermi-level at absolute zero (−273.15 °C) is called the Fermi energy and is a constant for each solid. At absolute zero there is no electron or hole available for conduction. The Fermi-level changes as the solid is warmed and as electrons are added to or withdrawn (compensated) from the solid. Thus, at elevated temperatures there will be electron-hole pair available for conduction. This results from the broken bond of the semiconductors. In this situation, concentration for both electrons and holes is referred to as intrinsic carrier concentration and denoted by n_i. For silicon, $n_i = 1.45 \times 10^{10}/cm^3$ at room temperature [1–6].

Each of the distinct energies with which an electron can be held within a solid is called an energy level. Each energy level can accommodate only a limited number of electrons. The Fermi-level is any energy level having the probability that it is exactly half-filled with electrons. Levels of lower energy than the Fermi-level tend to be entirely filled with electrons, whereas energy levels higher than the Fermi-level tend to be empty. When materials with different individual Fermi-levels are placed in intimate contact, some electrons flow from the material with the higher Fermi-level into the lower Fermi-level material. This transfer of electrons raises the lower Fermi-level material and lowers the higher Fermi-level material. When the transfer is complete, the Fermi-levels of the two materials are equal or become equal at the contact region as well as throughout the two material systems. This behavior is important in electronic devices that give junction devices.

The Fermi function $f(E)$ gives the probability that a given available electron energy state will be occupied at a given temperature. This is the probability of an electron in a particular state with energy E where T is the absolute temperature. The probability of a state at the Fermi-level has a 50% chance of being occupied by an electron which is obtained when the energy of the electron becomes equal to the Fermi energy, E_F. At higher temperatures a certain fraction, characterized by the Fermi function that comes from the Fermi-Dirac statistics $f(E)$ [1–6]:

$$f(E) = \left(\frac{1}{e^{\left[(E-E_F)/kT\right]} + 1} \right), \tag{2.6}$$

will exist above the *Fermi-level* designated by E_F with Boltzmann constant $k = 8.617 \times 10^{-5}\ eV/K$. Probability of available states of energy E being occupied is given by equation (2.6). For the doped semiconductors, p-type and n-type, the Fermi-level is shifted by the dopants or impurities. It is also referred to as the electron chemical potential in other contexts.

The basic nature of this function dictates that at ordinary temperatures, most of the levels up to the Fermi-level E_F are filled, and relatively few electrons have energies above the Fermi-level. The Fermi-level is on the order of electron volts (eVs), whereas the thermal energy kT is only about 0.0258 eV at 300 °K. If one puts those numbers into the Fermi function at ordinary temperatures, it is found that its value is essentially 1 up to the Fermi-level, and rapidly approaches zero above it. It is shown in Figure 2.3.

The illustration in Figure 2.4 shows the implications of the Fermi function for the electrical conductivity of a semiconductor. The band theory of solids gives the picture that there is a sizable gap between the Fermi-level and the conduction band of the semiconductor. At higher temperatures, a larger fraction of the electrons can bridge this gap and participate in electrical conduction. This is illustrated in Figure 2.5.

Although the Fermi function has a finite value in the band gap (between E_C and E_V), there is no electron population at those energies in the gap. The population depends upon the product of the Fermi function and the electron density of states. So in the gap there are no electrons because the density of states is zero. In the conduction band at °K, there are no electrons even though there are plenty of available states, but the *Fermi function* is zero. At high temperatures, both the density of states and the

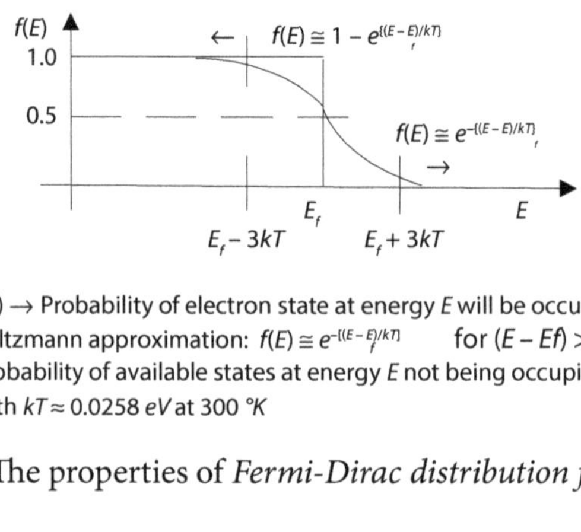

$f(E) \rightarrow$ Probability of electron state at energy E will be occupied
Boltzmann approximation: $f(E) \cong e^{-[(E-E_f)/kT]}$ for $(E - Ef) > 3kT$
Probability of available states at energy E not being occupied: $1 - f(E) = 1/[1 + e^{-[(E-E_f)/kT]}]$
with $kT \approx 0.0258\ eV$ at 300 °K

Figure 2.3 The properties of *Fermi-Dirac distribution function.*

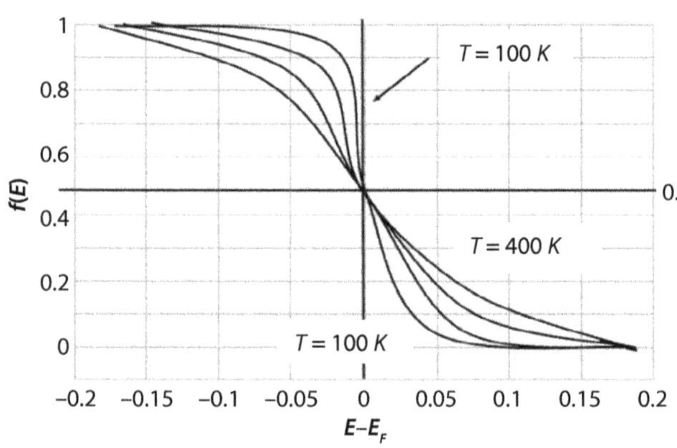

Figure 2.4 The Fermi-Dirac distribution functions at different temperatures.

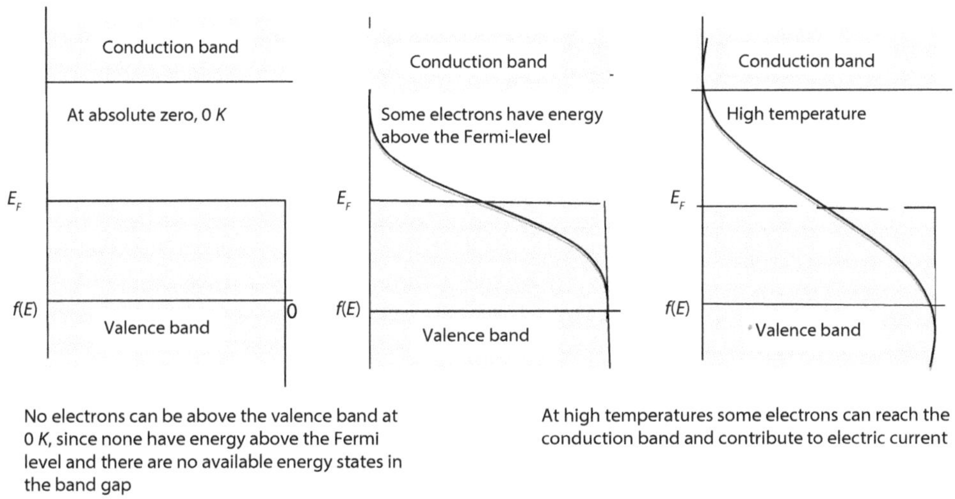

Figure 2.5 Electrons at the *conduction band* as temperature increases.

Fermi function have finite values in the conduction band, so there is a finite conducting population.

The Fermi function gives the probability of occupying an available energy state, but this must be factored by the number of available energy states to determine how many electrons would reach the conduction band. This density of states is the electron density of states, but there are differences in its implications for conductors and semiconductors. For the conductor, the density of states can be considered to start at the bottom of the valence band and fill up to the Fermi-level, but since the conduction band and valence band overlap, the Fermi-level is in the conduction band so there are plenty of electrons available for conduction. In the case of the semiconductor, the density of states is of the same form, but the density of states for conduction electrons begins at the top of the gap.

2.2 Basics of Semiconductors

The basic band diagram of an undoped or intrinsic semiconductor is shown in Figure 2.2. The band diagram of an *n*-type and a *p*-type semiconductor is shown in Figure 2.6. For enhanced doping concentration the Fermi-level pushes toward E_C for the *n*-type semiconductor whereas the Fermi-level pushes toward E_V for the *p*-type semiconductor. If the doping density or carrier concentration or doping concentration keeps increasing then the Fermi-level will ultimately push to the band edge, that is either to E_C or to E_V depending on the *n*-type or the *p*-type semiconductor,

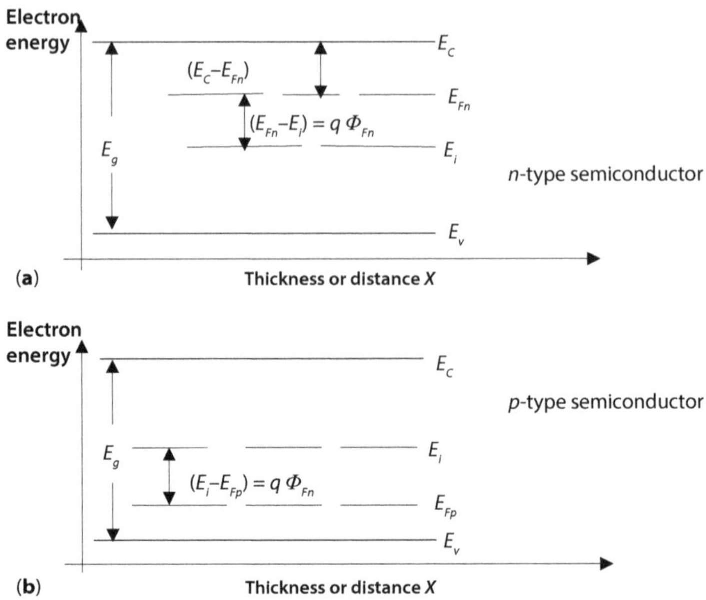

Figure 2.6 Energy band diagram for: (*a*) *n*-type semiconductor and (*b*) *p*-type semiconductor.

respectively. In that case the semiconductor becomes degenerated. Thus, regular doping gives non-degenerated semiconductor where Fermi-level resides within the band gap.

The doping concentration for the *n*-type and *p*-type semiconductors can be obtained by:

$$n = N_C e^{-\left[\left(E_C - E_f\right)/kT\right]} \text{ and } p = N_v e^{-\left[\left(E_f - E_v\right)/kT\right]}. \tag{2.7}$$

Both N_C (= 2.8 × $10^{19}/cm^3$ for Si) and N_V (= 1.04 × $10^{19}/cm^3$ for Si) are constants. The Fermi-level position for the *n*-type semiconductor with doping density N_d can be determined by

$$E_C - E_f = kT\ln\left(\frac{N_C}{N_d}\right), \tag{2.8}$$

and the Fermi-level position for the *p*-type semiconductor with doping density N_a can be determined by

$$E_f - E_v = kT\ln\left(\frac{N_v}{N_a}\right). \tag{2.9}$$

In the elemental crystalline semiconductors, the charge carriers are the *electrons* and *holes* for which the mobility is denoted as μ_n and μ_p, respectively. The conductivity of an intrinsic semiconductor containing both electrons and holes is given by

$$\sigma = nq\mu_n + pq\mu_p. \tag{2.10}$$

For the n-type semiconductor equation (2.10) can be approximated as

$$\sigma \approx nq\mu_n \tag{2.11}$$

with $n \approx N_d$ and $p = (n_i^2/N_d)$. For the p-type semiconductor equation (2.10) can be approximated as

$$\sigma \approx pq\mu_p \tag{2.12}$$

with $p \approx N_a$ and $n = (n_i^2/N_a)$.

2.3 Single-Crystal and Polycrystal Materials

Solid materials featuring crystalline solids refer to the periodic spatial arrangement of atoms. This systematic repeated arrangement of atoms forms a long-range order in three-dimensional pattern. The single-crystal materials possess periodic long-range order throughout the entire material system. This periodic arrangement of atoms obtained by the three-dimensional repetition of certain pattern ascribes a unit cell. When the periodicity of the pattern extends throughout a certain material system, then it is described as the single crystal. Often single crystal is described as monocrystal. A single crystal is formed by the growth of a crystal nucleus without another nucleation called secondary nucleation or impingement on other crystals. During the crystal growth process from solution, such as gas or liquid or solid, single crystal takes the shape of the container often having a well-defined form reflecting symmetry of the unit cell.

For a perfect crystal implying ideal situation provides single crystals free from internal demarcated regions or boundaries. The demarcated regions or boundaries are formed because of the discontinuities in the arrangement of the array of periodic atoms. That is why characteristic x-ray diffraction (XRD) pattern is obtained for a single crystal [7, 8]. Some types of single crystals exhibit anisotropy. This means that there is a variation of some of their physical properties according to the direction along which they are measured. As an example, the electrical resistivity or conductivity is the same in all directions. This anisotropy exists both for structure-sensitive properties and structure-insensitive properties. The structure-sensitive properties are strongly affected by the crystal imperfections, such as crystal cleavage during crystal growth. Other structure-sensitive properties of crystals, such as strength and diffusion coefficients seem to be governed by internal defects at the atomic scale. The structure-insensitive properties are not affected by the imperfections, such as in mechanical quantities depicting elastic coefficients.

The polycrystalline material possesses cluster of single crystals. Thus, polycrystalline material provides a long-range order only within the limited volume of the single crystal called grain. Also the separation or demarcation between the successive grains is called grain boundary. The grain boundaries are the interfaces where single crystals of different orientations meet. The orientation of each grain can be very different from each other. The grains are randomly shaped and connected to form the solid material containing grain boundaries. The size of the grains in three directions varies depending on the material system and method of its formation. However, there is no preferential extension of the single crystal within a grain in any direction. All single-crystal features are valid only within each individual grain [9].

If the crystal structure of all the grains is the same containing the same material or element(s) then this crystal structure refers to as single phase. When other crystal structure(s) exist within the same bulk material system then that phase refers to second phase. Likewise multiple phases can be present in a bulk material system depending on the type of material system prepared. In this situation the resulting arrangement of the grains, grain-boundaries, and other phases (when present) constitutes an integrated series-parallel network of conducting paths involving "m" junctions in parallel and "n" junctions in series between the electrode terminals [9–11]. The lumped behavior of this network indicates dominance of the electrical paths [9, 10] for a DUT. Each of these DUTs is subject to engineering applications depending on performance characteristics and appropriate needs.

In general, a solid material comprising of random orientation of crystalline regions is called crystallites, distinguished from a single crystal. These crystallites are very short-range crystals found in many glassy or amorphous systems. Sometimes the polycrystalline materials result from rapid solidification. Thus, crystallization commences at many sites, and the structurally ordered regions growing from each site intersect each other. The random arrangement of the boundaries between individual crystallites in a polycrystal causes them to scatter a beam of light instead of reflecting or refracting it uniformly, so that colorless polycrystals are opaque. Other mechanical or electrical or magnetic properties of single crystals are similarly altered by the absence of long-range order in polycrystals.

2.4 SCSJ and MPCHPH Systems

Engineering material systems adopt both single-crystal and polycrystal technologies for the applications of smart components and novel devices. The SCSJ system can be illustrated by either uniform or homogeneous configuration. Thus, to optimize performance of a DUT based on these technologies a thorough understanding is necessary on the structure-property-processing relationships of the associated material system. The single crystals are often treated in the clean-room for processing or fabrication to obtain necessary devices so that no contamination takes place for the junction devices. Modern integrated circuits are based on these junction devices using silicon (Si) or gallium arsenide (GaAs) single crystals. In many cases single junctions are made sequentially using state-of-the-art instruments. The resulting devices based on single crystals possess single junctions essentially involving single-crystal/single-junction (SCSJ) technology [9].

Using polycrystals for making devices often does not require clean-room environment to process or fabricate devices unless there is a desire for it to avoid so-called additional or extraneous unwanted contaminants. Polycrystals usually begin with the contamination having less high purity than silicon when starting with powdered materials. The resulting devices are invariably based on multi-junctions and may contain multi-phases depending on the type of complexity of the material systems besides adding other materials to it. In general, these devices are treated at high temperatures to form finite-sized polycrystals that may require liquid-phase sintering. The resulting devices possess random multiple junctions involving multi-phase/multi-component heterogeneous polycrystalline hybrid (MPCHPH) technology [9].

From the aforementioned description it is clear that both SCSJ and MPCHPH systems are in no way identical or similar when each presented as a single piece bulk material. Therefore, the factors determining performance characteristics of the DUT based on polycrystalline hybrid systems are not identical to the devices based on single junction of the single-crystal systems. Performing SCSJ-like data analysis for the MPCHPH system can lead to serious confusion in delineating simultaneously operative underlying phenomena [9] when "physical geometrical factors" are used in normalizing the as-measured electrical parameters or electrical quantities. These electrical quantities for the SCSJ systems can easily be normalized using physical geometrical parameters such as length or thickness and electrode area.

Treating MPCHPH system as the SCSJ system for the analytical approach can vitiate interpretation when polycrystalline microstructural inhomogeneity plays a key role in determining the operative electrical path(s) between the electrodes or terminals for the DUT. Therefore, normalization using physical geometrical parameters for the MPCHPH system leads to a misleading interpretation of the as-measured electrical data. The advantage of using as-measured electrical parameters or electrical quantities constituting the immittance function for the polycrystalline materials is explained herein. The state of normalization using physical geometrical factors can only be performed for a specific phenomenon when completely isolated from the total electrical response [9, 10]. This means that the lumped response must be delineated and unraveled in terms of underlying operative mechanisms. Then one parameter at a time should be identified or isolated corresponding to the associated mechanism which is subject to the state of normalization.

An MPCHPH system cannot be represented by either a uniform or homogeneous configuration. Under certain limiting conditions the MPCHPH can behave like the SCSJ system. But this does not mean that the MPCHPH system always can be used as the SCSJ system. A thorough understanding on these limiting conditions is necessary to identify the characteristics of the SCSJ-like response of the MPCHPH system. This response must be evaluated using the total contributions in the *frequency domain* under the explicit non-equilibrium experimental conditions. Such an evaluation may give overlapped mechanisms due to non-distinguishable successive semicircular relaxations and severe distortion in the relaxation processes in the complex plane representation of the immittance data [12–16].

The current flow in a SCSJ system that is very much involved with the straightforward electrical path(s) possibly in parallel manner across the opposite electrodes can be referred to as the shortest path(s). In an MPCHPH system the current flows through the easiest or shortest paths consisting of a network comprising of a series-parallel configuration. This means that the MPCHPH system constitutes "*m*" junctions in parallel and

"n" junctions in series. Such a configuration implies grains and other relevant phases that essentially produce lumped series effect with the grain boundaries as the interconnecting junctions that are parallel or nearly parallel or horizontal with respect to the physical electrode surface. This configuration is not necessarily to be the shortest electrical path(s) between the opposite electrodes. The series equivalent conducting path assumed to be perpendicular or nearly perpendicular or vertical with respect to the *physical electrode surface* must be operative between the two-terminal electrodes. For all practical purposes, possible interference from the cross-linked series-parallel *network* within the microstructure may be assumed to be negligible, and can have either no or very weak effect on the parallel conducting paths simultaneously operative between the two-terminal electrodes [8]. Only thorough understanding or complete comprehension of the operative electrical path(s) can delineate the MPCHPH system as the SCSJ system. Researchers must be careful in isolating each operative mechanism within the MPCHPH system before treating them as the SCSJ system. A ZnO-based varistor is a multi-component polycrystalline device. Multiple mechanisms simultaneously operative in this device are delineated using the frequency domain measurement [9–17].

The ultimate physical regions in the MPCHPH system contains a variety of domains such as grains or phases, grain boundaries, variation in the orientation of each grain and associated corresponding grain boundary, physical dimension (volume and/or grain size) of each grain, surface area of each grain, continuity of *lacy network* containing the grain boundaries, charge distribution at the surface of each grain implying interface states at the grain boundaries, types of charges (electron, hole, +ion, -ion) accumulation at each grain's surface, variation in grain donor or acceptor density, variation in the depletion region or the space charge layer, etc. The *lacy network* implies to the lace-like configuration for the *imaginary physical structure* without the grains [18–26]. This is similar to the consideration of the *lattice point* where an *atom* is supposed to reside in a continuous three-dimensional network making single crystal.

Figure 2.7 illustrates the total lumped representation of the MPCHPH system incorporating simultaneously operative competing phenomena. The lumped series effect is illustrated in Figure 2.7(a) where variation in grain-orientation provided three different series equivalent parallel or near parallel paths: (1) "A" comprises grain-to-grain contacts; (2) "B" comprises grain to interface to grain contacts; (3) "C" comprises the network region without the grains. The last path need not be the shortest between the two terminal electrodes reflecting a lacy network without the

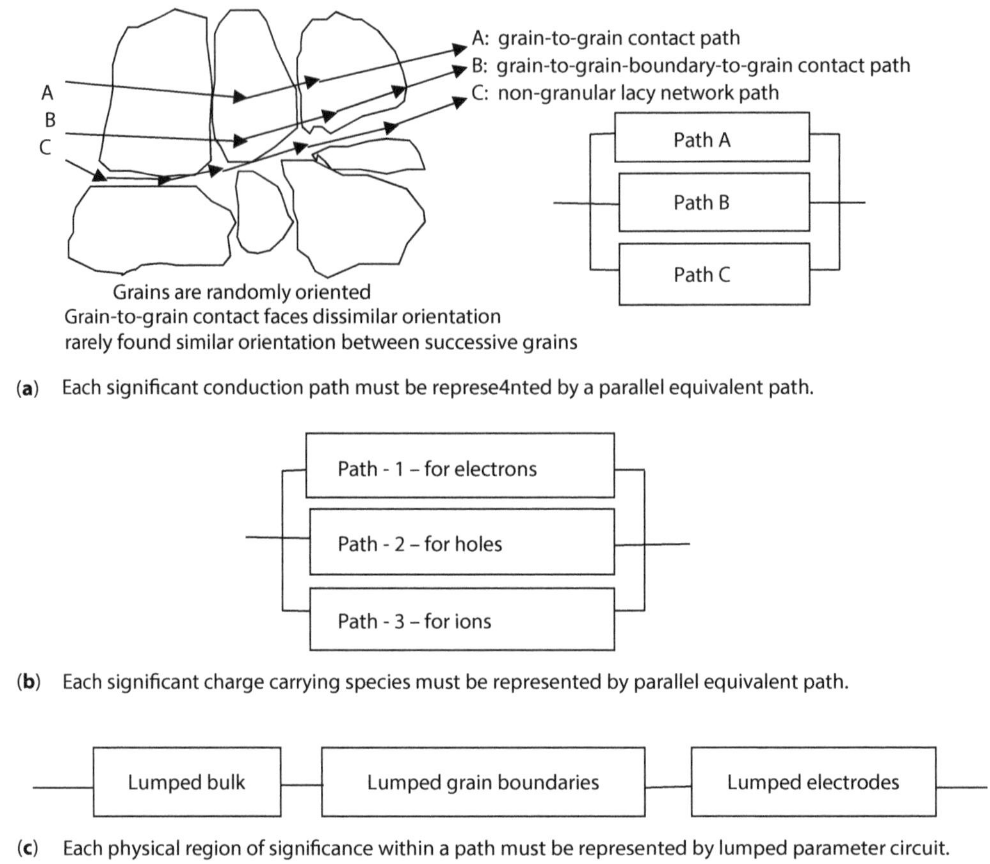

Figure 2.7 Schematic illustration of the lumped representation incorporating simultaneously operative competing phenomena in the electrically active MPCHPH systems.

grains. Each of these path configurations provide series equivalent conducting path presumed to be perpendicular or nearly perpendicular or vertical to the electrode surface which is operative between the two terminal electrodes. For the SCSJ system there will be only one electrical path associated with the bulk single crystal.

In order to assure a practical situation, possible interference from the cross-linked series-parallel network within the microstructures of the MPCHPH system may be considered as negligible. Overall, this can have either no or very weak effect on the parallel conducting paths simultaneously operative between the two terminal electrodes. Each contributing charge carrier within the lumped grains constitutes parallel combination to form the electrical path as shown in Figure 2.7(b). For the SCSJ system each contributing charge carrier within the lumped bulk single *crystal* constitutes parallel combination to form the electrical path. Additional contribution from the electrode material and its possible interaction with the MPCHPH system must be in series for the total lumped behavior of the grains, grain-boundaries, and other constituting phases in the microstructure. The physical location of each of these microstructural entities between the two terminals contributes to the net series effect as shown in Figure 2.7(c). For the SCSJ system lumped bulk contribution will appear in series with the effect of the electrodes at the surface.

A near idealized MPCHPH system is shown under electric field in Figure 2.8. It is illustrated with the type of possible operative mechanisms, type of charge carriers, and the type of the physical regions for the MPCHPH system. Based on the type of operative mechanisms in conjunction with Figure 2.7 a meaningful equivalent circuit can be developed for the MPCHPH system based DUT. Therefore, the investigator needs to be aware of the microstructural information for the material type as well as the tentative operative mechanisms.

Another perspective on the porosity of the MPCHPH system becomes a feature for the immittance data analyses. Often finite degree of porosity is desired for certain devices to perform gas sensitive behavior. Certain factors are the issues in determining the resulting performance. These are physical geometrical parameters, grain-to-grain contact area, total current-voltage response describing voltage per junction, response of the *dc* voltage superimposed with the *ac* small-signal voltage implying capacitance versus voltage or Mott-Schottky [16, 17] behavior, measurement frequency range of the immittance data, and the status of other non-equilibrium experimental variables like pressure.

As the degree of porosity vanishes, that is achieving nearly theoretical density of the material system the frequency domain response may differ greatly from that obtained in the porosity state of the DUT. This is expected because the physical contact area for the grain-to-grain junctions involving grain boundaries start to vary significantly. Evidently, physical processes causing net contributions to the underlying operative conducting paths can be influenced by the variation in the type of planer/non-planer contact area across each grain-boundary. A DUT of near theoretical density (99% or better) is less critical and, thus, its influence on the electrical measurements can be ignored. For a small change in the porosity, the frequency domain response may be reproduced as the operative mechanisms remain the same for either dense or porous system. The discussion on the solid state system does not exclude electrochemical-type DUT as the analyses of the immittance data are not different from those of the SCSJ and MPCHPH systems [8].

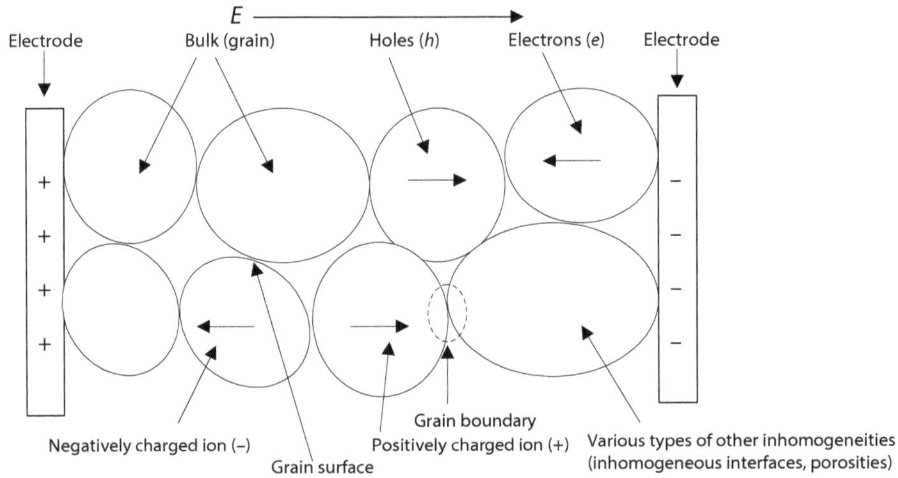

(**a**) type of charge carriers
1. electrons, *e*
2. ions, *h*
3. positively chargedions
4. negatively charged ions

(**b**) type of mechanisms
1. electron or hole conduction
2. ionic conduction
3. electronic polarization
4. lattice polarization
5. dipole polarization
6. barrier layer polarization
7. space charge polarization
8. Warburg or Faradaic polarization

(**c**) type of physical regions
1. electrodes
2. bulk (grains)
3. grain boundaries
4. grain surfaces
5. other inhomogeneous interfaces and/or porosities

Figure 2.8 Representation of the multi-phase/multi-component heterogeneous polycrystalline hybrid (MPCHPH) system showing physical regions and operative mechanisms.

2.5 Representation of the Competing Phenomena

Usually an equivalent circuit model of a DUT represents lumped behavior of the material system. This is a general practice for the SCSJ-type DUTs, such as *p-n* junctions or diodes, metal-semiconductor diodes or Schottky diodes, *p-n-p* and/or *n-p-n* transistors, MOS/MIS and MOSFET/MISFET devices, etc. [8]. A general illustration given in Figure 2.7 represents simultaneously operative competing phenomena occurring within a DUT. These phenomena can be used in developing an equivalent circuit model by following a set of rules [9, 10, 18, 19] summarized below:

1. each significant conducting path must be represented by a parallel equivalent circuit as shown in Figure 2.7(*a*);
2. each significant charge carrying species such as electrons or holes or ions must be represented by a parallel equivalent circuit path as shown in Figure 2.7(*b*);

3. each region of significance within an electrical path can be represented by a lumped parameter equivalent circuit with respect to the type of charge carrying species as shown in Figure 2.7(*c*);
4. each mechanism of polarization and associated phenomena must be represented by either series or parallel combination of resistors, capacitors, and inductors;
5. each of the components representing various conducting species and polarization processes has its own composition, microstructure, temperature, voltage, pressure, environment, and frequency dependence;
6. electrical paths of significance those yielding the least impediment to current flow;
7. regions of significance within the electrical paths are those presenting the largest impediment to current flow; and
8. analytical techniques such as exploitation of the complex plane formalisms in conjunction with the Bode plane analysis and spectroscopic plots can be used to elucidate an appropriate equivalent circuit from the measured immittance data when used concurrently with the information obtained from other methods.

With the application of the aforementioned rules, difficulties may arise for a complicated DUT system in which the electrical paths exhibit multiple current or voltage dividing lumped segments. Each of these lumped segments may not necessarily represent SCSJ-like behavior. As an example, a bulk phenomenon may suit SCSJ-like representation only when it is isolated from all other extraneous contributions or frequency-dependent components. Only then the state of normalization can be applied for the eventual extracted parameter. Should other simultaneously operative phenomena such as trapping within grain-boundaries and depletion regions, resonating behavior, charge accumulation effect at the DUT-electrode interfaces, electrode contribution, polarization from work-function mismatch, etc. demand SCSJ-like representations, then an adequate evaluation process of the total characteristics is necessary.

2.6 Effect of Normalization of the Electrical Parameters

A thorough understanding on the pattern of the dispersion of the immittance data with measurement frequency f is necessary for the

MPCHPH system before normalizing the as-measured electrical parameters or electrical quantities. This is conducive to resolve total electrical behavior of a DUT. Meaningful interpretation of the extracted parameters can be obtained following the identification of each contributing phenomenon of the DUT. Also careful examination warrants adequacy of the state of normalization using physical geometrical factors for the particular parameter(s). Usually physical geometrical factors imply to the electrode area and the distance between the two opposite electrodes.

The state of normalization using physical geometrical parameters can only be applied to those parameters that are truly associated with the physical processes within a DUT. In other words, the generalization of the state of normalization of a parameter by incorporating physical geometrical factors before performing analysis is inappropriate due to at least for the following reasons [9, 16, 17]:

1. total operative phenomena between the two electrode terminals unknown or not determined;
2. multiple relaxation processes often may not be distinct and existing possible distortion showing overlapped relaxation behavior;
3. multiple operative phenomena not distinguishable from a single lumped relaxation process – possibly requires further analytical tools/techniques to delineate;
4. limitation of the analytical tools/techniques and uncertainty of the window of visibility;
5. contribution of each phase in the microstructure to the total relaxation processes unknown;
6. contact area for each contributing electrical path between the successive grains unknown;
7. uncertain statistical variation in the contact area across a junction;
8. cross-section of effective contribution within each contact area unknown that is entire contact area not necessarily contribute to the total conduction due to planer or non-planer situation – not all of the contact areas or entire contact area involved in many DUTs;
9. distribution of spatial or localized traps unknown; and
10. distribution of electrical barrier voltages across a junction unknown.

These reasons may simultaneously exist or partly exist in the DUTs exhibiting multiple relaxations in any of the single complex planes (or more simultaneous complex planes) or a single relaxation in more than one complex plane. For multiple relaxations in a single complex plane are the [9]:

1. admixtures of Debye responses that influence each unknown operative electrical path;
2. admixtures of Debye and non-Debye responses that influence each unknown operative electrical path;
3. admixtures of multiple non-Debye responses that influence each unknown operative electrical path; and
4. non-Debye responses for each relaxation originating from the ordinate parameter where non-Debye behavior attributed to the depression of the ordinate parameter influencing unknown operative electrical path.

For a single relaxation obtained in any one complex plane [9] the:

1. range of frequency spectrum (measurement range as a function of frequency) is likely to be limited;
2. resulting Debye or non-Debye relaxation is likely to be lumped (possible for a Debye relaxation but common for a non-Debye relaxation) so that distinction may not be possible (described below in 3); and
3. ordinate parameter critical for the non-Debye response for both single and multiple relaxations.

Essentially a single-frequency measurement may be designated as the spot-frequency measurement. For the SCSJ system single-frequency immittance measurement may be meaningful as the frequency-dependent responses are eliminated at the ~1 MHz range. In this case the state of normalization using physical geometrical parameters may be useful. Conducting immittance measurement corresponding to an arbitrarily selected single-frequency corresponding to 10^6 Hz for the MPCHPH system invariably gives confusing information and thereby a subsequent erroneous interpretation. In this case incorporating physical geometrical factors may lead to either incorrect or physically meaningless conclusions. It is too much to speculate for the MPCHPH system about the mechanisms for the normalized real and imaginary immittance quantities corresponding to the single-frequency measurement. Formal assessment based on the single-frequency immittance data can be validated only when the entire frequency domain response is thoroughly examined and understood via a systematic evaluation process. This frequency domain response assures identification of the total dispersion characteristic as a function of measurement frequency. This dispersion can be significantly influenced by the presence of several types of distinguishable or non-distinguishable traps. These traps are characterized [9, 11, 16–26] by the energy depth, capture cross-sectional area, defect charge state such as donor-like or acceptor-like or neutral, capture-rate constituting elements such as resistance R and capacitance C of the relaxation time ($\tau = RC$), etc. When the frequency domain response of the simultaneously operative phenomena is not explored, the underlying physical processes of an unknown DUT remain unresolved.

Problems

1. Sketch the band diagram after finding the Fermi-level position in the Si wafer having donor doping density $10^{17}/cm^3$. What is the work-function of this Si wafer? If this Si wafer is used in fabricating a Schottky diode with Pt or W then what would be the forward breakdown voltage of each of the Schottky diodes? Explain why the forward breakdown voltage is equal or unequal for each diode.
2. Is a grain a single crystal? What is a polycrystal? If a Si single crystal and a Si polycrystal of the same geometry are used in measuring capacitance then what is likely to be observed? Are the two capacitances the same or different? If the two values are the same then explain why. If the two values are not the same then justify the observation.
3. A Si wafer of 0.5 cm thick formed a junction with a metal of 0.5 cm thick. This device is found to behave like an ohmic junction, not a Schottky diode. What could be the reason for such behavior? Could this junction be called as a device? Explain why or why not.
4. Determine the doping density for the n-type Si having $(E_C - E_{Fn}) = 0.15\ eV$. Determine the doping density for the n-type Si having $(E_C - E_{Fn}) = 0.01\ eV$. Compare the two values, and explain which one may exhibit larger capacitance. If there is a difference, then explain.

5. Calculate the conductivity for the n-type Si having $(E_C - E_{Fn}) = 0.1\ eV$. Take $\mu_n = 1280\ cm^2/V\text{-}sec$ and $\mu_p = 480\ cm^2/V\text{-}sec$.
6. A Schottky diode of platinum contact with silicon gives a forward breakdown voltage 0.4 V. Will the breakdown voltage increase or decrease if tungsten is used in lieu of platinum? Tungsten has a higher work function than platinum.

References

1. S. M. Sze and K. K. Ng, *Physics of Semiconductor Devices*, 3rd edition, Wiley Interscience, New Jersey, 2007.
2. S. M. Sze, *Semiconductor Devices: Physics, and Technology*, John Wiley & Sons, New York, 1985.
3. C. Kittel, *Introduction to Solid State Physics*, 4th Edition, Wiley Eastern Private Limited, New Delhi, 1974.
4. E. S. Yang, *Fundamentals of Semiconductor Devices*, McGraw-Hill Book Co., New York, 1978.
5. R. S. Muller and T. I. Kamins, *Device Electronics for Integrated Circuits*, John Wiley & Sons, Inc., New York, 1977.
6. D. A. Neamen, *An Introduction to Semiconductor Devices*, McGraw-Hill Book Co., Boston, 2006.
7. L. V. Azaroff, *Elements of X-Ray Crystallography*, McGraw-Hill Book Co., New York, 1968.
8. B. D. Cullity, *Elements of X-Ray Diffraction*, 2nd edition, Addison-Wesley Publication Company Inc., Reading, MA, 1978.
9. M. A. Alim; Electrical Characterization of Engineering Materials, *Active and Passive Electronic Components*, 19, 139–169, 1996.
10. M. A. Alim, S. Khanam, and M. A. Seitz, Immittance Spectroscopy of Smart Components and Novel Devices, *Active and Passive Electronic Components*, 16, 153–170, 1994.
11. M. A. Alim, Admittance-Frequency Response in Zinc Oxide Varistor Ceramics, *Journal of the American Ceramic Society*, 72, 28–32, 1989.
12. S. Trolier-McKinstry and R. E. Newnham, Sensors, Actuators, and Smart Materials, *MRS Bulletin*, 18, 27–33, 1993.
13. R. E. Newnham and G. R. Ruschau, Smart Electroceramics, *Journal of the American Ceramic Society*, 74, 463–480, 1991.
14. R. E. Newnham, Composite Electroceramics" *Annual Review in Materials Science*, 16, 47–68, 1986.
15. H.J. de Bruin and S. P. S. Badwal, Faradaic Impedance Dispersion of the Noble-Metal/Ceramic Electrolyte Interface, *Journal of the Australian Ceramic Society*, 14, 20–28, 1978.

16. M. A. Alim, M. A. Seitz, and R. W. Hirthe, Complex Plane Analysis of Trapping Phenomena in Zinc Oxide Based Varistor Grain Boundaries, *Journal of Applied Physics*, 63, 2337–2345, 1988.
17. M. A. Alim, An Analysis of the Mott-Schottky Behavior in ZnO-Bi_2O_3 Based Varistors, *Journal of Applied Physics*, 78, 4776–4779, 1995.
18. M. A. Seitz, Study of Heterogeneous Composite Materials via Lumped Parameter/Complex Plane Analysis, *International Journal of Hybrid Microelectronics* (*I.S.H.M.*), 3, 1–7, 1980.
19. J. J. Ackman and M. A. Seitz, Methods of Complex Impedance Measurements in Biologic Tissue, CRC Critical Review in Biomedical Engineering, 11, 281–311, 1984.
20. M. A. Alim and M. A. Seitz, Singular Nature of Preferential Conducting Paths at High Electric Fields in ZnO-Based Varistors, *Journal of the American Ceramic Society*, 71, C246–C249, 1988.
21. L. C. Sletson, M. E. Potter and M. A. Alim, Influence of Sintering Temperature on Intrinsic Trapping in Zinc Oxide-Based Varistors, *Journal of the American Ceramic Society*, 71, 909–913, 1988.
22. M. A. Alim, S. Li, F. Liu, and P. Cheng, Electrical Barriers in the ZnO Varistor Grain Boundaries, *Physica Status Solidi* (*A*), 203, 410–427, 2006.
23. M. A. Alim, Influence of Intrinsic Trapping on the Performance Characteristics of ZnO-Bi_2O_3 Based Varistors, *Active and Passive Electronic Components*, 17, 99–118, 1994.
24. M. A. Alim, Influence of Multiple Trapping Phenomena on the Applications Criteria of ZnO-Bi_2O_3 Based Varistors, *Active and Passive Electronic Components*, 17, 57–65, 1994.
25. M. A. Alim, High-Frequency Terminal Resonance in ZnO-Bi_2O_3-Based Varistors, *Journal of Applied Physics*, 74, 5850–5853, 1993.
26. M. A. Alim, M. A. Seitz and R. W. Hirthe, High-Temperature/Field Alternating Current Behavior of ZnO-Based Varistors, *Journal of the American Ceramic Society*, 71, C52–C55, 1988.

3

Dielectric Representation and Operative Mechanisms

3.1 Dielectric Constant of Materials: Single Crystals and Polycrystals

The general field of dielectrics did not advance much in the past century except in the exploration of large scale semiconducting materials. Variety electroceramic or ceramic materials, composites, piezoelectrics, ferroelectrics, etc. advanced with the latest state-of-the-art processing technologies for controlled electrical and electronic applications. Composites based on organic-organic, organic-inorganic, and inorganic-inorganic materials are not exceptions. Microcircuits or integrated circuits advanced rapidly satisfying technological demand. Economic considerations gradually diverted advancement toward the polycrystal technology though single-crystal technology is retaining most of the prime applications. All these material systems require thorough investigation for achieving improved devices, utilizing knowledge in dielectrics.

Relative permittivity is presented as permittivity or dielectric constant, ε_r, for a material system. The same parameter is often termed as static relative permittivity. Too many names exist for a single parameter. The dielectric constant determines characteristics of the material system as the ratio of the permittivity of the medium and the permittivity of free space [1–17]. For a parallel-plate capacitor [1] containing a dielectric of permittivity (or relative permittivity or dielectric constant) ε_r, neglecting the edge effect or electrode effect, the capacitance C can be expressed as:

$$C = \frac{\varepsilon_o \varepsilon_r A}{d}, \tag{3.1}$$

where ε_o is the permittivity of free space with a value 8.854×10^{-12} F/m, A is the electrode area (of each electrode and expected to be equal for homogeneous measurement), and d is the thickness or distance between the two opposite electrodes. The exponent changes to 10^{-14} when the unit is expressed in F/cm. Equation (3.1) satisfies single-crystal systems for which dc and ac measurements for the terminal capacitance may coincide to obtain dielectric constant ε_r. Invariably electrode-to-surface contact for the dielectric material is ohmic or linear so that there is no edge effect. Thus, neglecting the edge effect implies ohmic contact. The product representing the quantity $\varepsilon_o\varepsilon_r$ is often termed as permittivity for which the unit depends on the exponent value.

The type of capacitance expressed in equation (3.1) follows Ohm's law under dc voltages whereas under ac voltages it will impede the flow of current by an amount that varies with applied frequency f. This equation is valid for a single crystal because it represents a continuous single phase where the electric field distribution is homogeneous across the electrodes. The electrode effect is neglected when the contact is considered as ohmic or linear satisfying relevant matching condition for work function of the metal [18] as electrode material and the dielectric. Using equation (3.1) the dielectric constant or permittivity ε_r of a dielectric is given by:

$$\varepsilon_r = \frac{Cd}{\varepsilon_o A} \tag{3.2}$$

where (d/A) is the geometry-related normalizing factor. In the case of a single crystal of a material ε_r is the small dimensionless number, whereas for a polycrystal of the same material it is no surprise that it could be a hundred or thousand. It is also true that despite both materials have the same geometry, a polycrystal yields often a hundred or thousand times larger value for ε_r than the single crystal. This means an added quantity or term appears in the terminal capacitance of the polycrystal. Therefore, equation (3.1) or (3.2) cannot be used in determining the so-called dielectric constant of a polycrystal unless the origin of this added quantity of the capacitance is thoroughly investigated and understood. Similarly, determining conductivity also may not be useful referring to the grains or lumped grains. Invariably the utterance of the dielectric constant of a polycrystal in a strict sense is misleading and a simple misnomer.

The dielectric constant ε_r is found to be about 11.7 for Si or 8.5 for ZnO single crystals. The permittivity of a single crystal indicates the degree to which the medium can resist the flow of electric charges and is always greater than unity [1]. The dielectric materials can be classified into active dielectric and passive dielectric materials. When a dielectric material is kept in an external electric field, if it actively accepts the electricity, then it is known as active dielectric material. Active dielectrics can easily be adapted to store the electrical energy whereas the passive dielectrics restrict the flow of electrical energy in them. Thus, the active dielectrics act as the semiconductors like Si, Ge, ZnO, SnO_2, etc. and the passive dielectrics act as the insulators like mica, rubber etc. [7–17, 19–21].

In polycrystals the dc conductivity is usually dominated by the grain boundaries [22–28]. It is affected by the development of the interfacial charges. Thus, nonlinear resistance can arise for a polycrystal. Such a DUT may contain multiple dc conduction paths implying possible parallel operative electrical paths. A large value of ε_r generally originates from the presence of multi-phases demarcated with grain boundaries beside the contributions of the defect states within the electrical field falling regions. The electrical field is experienced in the resistive regions of the material system. As an example, bulk and the depletion region across the grain boundary experience the drop of ac small-signal voltage reflecting electrical field in these physical regions. Therefore, thorough understanding of the microstructure of the polycrystal is essential in order to understand the origin of the increased amount of permittivity or corresponding capacitance when electrical field falling regions are present [26].

For a polycrstal, the value of ε_r above the usual range of the single crystal can be considered as the added quantity. This added quantity exhibits frequency dependence which is associated with one or more underlying mechanisms constituting the terminal capacitance. Thus, the investigation of polycrystals needs clarity with the underlying operative mechanisms.

After the elimination of the added quantity from the terminal capacitance the left out quantity is essentially associated with the bulk which is similar to that of the single crystal-like behavior emphasizing geometric contribution. The bulk geometric capacitance yields dielectric constant or relative permittivity ε_r. In this case the depletion region is assumed to be non-existent.

If another geometric quantity similar to that of the bulk exists (behaving like a geometric quantity) then this term is likely associated with the depletion region across the grain boundaries. Thus, two geometric terms may be identified from the left out quantity which is frequency-independent. The eliminated quantity is a frequency-dependent associating variation of charges with respect to the sinusoidal signal-voltage at applied frequencies. The frequency-dependent terminal capacitance [22–29] can be written by:

$$C_P = \left(\frac{\varepsilon_o \varepsilon_r A}{d}\right)_{Bulk} + \left(\frac{\varepsilon_o \varepsilon_r A}{d}\right)_{Depletion-Region} + \frac{dQ}{dV_{ac}} \tag{3.3}$$

or

$$C_P = (C_{Geometric})_{Bulk} + (C_{Geometric})_{Depletion-Region} + C_{Frequency-Dependent} \tag{3.4}$$

or

$$C_P = C_{Geometric} + C_{Frequency-Dependent} = C_{High-Frequency} + C_{Frequency-Dependent} \tag{3.5}$$

with

$$C_{Geometric} = C_{High-Frequency} = \frac{\varepsilon_o \varepsilon_r A}{d} \tag{3.6}$$

and

$$C_{Frequency-Dependent} = \frac{dQ}{dV_{ac}}, \tag{3.7}$$

where the symbol C_P in equation (3.3) or (3.4) or (3.5) implies terminal capacitance measured in the form of admittance or R-C parallel circuit mode which is always the case when determining the dielectric constant ε_r of a DUT. The geometry related term expressed in equation (3.6) can be recognized at the high frequency range. It may comprise of one or two contributing elements as found in many single junction or polycrystal devices as noted in equation (3.3) or (3.4). For ZnO-based varistors two geometric quantities are observed in the high frequency domain one after the other, preferably barrier layer capacitance followed by the bulk capacitance [22–26]. The frequency-dependent term is the added term above the geometric term only recognized below the high frequency domain. Usually MPCHPH-based DUTs exhibit the frequency-dependent term [22–29].

For the single-crystal based single-junction devices there is rare domination of traps as the semiconductor is maintained almost free from point defects or structural flaws or any extraneous contamination. Invariably near perfection of the crystal structure of the semiconductor hardly exhibits frequency dependence in the measurement of the terminal capacitance. That's why as a frequent practical rule of thumb 1 MHz is often used in the measurement, because of the flat value of the terminal capacitance for the trap-free condition.

3.2 Dielectric Behavior of Materials: Single Crystals and Polycrystals

For a perfect or ideal dielectric $R \rightarrow \infty$ or $R = \infty$, and the associated reactance X_C with $\omega = 2\pi f$ is given by:

$$X_C = \frac{1}{\omega C}. \tag{3.8}$$

In the real world no capacitor is ideal or perfect as it is somewhat leaky implying conducting electricity under the dc voltage. Therefore, R_P is finite possessing conductivity σ or resistivity ρ for the dielectric. It can be described by an equivalent circuit comprising of a parallel combination of resistance R_P and of capacitance C_P as shown in Figure 3.1. The admittance [30] of this R-C parallel combination is given by:

$$Y = \frac{1}{R_P} + j\omega C_P = G_P + j\omega C_P. \tag{3.9}$$

The dimensionless parameter ε_r known as dielectric constant corresponds to the R-C parallel mode measurement implying admittance at applied frequency f. The corresponding impedance [30] for the R-C parallel combination is given by:

$$Z = \frac{1}{Y} = R_S + \frac{1}{j\omega C_S} = R_S - j\left(\frac{1}{\omega C_S}\right). \tag{3.10}$$

The real and imaginary quantities of admittance are given in terms by the impedance parameters as:

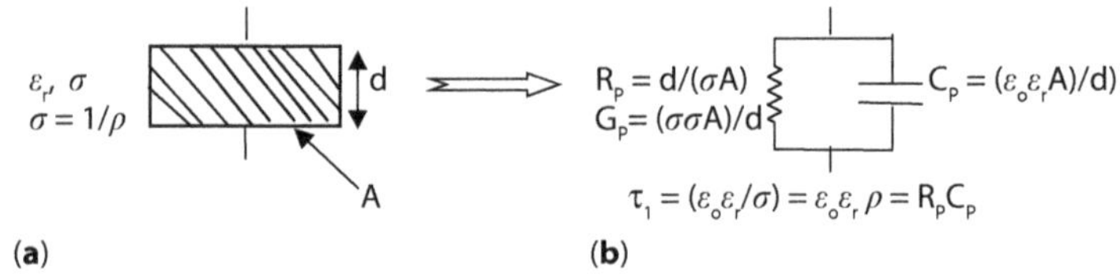

Figure 3.1 Representation of a dielectric material: (a) possessing dielectric constant ε_r, conductivity σ or resistivity ρ, thickness d, and electrode area A; (b) equivalent circuit of the same dielectric with time constant τ having resistance R_p and capacitance C_p.

$$G_P = \frac{R_S}{R_S^2 + \left(\frac{1}{\omega^2 C_S^2}\right)} \tag{3.11}$$

and

$$C_P = \frac{\frac{1}{\omega C_S}}{R_S^2 + \left(\frac{1}{\omega^2 C_S^2}\right)}. \tag{3.12}$$

The real and imaginary quantities of impedance are given in terms by the admittance parameters as:

$$R_S = \frac{G_P}{G_P^2 + \omega^2 C_P^2} \tag{3.13}$$

with

$$\frac{1}{\omega C_S} = \frac{\omega C_P}{G_P^2 + \omega^2 C_P^2} \tag{3.14}$$

and

$$C_S = \frac{G_P^2 + \omega^2 C_P}{\omega^2 C_P}. \tag{3.15}$$

Summarizing the behavior of a real world dielectric material per Figure 3.1 under the dc condition:

$$R_P = \frac{d}{\sigma A} \quad \text{or} \quad G_P = \frac{\sigma A}{d} \tag{3.16}$$

and

$$C_P = \frac{\varepsilon_o \varepsilon_r A}{d}. \tag{3.17}$$

The current in this dielectric material does not reach maximum and minimum at the same time under the application of the ac voltage. The phase angle θ represents the amount by which the current leads the voltage [30]. For $G_p = 0$ implies $R_p = \infty$ and $C_p = C_p$, then $\theta = 90°$. Similarly when $C_p = 0$ as found in a resistor R_p, then $\theta = 0°$. These conditions satisfy equation (3.6) and can be depicted in Figure 3.2 as vector diagram with real part conductance G_p $(=1/R_p)$ on the x-axis and imaginary part susceptance ωC_p on the y-axis [30]. The angle δ $(= 90° - \theta)$ in Figure 3.2 determines the loss tangent defined [30] as tan δ which is given by:

$$\tan\delta = \frac{G_P}{\omega C_P} = \frac{\sigma A d}{\omega d \varepsilon_o \varepsilon_r A} = \frac{\sigma}{\omega \varepsilon_o \varepsilon_r} = \frac{1}{\varepsilon_o \varepsilon_r \omega \rho}. \tag{3.18}$$

Now the real current flows through R_p $(= 1/G_p)$ and the imaginary current flows through C_p. Thus, from Figure 3.1 the total current flowing in the circuit is given by the sum of these two currents:

$$I_{Total} = I_{Real} + I_{Imaginary}. \tag{3.19}$$

This means that at constant voltage for the circuit in Figure 3.1 using Ohm's law:

$$I_{Real} \propto 1/R_P \quad \text{or} \quad I_{Real} \propto G_P \quad \text{and} \quad I_{Imaginary} \propto \omega\, C_P. \tag{3.20}$$

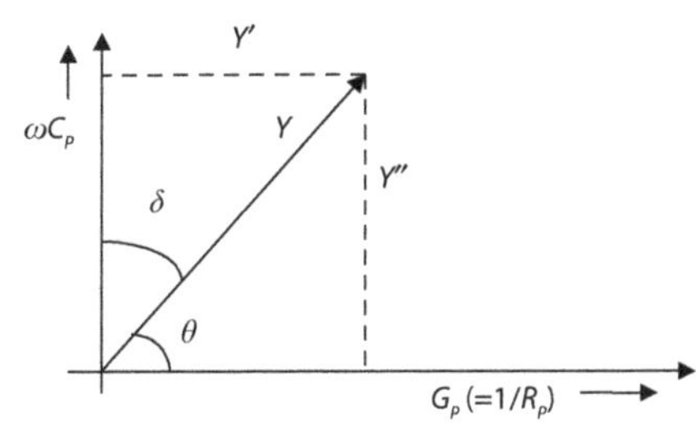

Figure 3.2 Electrical response in the form of admittance of a dielectric material.

The ratio of these two currents gives:

$$\frac{I_{Real}}{I_{Imaginary}} = \frac{Y'}{Y''} = \frac{G_P}{\omega C_P} = \tan\delta. \tag{3.21}$$

In other words the loss tangent or dielectric loss or tan δ represents the ratio of the real current to the imaginary current through the DUT when the mode of measurement reflects an R-C parallel circuit. It may be noted here that G_p/ω when normalized with the geometrical factor d/A is associated with the imaginary part of the dielectric constant (σ A)/(ω d) which is ε′, and C_p when normalized with the geometrical factor d/A is associated with the real part of the dielectric constant (ε_O ε_r)(A/d) which is ε″. Thus, equation (3.18) or (3.21) becomes:

$$\tan\delta = \frac{I_{Real}}{I_{Imaginary}} = \frac{Y'}{Y''} = \frac{G_P}{\omega C_P} = \frac{\varepsilon'}{\varepsilon''}. \tag{3.22}$$

When the relaxation time τ and the frequency of the applied electric field are identical or similar then a phase lag occurs and energy is absorbed causing the dielectric loss. In the classic Cole-Cole plot [31] representing ε′ versus ε″ the semicircular arc provided maximum value of tan δ at the peak where the relaxation time τ becomes equal to 1/ω or 1/(2πf). This is the dielectric loss. Depending on the direction of the applied frequency f or angular frequency ω, left (anti-clockwise) or right (clockwise) along the semicircular arc, the relaxation time τ may be referred to as either faster or slower than the frequency of the applied electric field. Polarization occurs instantaneously when the relaxation time is faster than the frequency of the applied electric field, while no polarization occurs when the relaxation time is slower than the frequency of the applied electric field. Polarization is the charge storage or charge separation process may be attributed to one or more associated mechanisms existing in various material systems having continuous uniform bulk structures in single crystals, and discontinuous non-uniform heterogeneous structures in polycrystals [32, 33]. Discussion on polarization is available in most texts [7–17, 19–21].

For a polycrystal, misleading information or interpretation [22–29] can easily be obtained using equation (3.1). It is essential to isolate geometric capacitance from the observed frequency-dependent contributions or mechanisms contained in the terminal capacitance [22–29]. This is because the electric field distribution may not be homogeneous due to the presence of the single phase grains or possible second phase grains or multi-phase grains or grain boundaries or possible effect of orientation of the grains, etc. between the electrodes. Strong frequency-dependent behavior may be noticed for the polycrystalline materials. Once the geometric contribution is isolated from the terminal capacitance only then the frequency-dependent portion can be attributed to the type of the operative mechanisms [22–29].

If simultaneously multiple operative competing mechanisms existed then each operative mechanism is expected to be visible in the complex plane rather than in the Bode plane. This is because the resolution in the frequency domain is likely to be poor or inferior via Bode plane than the complex plane unless the lumped behavior is dominant in the DUT. Once the behavior is ascertained in the complex plane then physical regions responsible can be investigated. Thus, microstructural information or structural configuration of the DUT is absolutely necessary for a better viewpoint.

In general, the polycrystalline materials are slightly lossy systems because of the interfacial polarization at the crystallites or grains and charge transport across the grain boundaries. For space charge or barrier layer polarization, an added geometric capacitance arises beside the bulk contribution of the DUT. In the presence of the space charge or barrier layer capacitance the DUT exhibits geometric contribution in addition to the bulk single crystal-like lumped behavior for the rest of the physical regions. Thus, two geometric contributions may be observed in succession at high frequencies in the frequency domain as these two capacitances are always in succession once the rest of the frequency-dependent terms are eliminated. These two capacitive elements physically must be operative in parallel within the DUT. Thus, the equivalent circuit configuration [7–17, 19–21] must include these two capacitances in parallel.

The electrical thickness essentially sustains the applied electrical field under a non-equilibrium condition. The lumped effect of the grains named as bulk and the lumped barrier layers named as electrical thickness constitute a single out-of-phase geometric component (i.e., capacitance) operative in parallel to the trapping response existing within the electrical thickness [22–29]. In general, this behavior is also observed in the p-n junction where its depletion region and bulk behavior dominate as point defects or defect states in the bulk are much less dominant or not significant or may be negligible for the terminal capacitance. It is important to note that the electrical field falling regions account for the conduction as well as capacitance for the DUT. Sometimes the electrical field falling region is referred to as the electrical length or electrical layer within the DUT [22–29].

3.3 Origin of Frequency Dependence

Essentially frequency dependence satisfies equation (3.3) or (3.7) where the $C_{Frequency\text{-}Dependent}$ term becomes influential or dominant while measuring terminal capacitance. In general, all kinds of material systems exhibit some sort of frequency dependence. In order to understand the origin of frequency dependence it is necessary to comprehend polarization mechanisms operative within the dielectric material. The charges in a dielectric material do not experience bulk motion in the presence of an applied external electric field but the charges rearrange in some way. This is commonly referred to as the polarization, and thereby the DUT so stressed is said to be polarized [34]. A variety of polarization mechanisms is observed in nearly all material systems and thereby frequency dependence emerges.

Due to the focus on the solid state materials, semiconductors and dielectric materials are in constant consideration for searching variety devices. In general, dielectric materials are insulators where all the electrons are tightly bound to their parent molecules and there are no free charges. Usually a band gap exceeding 3 eV tends toward a dielectric material. Of course, there are some exceptions where band gaps above 3 eV are known as semiconductors. Nevertheless, all semiconductors considered dielectric require thorough investigation for potential applications. Most dielectric materials exhibit a large specific resistance and negative temperature coefficient of resistance (NTCR) [1–17]. For the insulating materials it is not possible for the electrons in the valence band to excite to the conduction band by crossing the energy band gap with normal application of voltages and thermal energy.

Dielectric dispersion is the dependence of the permittivity of a dielectric material on the frequency of an applied sinusoidal electric field [7–17, 19–29, 31]. The electric field is assumed to consist of electric lines of force. These lines of force originate from the positive charges and terminate to the negative charges. Since there is a lag between the changes in polarization and the changes in the electric field, the permittivity of the dielectric is a complicated function of frequency of the electric field. Dielectric dispersion is very important for the applications of dielectric materials and for the analysis of polarization systems. Often the application of voltage stress causes polarization in the DUT. When the electric field E is increased the strength of the induced dipole is also increased. Thus, the induced dipole moment is proportional to the intensity of the electric field.

Dielectric relaxation is the momentary delay or lag in the dielectric constant of a material. This is usually caused by the lag in molecular polarization with respect to the changing electric field in a dielectric medium inside the parallel-plate capacitor [1, 7–17]. Relaxation, in general, is a delay or lag in the response of a linear system. Therefore, dielectric relaxation is measured relative to the expected linear steady state or equilibrium dielectric values. Dielectric relaxation refers to the relaxation response of a dielectric medium to an external sinusoidal electric field. This relaxation is often described in terms of permittivity as a function of frequency. For ideal systems this response can be described by the Debye equation [35].

Dielectric polarization appears whenever charges in a material system are somewhat displaced with respect to one another under the influence of an electric field [7–17]. This displacement or distortion related to ionic polarization and electronic polarization shows behavior of the resonance [7–17]. The character of the distortion process depends on the structure, composition, and surrounding of the DUT. In a capacitor, the negative charges are displaced within the dielectric toward the positive electrode whereas the positive charges shift in the opposite direction. Since charges are not free to move in an insulator, restoring forces are activated, causing transfer of energy. On charging a capacitor, the polarization effect opposing the applied electric field draws charges onto the electrodes storing energy. On discharge this energy is released.

An atom in its ground state has no dipole moment. However, in the presence of an electric field E a dipole moment results from the relative displacement of the positive and negative charges within the atom. This is termed as polarization P which is proportional to the applied electric field E, and it is in the same direction of E. It can be written as:

$$P \propto E \qquad \text{or} \qquad P = \alpha E = \varepsilon_o \chi_e E, \tag{3.23}$$

where α $(= \varepsilon_o \chi_e)$ is called the atomic polarizability with χ_e as electric susceptibility and ε_O is the permittivity of free space. The electric susceptibility of a dielectric material is a measure of how easily it polarizes in response to the electric field E [7–17]. This, in turn, determines the permittivity of the material. The electric susceptibility of a medium or a DUT is related to its relative permittivity ε_r by:

$$\chi_e = \varepsilon_r - 1. \tag{3.24}$$

For a vacuum, $\varepsilon_r = 1$. Thus, $\chi_e = 0$. Using equation (3.23) polarization can be expressed as:

$$P = \varepsilon_o (\varepsilon_r - 1) E. \tag{3.25}$$

Dielectric polarization is the displacement of charged particles under the action of the externally applied electric field. Several mechanisms are responsible for the electric polarization [7–17]. The total polarization of a dielectric arises usually from four different sources of charge displacement: (a) electronic; (b) ionic; (c) orientation of permanent dipole; and (d) space charge.

The total contribution of polarization [7–17, 19–21] to the dielectric constant is, therefore, a summation of these sources given by:

$$P_{total} = P_{electronic} + P_{ionic} + P_{orientation} + P_{space\text{-}charge}. \tag{3.26}$$

In the R-C parallel mode measurement each of these contributions appears as the added term beside the geometric term as a function of applied frequency. This is one of the cases of a general phenomenon known as dispersion occurring in the DUT which is termed as the frequency-dependent response for the ac small-signal voltage. The frequency dependence of the added term requires identification and interpretation for the type of DUT.

3.4 Effect of Polarization

An atom is electrically neutral, possessing negatively charged electrons in the shells around the positively charged nucleus. The center of the negatively charged electrons is a hypothetical concept that coincides with the positive charge of the nucleus implying that the atom has no net dipole moment. However, when the atom is placed in an external electric field, the center of the positive charge is displaced along the field direction while the center of the negative charge is displaced in the opposite direction. This effect is common to all materials as it involves distortion of the center of charge symmetry of the basic atom. Under the influence of an applied electric field the nucleus of an atom and the negative charge center of the electrons shift. This shift or displacement creates a dipole or dipole moment. Thus, when a dielectric material is placed in an electric field, such dipoles are created in each atom.

An electrical field will always displace the center of charge of the electrons with respect to the nucleus and, thus, induce a dipole moment [7–17]. Thus, the dipole polarization can occur wherever dipole exists [32, 33]. For a molecule which has centers of positive and negative charge that are separated each other is referred to a dipole [32, 33]. If the dipole is free to orient in electric field then dipole polarization can occur. Under the application of thermal energy dipole orientation becomes random. Thus, complete alignment of dipole in an electric field cannot occur. The thermal energy randomizes the alignment which causes temperature dependence when temperature is increased. In this case the material system usually liquid or gas must have natural dipoles or intrinsic dipoles which can rotate freely. In thermal equilibrium, the dipoles will be randomly oriented and, thus, carry no net polarization. The external applied field aligns these dipoles to some extent and induces a polarization for the material.

The magnitude of dipole polarization depends on the nature and the concentration of dipoles depending on the stability of ambient temperature. At low frequencies dipoles reorient in an electric field and lead to polarization. As reorientation occurs, the charge is transported and transient conduction results. No further conduction occurs until the electric field is reversed. At high frequencies, dipoles are not able to follow changes in the electric field, and no polarization or conduction occurs. In the intermediate frequency range the dipole is reoriented in the electric field before the field is reversed. The time required for reorientation at this frequency is referred to as the dipole relaxation time, τ_d. Corresponding to this frequency conduction current is a maximum, and the polarization process is dependent on the size of the dipole and the nature of the media. The dipole polarization process can be represented by a capacitor in parallel with a series R-C combination which bears the concept of a blocking circuit. Materials exhibiting such behavior are temperature-dependent [32, 33].

The dipole orientation is a phenomenon involving the rotation of permanent dipoles under an applied electric field. Although permanent dipoles exist in ceramic type materials such as silica or SiO_2, which has no center of symmetry for positive and negative charges, dipole orientation is not found to occur. Since the dipole is restricted from shifting by the rigid crystal lattice, reorientation of the dipole is precluded as destruction of the lattice would ensue. Dipole orientation is more common in polymers in which by virtue of their atomic structure permit reorientation. This mechanism of permanent dipoles is not the same as that of the induced dipoles of ionic polarization [7–17].

The electronic polarization occurs due to the displacement of the positively charged nucleus and negatively charged electrons under the application of an external electric field. This means the center of the negative charge of the electron cloud surrounding the nucleus of an atom is displaced from the center of the positive charge of the nucleus while the atom is subjected to an electric field [7–17, 32, 33]. This polarization effect exists in almost all materials, though too small, despite the large number of atoms within the material because the moment of the dipoles is very short. The electronic polarization can respond to the variation of the electric field at frequencies approaching to that of the ultraviolet region of the spectrum in the vicinity of 10^{15} Hz, and leads to a constant level of polarization at lower frequencies. Thus, this polarization process represents capacitance at frequencies less than that of the optical region of the spectrum. Furthermore, this polarization has very little or no temperature dependence. This polarization responds at frequencies approaching the ultraviolet region of the spectrum, and leads to a constant level of polarization at lower frequencies. The electronic polarization has little to no temperature dependence [32, 33]. This polarization is also sometimes called atom polarization or atomic polarization as a misnomer term. The lattice polarization is termed as the atomic polarization that exists in crystalline materials having interpenetrating positive and negative ion sublattices [32, 33].

If a crystal or molecule consists of atoms of more than one kind, then the distribution of charges around an atom in the crystal or molecule leans toward a positive or negative corner. As a result, when lattice vibrations or molecular vibrations induce relative displacements of the atoms, the centers of positive and negative charges are also displaced. The locations of these centers are affected by the symmetry of the displacements [1–17]. When the centers don't correspond, then polarization arises in molecules or crystals. This polarization is called ionic polarization. It is caused by relative displacements between positive and negative ions in ionic crystals such as NaCl. In this case a solid material must have some ionic character. Thus, ionic polarization arises due to the displacement of the negative (-ve) ions and positive (+ve) ions in opposite directions, and it occurs in ionic solids in the presence of applied electric field [7–17, 19–21]. The displacement is independent of temperature like electronic polarization. It then has internal dipoles. But these built-in dipoles cancel each other exactly and are unable to rotate. The external field then induces net dipoles by slightly displacing the ions from their rest position. It causes the ferroelectric effect as well as dipolar polarization.

When the applied frequency f increases ionic polarization can no longer track the electric field around 10^{13} Hz, and electronic polarization loses its response in the ultra-violet region around 10^{15} Hz. As expected, ions that are larger are less mobile and have a less rapid response. The polarization effect of ionic displacement decreases at 10^{13} Hz. At this frequency, the ionic displacement begins to lag the field reversals and thereby increasing the loss factor and contributing less to the dielectric constant. Extensive research experience handling various material systems and variety instruments can aid in visualizing the foregoing frequencies for potential measurement.

The orientation polarization arises with the polar molecule [35] as it is caused by the permanent dipole moment. In the case of polar molecule the positive and negative charges do not coincide. Only molecules possessing a permanent dipole moment exhibit orientation polarization. When one element or atom has more electronegativity than the other element or atom, then the former element or atom pulls the bonded electrons toward itself more strongly than the other element or atom. Thus, in the absence of an electric field a net dipole moment exists. When the field is applied then the positive portion aligns along the direction of electric field and the negative portion aligns in the opposite direction. This kind of polarization is referred to as the orientation polarization [7–17].

The space charge polarization occurs due to the diffusion of ions or migration of charges along the field direction and thereby giving rise to the redistribution of charges [7–17, 19–21] in the DUT. This mechanism is extrinsic to any crystal lattice. It arises due to the charges that attribute to the contaminants or dopants or irregular geometry of the interfaces of the polycrystalline systems and is, therefore, an extraneous contribution. Essentially the dopant is referred to the deliberately added foreign atom to the host material. These charges are partly mobile and migrate under the applied electric field though the charges achieve equilibrium.

Sometimes the depletion region is referred to as the space charge region that is found in the p-n junction shown in Figure 3.3, where charges of opposite nature migrate across the demarcation. This demarcation is the physical boundary for each doped region. Far away from the junction the bulk is neutral, carrying p-type carrier in the p-semiconductor or n-type

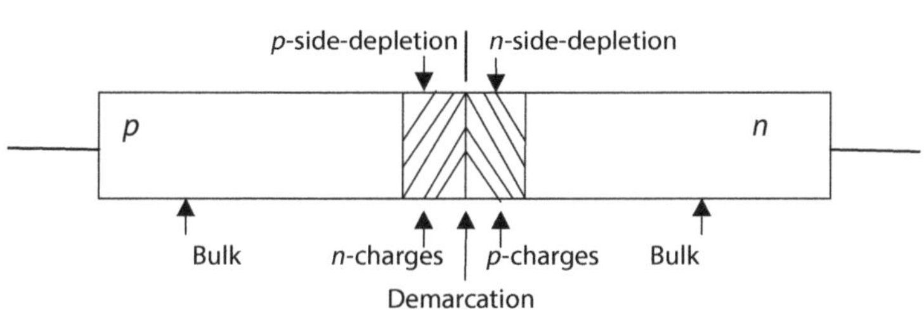

Figure 3.3 Depletion region in the p-n junction where far away from the junction is the bulk region.

carrier in the n-semiconductor. Therefore, the depletion region only exists across the physical boundary of the two distinct regions in a narrow range. For a trap-less depletion region, frequency response provides the concept of parallel R-C combination in series with the bulk contribution as it can be visualized from Figure 3.3. The values of the capacitor C and the resistor R are the functions of frequency for the space charge polarization.

Surfaces, interfaces, demarcation in polycrystals as grain boundaries, inter-phase boundaries for the multiple phases, etc. may carry charges. These charges respond to the external electric field indicating interfacial polarization of the DUT. Thus, interfacial polarization can exist usually in heterogeneous microstructures where a region of high conductivity (low resistivity grain) is surrounded by a matrix region of low conductivity (high resistivity grain-boundary). Interfacial polarization is caused at the electrodes preferably for the liquid systems per Figure 3.4(a) whereas the same is caused at the crystallites or grains for the solid systems per Figure 3.4(b). This leads to an accumulation of charge resulting in space charge polarization [32, 36–39] similar to that of the p-n junction. This polarization gives the concept of a parallel R-C combination in series with the bulk contribution where the values of the capacitor C and the resistor R are constant. High resistivity interfacial layers can block or partly block the transport of charges.

The Faradaic polarization can occur with the accumulation of space charge at interfaces leading to a chemical potential to drive a chemical reaction. The space charge polarization cannot occur without the existence of interfacial polarization. Similarly Faradaic polarization cannot occur without first there existing space charge polarization. Thus, space charge polarization is noticeable at lower frequencies than the interfacial polarization. Faradaic polarization becomes evident at lower frequencies than the space charge polarization. In some cases, the free energy of reaction for the Faradaic polarization is so low that immeasurable space

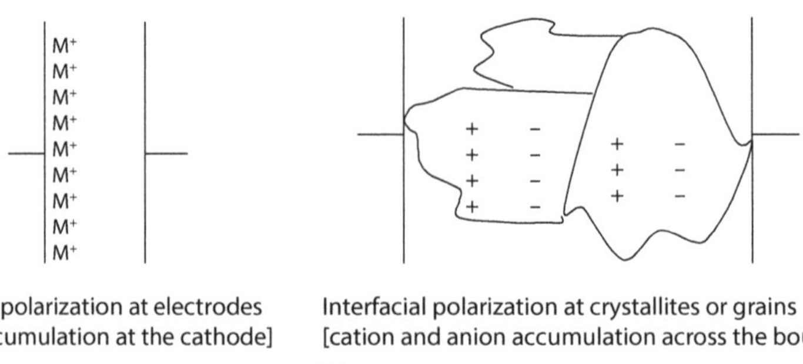

Figure 3.4 Interfacial polarization at (a) electrodes; and at (b) crystallites or grains.

charge polarization occurs before the Faradaic polarization becomes a dominant process. These three polarization processes are highly dependent on charge transport, and will reflect the temperature dependence of the dominant conduction process in the material under consideration. The Faradaic process would be expected to have an additional dependence on temperature due to the reaction processes involved [32]. This polarization process gives the concept of a R-C parallel combination circuit as it is termed as the Faradaic impedance or admittance which is often difficult to resolve or comprehend. The same circuit configuration is designated for the Warburg polarization which is also difficult to resolve or comprehend. The values of the capacitor C and the resistor R are the functions of frequency for the Faradaic or Warburg polarization processes.

The Faradaic process is dictated by the Faraday's Law. This polarization mechanism can be modeled as an equivalent circuit using frequency-dependent components [32, 33, 40]. Sometimes interfacial polarization at electrodes may appear as Faradaic polarization when the accumulation of charges is related to the reaction. Warburg polarization is a special case of Faradaic polarization when the Warburg impedance components can be written as:

$$Z_{Warburg} = R_{Warburg} + X_{Warburg} = A\,\omega^{-0.5} + A\,\omega^{-0.5}, \tag{3.27}$$

where ideal Faradaic impedance is given by:

$$Z_{Faradaic} = R_{Faradaic} + X_{Faradaic} = A\,\omega^{-\beta} + A\,\omega^{-\beta}. \tag{3.28}$$

An ideal Faradaic polarization refers to a reaction related transport mechanism yielding two exponents of equal value as in equation (3.28). Deviation from the nature of equation (3.28) causes non-Faradaic polarization giving rise to the impedance of the form:

$$Z_{Non-Faradaic} = R_{Non-Faradaic} + X_{Non-Faradaic} = A\,\omega^{-\beta_1} + A\,\omega^{-\beta_2} \tag{3.29}$$

where β_1 and β_2 are unequal. Overall, equation (3.29) indicates complicated type non-Faradaic impedance because of the inequalities in the exponents of β_1 and β_2. The non-Faradaic impedance becomes more complicated when the coefficient of $\omega^{-\beta_1}$ and $\omega^{-\beta_2}$ are unequal which is expressed as:

$$Z_{Non-Faradaic} = R_{Non-Faradaic} + X_{Non-Faradaic} = A\,\omega^{-\beta_1} + B\,\omega^{-\beta_2} \tag{3.30}$$

implying A and B are unequal. Possible theoretical treatment to delineate the coefficients A and B along with the exponents β_1 and β_2 is not available to date.

The ferroelectric transition caused by the lining up of the orientations of permanent dipoles along a particular direction which is called an order-disorder phase transition. The ferroelectric polarization process can exist in certain crystalline materials having non-centro-symmetric crystal structures, and permanent dipoles which can reorient cooperatively in electric field [7–17, 32, 33]. Ionic displacement is common in ceramic type materials that are consisting of crystal lattices occupied by cations and anions. Under the influence of an electric field dipole moments are created by the shifting of these ions towards their respective opposite polarity of the field. The displacement or moment arm of the dipoles can be relatively large in comparison to the electronic displacement and, therefore, can give rise to enhanced dielectric constant in some ceramics.

The variation of the polarization mechanisms with frequency is reflected in Figure 3.5 in the frequency domain when measuring dielectric constant of a capacitor. The total net effect of frequency on the dielectric measurement is illustrated in the tentative generic form. As expected for the value of capacitance, the dielectric constant always decreases with increased frequency for all DUTs depending upon the dominance of the polarization type. However, much complexity in the dielectric response of the DUT may be observed below 1 GHz for many MPCHPH systems. Sometimes the sub-effects of the mechanisms are predominant for the DUT. Thus, variation of the generic form of Figure 3.5 depends on the type of DUT. Dipole orientation and space charge polarization have little slower frequency responses.

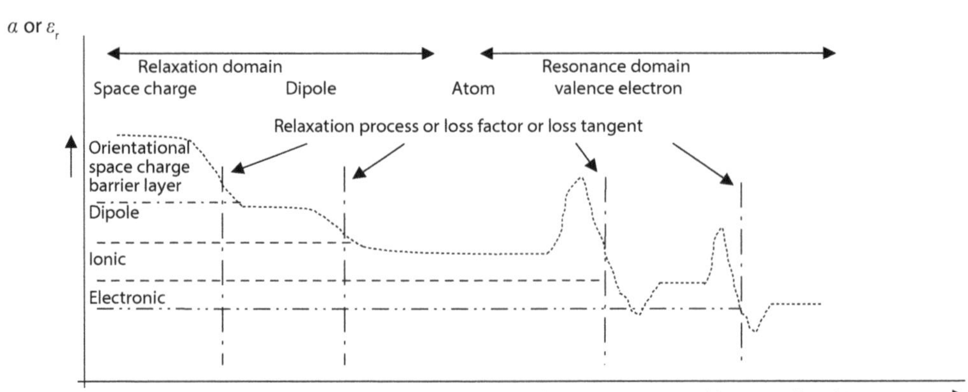

Figure 3.5 Frequency dependence of polarizability or dielectric constant showing several contributing mechanisms.

The peaks which occur near the limiting frequency for ionic and electronic polarization are due to the resonance points, where the applied frequency equals the natural frequency of the material.
The regular instruments for measuring immittance at extremely high frequencies are rarely available. In the frequency region above ultra-violet region, permittivity approaches ε_0 in every substance where ε_0 is the permittivity of the free space. Since permittivity indicates the strength of the relation between electric field and polarization then the polarization process loses its response resulting in decrease in permittivity. The mechanisms of polarization have varying time response capability to an applied frequency f of the sinusoidal stress, and the net contribution of polarization to the dielectric constant ε_r is, therefore, frequency-dependent. Electronic displacement responds rapidly to the field reversals, and no lag of the polarization contribution occurs up to 10^{17} Hz.

3.5 Equivalent Circuit Representation of the Mechanisms and Processes

Every material system experiences conduction through electron, hole, and ion transport across the electrodes. A conduction process can be represented as a simple resistance R or conductance G. Domination of the charge carrier depends on the type of material system. In a p-type semiconductor, the hole is transported while in the n-type semiconductor electron is transported. Ions are transported in electrolytes or in biological materials. Thus, conductivity depends on the charge carriers implying concentration. Ions are large in size and basically resemble the atom in mass. Therefore, the transport of ions requires massive energy involving the diffusion coefficient, D, of the ionic species. That is why aqueous media involves ionic transport accounting for atomic or molecular size of the ionic species and the viscosity of the media [32, 33]. Certainly note that the viscosity is highly temperature-dependent for the liquid systems.

In many DUTs electrons or holes, and ions are the major charge carrying species, which can be due to electronic, dipole, interfacial, space charge, and Faradaic polarization processes [32, 33]. Each underlying operative mechanism or process occurring within the DUT can be represented by the concept of an equivalent circuit. Each circuit is essentially based on valid logic, an underlying operative mechanism, and intuition utilizing common sense or wisdom for the observed functions or responses of the DUT. Upon identifying operative mechanisms it is possible to develop an eventual equivalent circuit as noted earlier [32, 33]. Complex plane plots in conjunction with the Bode plots can confirm the equivalent circuit model for the DUT [22–29].

For visualizing or apprehending a potential situation for each operative mechanism or process at zero electric field (i.e., no bias or zero-bias) and under electric field (i.e., applied bias), the list of equivalent circuits is summarized in Figure 3.6. It involved consistent observation, handling variety of material systems, extensive effort having patience, track record of the diagnosis of the mechanisms, intuition, instrumentation, measurement procedures, data acquisition and eventual analysis, etc. It is understood that a wide range of texts [7–21] and numerous unlimited sources aided to provide this listing besides the guidance of initial work [32, 33].

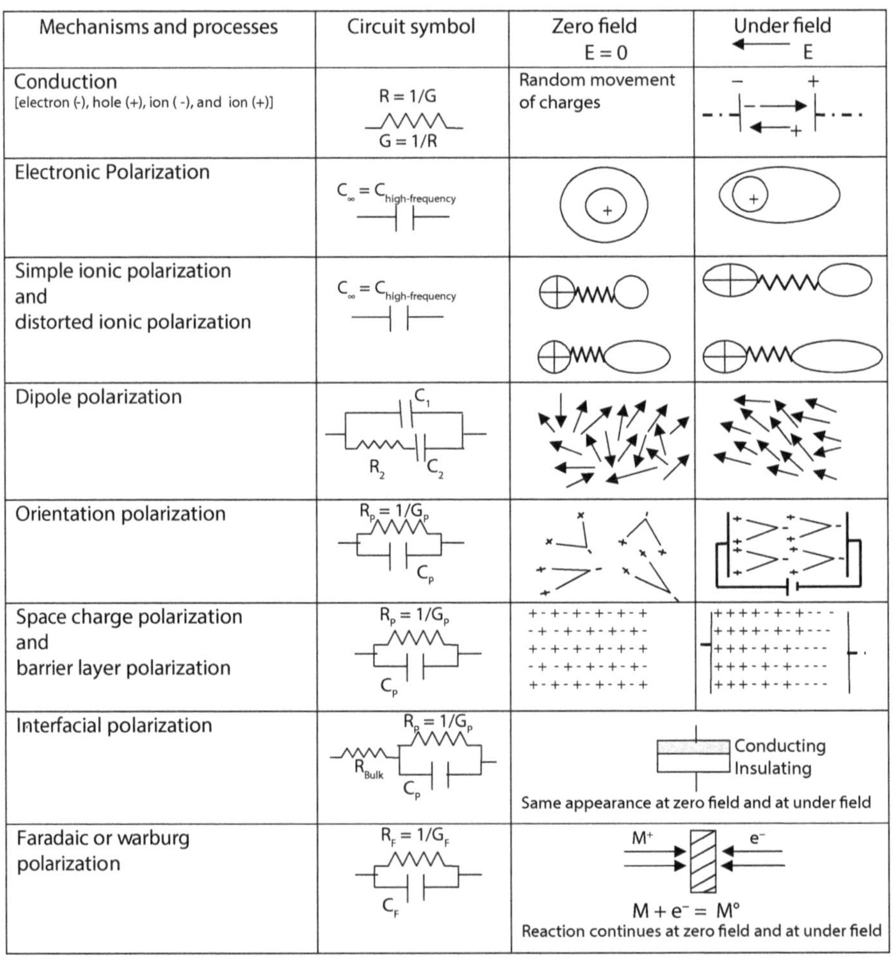

Figure 3.6 Equivalent circuit model for potential operative mechanism and process both at zero bias and under bias occurring within the DUT.

The concept of electrical circuits for each potential mechanism listed in Figure 3.6 requires identification. Each of these circuits can be delineated via any of the four complex plane plots upon using the ac small-signal electrical data, acquired in a broader range or window of the measured frequencies, besides other supporting investigations such as Bode plane analysis and spectroscopic analytical approaches. Furthermore, each of these circuits can unravel the identity of the tough operative mechanisms within the DUT.

3.6 Defects and Traps

There are many concurrent names for defects and traps such as recombination centers, generation centers, shallow levels, shallow impurities, shallow defect states, deep impurities, deep levels, deep defect states, etc. Sometimes these are synonyms and sometimes each is different from the other depending on the situation or context or occurrence or explanation or interpretation or understanding. These were discussed for decades in various ways though lack of clarity causes confusion to the investigators [41–62]. Often defects and traps are intermixed and interchanged freely with respect to each of the foregoing nomenclatures. Detailed explanation and clarification are necessary for the role of the defects and the traps in any site within the semiconductors.

The role of defects and traps is more complicated in the polycrystalline systems than in the single-crystal based devices. Compound semiconductors are more complicated regardless of the crystal status, single crystal or polycrystal. The general tendency of the investigators is to use simplistic routes as found in the case of elemental single crystal. This is because it may aid in resolving complicated trapping and de-trapping processes of the polycrystalline materials. But this is not the viable way to obtain meaningful solutions or interpretations for the operative underlying problems of the DUT. Essentially trapping of charges is caused by a defect site, and then the same site releases the same charges in moments. This release of the charges is the de-trapping which leads to a series event resulting in a justified R-C series circuit. This means that the mobile charges are in-phase for conduction while trapped charges are in out-of-phase conduction. In this way the concept of the R-C series circuit arises. Thus, the measured ac small-signal electrical data must display logical relaxation for the R-C series circuit either in the Y^*-plane or in the C^*-plane, if the process is unmasked from the dc resistance and the geometric capacitance of the DUT. If the dc resistance and the geometric capacitance mask the R-C series circuit feature in terms of relative values, then the ac small-signal electrical data may appear only in the Z*-plane or in the M*-plane relaxations. That is why often the measured data reveal the Z*-plane plot to have two consecutive semicircles in the low frequencies and high frequencies, respectively, referring to the grain-boundary behavior and the grain behavior, respectively, for the polycrystalline systems. In this way no trapping response is evident in the Z*-plane. The Y*-plane or the C*-plane can resolve the masking nature of the R-C series circuit by displaying the dc resistance and the geometric capacitance. This implies that when the same data is used either in the Y*-plane or in the C*-plane then it is likely that a relaxation may be obtained with intercept, along with possible distortion in the semicircular behavior. This distortion is attributed to the influence of the dc conductance or geometric capacitance that Grant [63] and Coelho [7] suggested.

The duration for trapping until de-trapping in the site may be referred to as the relaxation time τ (= R-C) for that particular trapping process. Another concept with the duration of mobility following de-trapping until the same charges getting trapped by the next site. This means that the existence of the de-trapped charges as the conduction charges for the duration of mobility before being trapped by the next site may be referred to as the relaxation time τ (= R-C). It is a self-logic combined with strong intuition that determines the meaning of the τ (= R-C). Thus, the relaxation time τ (= R-C) may have various meanings.

For semiconductors there exist allowed energy states for the defects in the band gap [2–17, 41–47]. Figure 3.7 shows donor and acceptor energy levels along with the deep trap energy level E_T located near towards the middle of the band gap. A donor impurity is called shallow if the donor energy level E_D is close to the conduction band E_C. Likewise an acceptor impurity is called shallow when the acceptor energy level E_A is close to the valence band E_V. For a particular energy level, E_D, depending on the situation and type of ionization the following three mechanisms is noted:

1. charge trapping center,
2. recombination center, and
3. generation center.

The symbol '—' represents electrons whereas '+' represents holes. Depending on the type of impurities additional energy levels, E_A and E_T are initiated in the band gap. The same three mechanisms can apply to

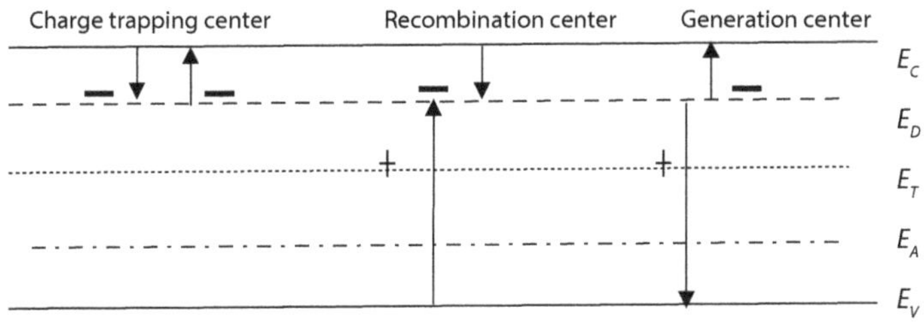

Figure 3.7 Band diagram of a semiconductor showing conduction band edge E_C, valence band edge E_V, donor energy level E_D, acceptor energy level E_A with E_T as the deep trap energy level for charge trapping center, recombination center, and generation center for the specific energy level E_D.

each of these energy levels. Thus, the complex nature of the impurities gives rise to different energy states for the impurities or traps or defect states in the band gap [41–62].

In general, a shallow impurity represents an atom either having at least one additional valence electron implying donor (n-type impurity) type or having at least one less valence electron implying acceptor (p-type impurity) type than the atom of the host elemental semiconductor (or usually cation of the compound semiconductor) that would normally occupy the lattice site. If the impurity is a complex clustered species regardless of shallow or deep as a defect then the trapping and de-trapping situation becomes highly complicated as often found in multi-component oxide materials such as varistors or doped perovskites. Usually a deep level is trapping charge carrier that acts as a recombination center. The same deep level can have different charge states depending on the occupancy of levels on the states. Thus, the energy level varies with the charge state as it can be influenced by the lattice relaxation depending on the charge state [41–47]. This kind of variation in energy level for the same type defect complex acting as trap can give rise to non-Debye response. This means that the ac small-signal electrical data can yield depressed semicircular relaxation in any of the complex plane plots.

The deep trap energy level may be referred to the middle of the band gap, although donor and acceptor levels may be possible trap centers. Overall, this means that the ionization energy of any species for the host material is smaller than the band gap. An acceptor impurity has the capability of capturing an electron from the bulk (crystal or grain) which is a trapping behavior. Similarly a donor impurity has the capability of giving up an electron to the bulk leaving a trapping behavior. This is appropriate when the dopant becomes donor or acceptor by occupying the lattice site of the host semiconductor.

Transfer or movement of electron between the E_C and E_V or vice versa refers to the:

1. generation, and
2. recombination.

The generation is the transfer of an electron from E_V to E_C, and thereby a hole is created, whereas the recombination is the movement or transfer of an electron from the E_C to E_V and thereby causes destruction or removal of a hole. The destruction of a hole implies filling or recovering the valence orbital. In other words, the transfer of electrons from the E_V to the E_C is referred to as the generation of the electron-hole pair. Thus not only a free electron is created in the E_C, but also a hole in the E_V can contribute to the conduction processes, i.e., charge current. The inverse process is termed recombination of the electron-hole pair. This is shown in Figure 3.8.

A large amount of energy is necessary for band-to-band, i.e., between E_V and E_C, transition. The presence of trap levels within the band gap caused by the impurities facilitates this transition process [41–48]. These processes are depicted in Figure 3.9 as four plausible mechanisms. Each of these is described vividly as:

1. hole emission implying capture of a hole by the E_V (it means an electron jumps from the E_V to the trapping level as in hole emission of Figure 3.9),
2. hole capture implying emission of a hole to the E_V (it means an electron moves from an occupied trap to the E_V as a hole disappears or destroys as in hole capture of Figure 3.9),

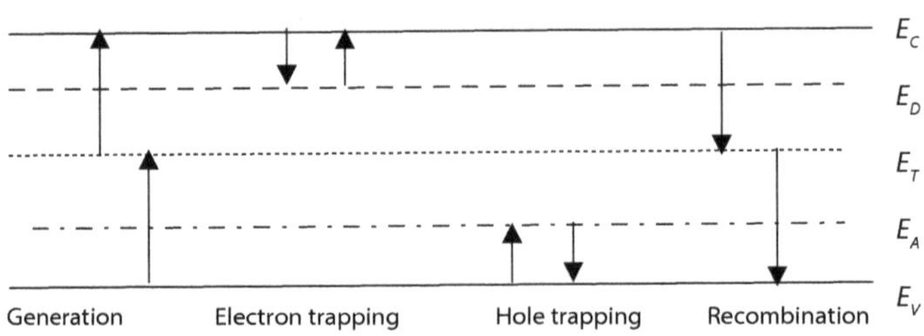

Figure 3.8 Band diagram of a semiconductor showing conduction band edge E_C, valence band edge E_V, donor energy level E_D, acceptor energy level E_A, and E_T is the deep trap energy level. Charge trapping center, recombination center, and generation center are shown from a specific energy level E_T.

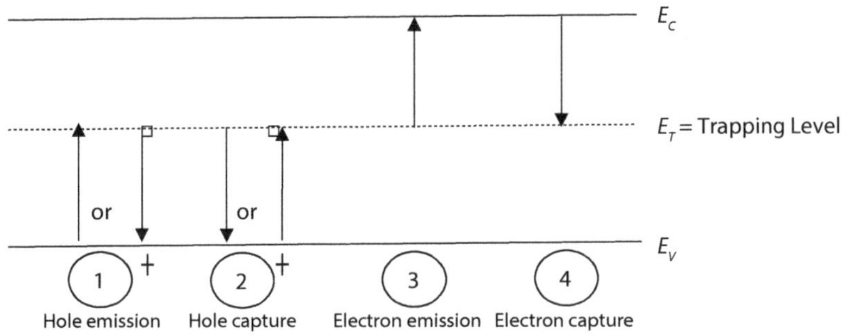

Figure 3.9 Processes of emission and capture for a single trap center E_T.

3. electron emission implying emission of an electron to the E_C (it means an electron jumps from trapping level to the E_C as in electron emission of Figure 3.9), and
4. electron capture implying capture of an electron from the E_C (it means an electron moves from the E_C to an unoccupied trap at the trapping level as in electron capture of Figure 3.9).

The Shockley-Read-Hall (SRH) model [51, 52] was introduced in 1952 to describe the statistics of the recombination and the generation of holes and electrons in semiconductors occurring through the mechanism of trapping. The transfer of electrons from the E_V to the E_C is referred to as the generation of electron-hole pairs (or pair-generation process). The inverse process is termed recombination of electron-hole pairs. A free electron is created in the E_C, and also a hole is created in the E_V which can contribute to the charge current. The band gap or forbidden band between the upper edge of the E_V and the lower edge of the E_C is large in semiconductors. It means that a large amount of energy is necessary for the direct band-to-band generation event. The presence of trap levels within the band gap caused by the crystal impurities facilitates this process [2–6, 41–50, 64–66]. These mechanisms are illustrated in Figure 3.9.

The intrinsic point defects remain intact in the bulk regardless of doping. A donor causes an impurity center in the semiconductor leaving a positive charge by causing the donation of an electron in the E_C. Meanwhile, an acceptor causes an impurity center in the semiconductor leaving a negative charge by causing the absence of an electron in the E_V. The absence of an electron is termed as the presence of a hole, which is considered as the same as an electron except for it having a positive charge. Any standard text describes [2–6, 41–50, 64–66] in detail about the hole. This means that the impurity center becomes positively charged while electron appears in the E_C, and negatively charged while the hole appears in the E_V. In this way n-type conductivity is achieved attributing to the electrons, whereas p-type conductivity is achieved attributing to the holes. These electrons and holes are capable of participating in the transport processes.

A trap is a center or a site in a crystal capable of holding a charge either negative (electron) or positive (hole). This center originates from the type of the defect. Because of the charge holding nature of the center, the defect type is called a trap. Thus, trap is the defect state that is essentially the energy structure representation or implication or insinuation of the defect. Often defects are referred to as the defect states. Some defects have little or no energy structure impact so that not all defects necessarily create traps or act as traps. Traps can be shallow or deep depending on the relative energy depth of the energy well defining the trap. Therefore, a trap is a confining situation, hence a trap, because the energy well can capture or release a carrier electron or hole [2–6, 41–50, 64–66].

In general, time constant τ, energy depth E, capture rate c, capture cross-section σ, and emission (or generation) rate e are the important parameters characterized for the traps. The capture cross-section ranging between 10^{-22} cm^2 to 10^{-12} cm^2 is well established in the text [41–62]. The capture cross-section is usually found to be much less than 10^{-14} cm^2 for the well-formed functional devices. This suggests that the trap is an attractive center, and therefore, referred to as the donor-like trap. Usually large capture cross-section suggests repulsive center, and therefore, referred to as the acceptor-like trap.

The presence of deep level traps affects the electrical characteristics of the DUT severely causing low frequency noise. Characterization of deep levels may be important for performance issues of the DUT. This can involve energy depth (level), capture cross-section, capture rate, emission or generation rate charge states (+ or −), carrier density or concentration, degeneracy ratio, and possible chemical species. Several types of defects can be originated from the nature of the crystal or polycrystal. Defects in crystals or polycrystals can be classified into the following categories [67–72]:

1. zero-dimensional defect,
2. one-dimensional defect,
3. two-dimensional defect, and
4. three-dimensional defect.

All kinds of point defects are referred to as zero-dimensional defect. Thus, point defects include vacancy, interstitial, self-interstitial, anti-site, substitutional, variety combination of vacancy and interstitial such as the Schottky defect, Frenkel defect, etc. These point defects are often mentioned as intrinsic defects [67–72]. Certain complexes can cause extrinsic defects. Often substitutional impurities referred to as dopant or foreign atom in the host material belong to this defect type. One-dimensional defect usually refers to line type defect such as dislocation (line or edge or screw or threading edge), separation or demarcation between two successive grains or poly-grains or multi-grains often noted as grain boundary, stacking fault, etc. [67–72]. Two-dimensional defect usually refers to the defect associated with area or planar type such as twin boundaries, sub-boundaries, surfaces with dangling bonds, etc. However, certain types of stacking faults and grain boundaries belong to this defect. Many standard texts [41–47, 67–72] have discussed these defects. Three-dimensional defect is often termed as volume defect referring to the voids or air-tight cracks implying agglomerates of continuous vacancies of atoms. Several standard texts have discussed these defects in detail [67–72].

In general, an intrinsic crystal contains some sort of point defect in the form of impurity and vacancy in the lattice site beside interstitial [41–47, 67–72]. An example of covalent bond in Si is considered, illustrating the role of a vacancy as a point defect. An intrinsic defect often causes dangling bonds and relaxation of lattice to minimize energy. Obviously the same point defect as vacancy can reflect two kinds of responses as a function of applied frequency. These two responses are illustrated as the:

1. regular vacancy involving four non-relaxed dangling bonds showing dangling in Figure 3.10(a), and
2. regular vacancy involving relaxed dangling bonds showing no dangling in Figure 3.10(b).

The presence of non-relaxed dangling bond in a Si crystal shown in Figure 3.10(a) is the source of a trap attributing to the type of the vacancy defect state, whereas the relaxed dangling bond in Figure 3.10(b) is not likely to act as a trap as if no defect exists. In this way each point defect is not a source of a trap as it may not have energy structure impact. In reality, compound semiconductors possessing a vacancy defect state may constitute the non-relaxed dangling bond for the cations where anions are ignored. This type of non-relaxed dangling bond may form a complicated defect site for the trap actions, possessing varying time constant τ, energy

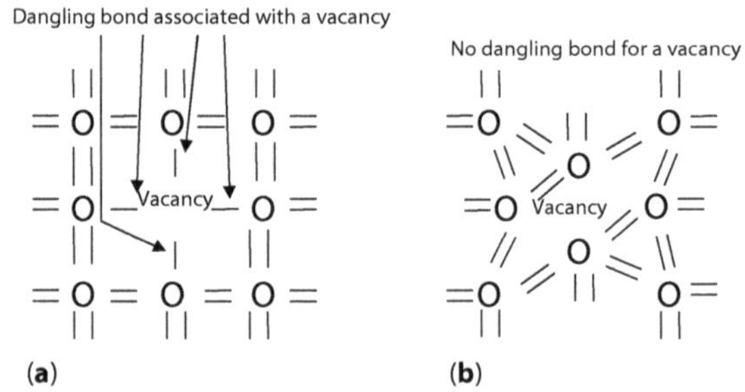

Figure 3.10 Representation of a covalent semiconductor where O is the Si atom in the crystal having (a) a regular vacancy involving four non-relaxed dangling bonds; and (b) a relaxed vacancy with no dangling bond.

depth E, capture rate c, capture cross-section σ, and emission (or generation) rate ε. This variation causes the origin of the depression angle for the relaxation process as the semicircle. Perhaps relaxed dangling bond in compound semiconductor is rare.

The foregoing situations with the dangling bond can be extended to the grains of the polycrystal system for point defects and doping. In addition to these situations, the grain boundary demarcation as discontinuity of the crystal (i.e., grain) plays an important role for the polycrystals. The dangling bonds caused by the grain boundary surfaces can act as the defect states. Needless to say, all these are the subject to some sort of polarizations [34] causing frequency dependence of the DUT beside non-relaxed dangling bond associated with the point defects. The non-relaxed dangling bond can cause complicated clustered species possessing different charge state with varying energy level to give rise as a defect site. Trapping role for such a defect is certainly complicated to delineate the exact behavior and, thus, originates complicated non-Debye behavior. Perhaps ZnO-based varistor may possess this kind of defect states [24] where trapping capacitance exhibited partial thermal activation instead of 100% thermally activation for the trapping resistance.

For interstitial impurities in the elemental bulk semiconductor ionization may be straightforward whereas for the compound semiconductor the situation may be slightly complicated depending on cation and anion issues at a given temperature. In many cases donor or acceptor for the host semiconductor is determined from the location of elemental cation or anion in the periodic table. This includes the occupation of interstitial cation or anion of the constituting host compound semiconductor besides referring to metal excess (donor) or metal deficit (acceptor) situations [67–72]. It may be noted that the type of element for certain compound semiconductors does not prompt the nature of the doping. As an example, doping GaAs (III-V compound) with Ge or Si (Group IV) can produce either a donor when occupying Ga site or an acceptor when occupying As site. In the case of interstitial Ge or Si could be either a donor or an acceptor. Thus, Ge or Si behaves as amphoteric dopant for GaAs. The probability of occupation of a certain type of site depends essentially on the chemical potential or Fermi level of the atom at that site depending on attractive (binding) and repulsive forces (size effects) [41–47]. Steady characterization of GaAs concerning this situation ascertain the type of charges for the dopant.

Major role of traps is to reduce the majority concentration in the crystal or grain. Traps act as recombination sites for light generated minority carriers. Several aspects of traps and trapping effects are discussed for the single crystal in the text for decades [41–62, 64–72]. Upon careful examination and systematic evaluation of the DUT using complex plane formalisms and Bode plane analysis in conjunction with several other trap characterization tools and techniques [41–62], the ultimate device behavior of the polycrystal based materials can be ascertained. Invariably characterization of traps in polycrystals is a challenge prior to anticipating potential applications due to better control of the devices.

3.7 Point Defects and Stoichiometric Defects

The point defects are considered for both single-crystal and polycrystal materials regardless of elemental or compound semiconductor. For compound semiconductor, cation and anion provide a fixed ratio in order to ascertain perfect chemistry which is referred to as stoichiometric crystal. When this ratio deviates from the expected fixed value then the compound can be termed as non-stoichiometric [41–47, 67–72].

For GaAs as a perfect crystal implying no point defect as if entire lattice points are filled with appropriate cation and appropriate anion then the ratio of cation to anion must be unity. In the presence of a few Ga vacancies in the lattice sites having no defect for As then the ratio of cation to anion is less than unity. This non-stoichiometric chemical formula would be $Ga_{1-x}As$, where "x" refers to the amount of the Ga vacancies in the lattice sites. In the presence of a few As vacancies in the lattice sites having no defect for Ga then the ratio of cation to anion is greater than unity. This non-stoichiometric chemical formula would be $GaAs_{1-y}$, where "y" is the amount of As vacancies in the lattice sites [41–47, 67–72].

In another situation if "x" is the amount of Ga interstitials in the crystal without point defect for As then the non-stoichiometric chemical formula $Ga_{1+x}As$ yields the ratio of cation to anion greater than unity. In the case of "y" as the As interstitials in the crystal without point defect for Ga then the non-stoichiometric chemical formula $GaAs_{1+y}$ yields the ratio of cation to anion less than unity. The foregoing scenario of GaAs is too complicated when all these point defects simultaneously exist though temperature variation and subsequent ionization of the interstitial atoms are not considered or taken into account. Because of this realistic situation, binary compound semiconductors are too complicated to handle. Production of single crystal of a compound semiconductor assuring minimized point defects is certainly a tedious approach [41–47, 67–72].

When the compound semiconductors are polycrystalline then the real world becomes ultra complicated. That is why polycrystal compound semiconductors are usually dealt in a non-clean-room environment. Moreover these compound semiconductors are often treated as ceramic materials, if oxides. Due to the electronic nature of the compound semiconductors numerous binary compounds are treated as electroceramic materials. In order to grow a polycrystalline structure, employing electroceramic route for processing includes high temperature sintering. Often stages of processing steps are used in controlled manner to ascertain the quality of the end-product [41–47, 67–72].

Some definitions can be straightened out for stoichiometric and non-stoichiometric compound semiconductors using standard notation of point defects comprising of cation as M, anion as X, vacancy as V, and interstitial as I. Thus, $V_M \rightarrow$ cation vacancy, $V_X \rightarrow$ anion vacancy, $I_M \rightarrow$ cation interstitial, $I_X \rightarrow$ anion interstitial, $M_X \rightarrow$ cation substituted by anion, $X_M \rightarrow$ anion substituted by cation, $M_M \rightarrow$ cation substituted by cation, X_X anion substituted by anion, and combinations thereof provide a list of defect types [67–72] in Figure 3.11 for the polycrystal systems. The symbols, terms, and nomenclatures therein are self explanatory.

3.8 Leaky Systems

In general, alternately capacitive path turning to resistive path and resistive path turning to capacitive path for a DUT is referred to as a leaky system. To achieve this kind of periodic or single-sided behavior it is more than likely that the solid-state DUT involves consistent electrochemical processes. The electrochemical processes are presumed to be

Defects	Polycrystal
Vacancy (vacancy disorder)	(1) V_M (2) V_X (3) $V_M + V_X$ (n = m) (4) $nV_M + mV_X$ (n ≠ m) (n and m are the displaced atoms from usual lattice sites/points)
Interstitial (interstitial disorder)	(1) I_M (2) I_X (3) $I_M + I_X$ (n = m) (4) $nI_M + mI_X$ (n ≠m) (n and m are the displaced atoms from usual lattice sites/points)
Schottky defect:	$V_M + V_X$ Cation vacancy + anion vacancy
Frenkel defect	(1) Cation frenkel → $I_M + V_M$ (2) Anion frenkel → $I_X + V_X$
Substitutional defect	$M_X + X_M$ (anti-structure disorder → i.e., wrong placement of atoms in the respective lattice sites/points)
Simple substitutional defect	M_M or M_X X_M or X_X (occurs during doping)
Substitutional + interstitial	(1) $M_X + I_X$ (2) $X_M + I_M$
Substitutional + vacancy	(1) $M_X + V_M$ (2) $X_M + V_X$

Figure 3.11 Notations of point defects for the crystal and polycrystal systems.

time-dependent. Consistent occurrences of charge capture and charge release can lead to Faradaic or Warburg response. Certainly this kind of response is complicated to resolve the exact underlying operative mechanism. For ZnO varistors [22–26, 73–82] leaky behavior is obvious but remains too tedious to resolve the issues.

Problems

1. The capacitance of a device is measured by an impedance analyzer and found to be 300 pF at 100 Hz and 200 pF at 10 MHz. The capacitance remained the same above 10 MHz. What is the geometric capacitance for the device? Is it possible to extrapolate the capacitance at the dc (f → 0 Hz) condition? Explain and justify the answer.
2. The capacitance of a device is measured by an impedance analyzer and found to be 200 pF at 10 MHz and 300 pF at 100

Hz. The capacitance remained the same above 10 MHz. If this is a Schottky diode then is it possible to get a C_p^{-2} versus V_{dc} plot having a straight line? What frequency should be used for the measurement to get only one straight line? Explain and justify the answer. Consult the Mott-Schottky plot to obtain an explanation.

3. Is it possible to obtain a relaxed vacancy with no dangling bond at the surface of a single-crystal Si wafer? Is it possible to obtain a relaxed vacancy with no dangling bond at the surface of a Si polycrystal grain? Explain the answer.
4. Can an interstitial point defect be a source of a non-relaxed dangling bond? How complicated can a trap site be, considering all kinds of point defects in a simultaneous fashion? Explain and justify the answer.
5. How many types of point defects can be visualized for the compound semiconductor, GaAs? List these defects with justification. Why can Si or Ge be the amphoteric dopant for GaAs? Explain and justify the reasons.

References

1. R. Resnick and D. Halliday; *Physics I & II*, John Wiley & Sons, New York, 1966.
2. C. Kittel; *Introduction to Solid State Physics,* 4th edition, John Wiley, New York, 1971.
3. A. J. Dekker; *Solid State Physics*, Macmillan and Company Ltd., London, 1962.
4. A. J. Dekker; *Electrical Engineering Materials*, Prentice-Hall Inc., Englewood Cliffs, N. J., 1959.
5. N. W. Ashcroft and N. D. Mermin; *Solid State Physics*, Holt, Rinehart and Winston Publisher, New York, 1976.
6. M. Ali Omar; *Elementary Solid State Physics*, Addison-Wesley, New York, 1975.
7. R. Coelho; *Physics of Dielectrics for the Engineer,* Elsevier Scientific Publishing Co., Amsterdam, 1979.
8. A. A. Zaky and R. Hawley; *Dielectric Solids,* Dover Publications, New York, 1970.
9. V. V. Daniel; *Dielectric Relaxation*, Academic Press, New York, 1967.
10. J. C. Anderson; *Dielectrics*, Reinhold Publishing Corporation, New York, 1964.

11. C. P. Smyth; *Dielectric Behavior and Structure*, McGraw-Hill Book Company, New York, 1955.
12. A. K. Jonscher; *Dielectric Relaxation in Solids,* Chelsea Dielectrics Press, London, 1983.
13. H. Frohlich; *Theory of Dielectrics*, 2nd edition, Oxford (Clarendon) Press, London, 1958.
14. P. Debye; *Polar Molecules,* Dover Publications, New York, 1929.
15. A. R. von Hippel; *Dielectrics and Waves,* John Wiley & Sons, New York, 1954.
16. A. R. von Hippel; *Dielectric Materials and Applications,* John Wiley & Sons, New York, 1954.
17. A. R. von Hippel; *Molecular Science and Molecular Engineering,* John Wiley & Sons, New York, 1959.
18. E. H. Rhoderick; *Metal-Semiconductor Contacts,* Clarendon Press, Oxford, U.K., 1978.
19. J. D. Jackson; *Classical Electrodynamics*, 2nd edition, John Wiley & Sons, New York, 1975.
20. D. M. Smyth; "Dielectric-Loss Mechanism in Amorphous Oxide Films, in *Oxide and Oxide Films, Volume II,* J. W. Diggle edited, Marcel Dekker, Inc. New York, 1973.
21. J. Volger; "Dielectric Properties of Solids in Relation to Imperfections," in *Progress in Semiconductors, Volume IV* A. F. Gibson edited, John Wiley and Sons, Inc. N. Y., 1960.
22. M. A. Alim, M. A. Seitz and R. W. Hirthe; "Complex Plane Analysis of Trapping Phenomena in Zinc Oxide Varistors," *Journal of Applied Physics,* 63(7), 2337–2345, 1988.
23. M. A. Alim, S. Li, F. Liu, and P. Cheng; "Electrical Barriers in the ZnO Varistor Grain Boundaries," *Physica Status Solidi* (A), 203(2), 410–427, 2006.
24. M. A. Alim; "Admittance-Frequency Response in Zinc Oxide Varistor Ceramics," *Journal of the American Ceramic Society*, 72(1), 28–32, 1989.
25. L. C. Sletson, M. E. Potter and M. A. Alim; "Influence of Sintering Temperature on Intrinsic Trapping in Zinc Oxide-Based Varistors," *Journal of the American Ceramic Society*, 71, 909–913, 1988.
26. M. A. Alim; "Electrical Characterization of Engineering Materials," *Active and Passive Electronic Components,* 19, 139–169, 1996.
27. M. A. Alim; "Immittance Data-Handling/Analyzing Criteria for Heterogeneous Systems," Materials Research Society Symposium Proceedings: *Electrically Based Microstructural Characterization*, R. A. Gerhardt, S. R. Taylor, and E. J. Garboczi edited, 411, 13–23, 1996.
28. A. M. Azad, L. B. Younkman, S. A. Akbar and M. A. Alim; "Characterization of TiO_2-Based Sensor Materials Using Immittance Spectroscopy," *Journal of the American Ceramic Society*, 77(2), 481–486, 1994.

29. P. Guggilla, A. K. Batra, J. R. Currie, M. D. Aggarwal, M. A. Alim and R. B. Lal; "Pyroelectric Ceramics for Infrared Detection Applications," *Materials Letters*, 60, 1937–1942, 2006.
30. J. D. Irwin and R. M. Nelms; Basic Engineering Circuit Analysis, 10th Edition, John Wiley & Sons, Inc. New York, 2011.
31. K. S. Cole and R. H. Cole; "Dispersion and Absorption in Dielectrics I.: Alternating Current Characteristics," *Journal of Chemical Physics*, 9, 341–351, 1941.
32. J. J. Ackmann and M. A. Seitz; 'Methods of Complex Impedance Measurements in Biologic Tissue," *CRC Critical Reviews in Biomedical Engineering,* 11(4), 281–311, 1984.
33. M. A. Seitz; "Study of Heterogeneous Composite Materials via Lumped Parameter/Complex Plane Analysis," *International Journal of Hybrid Microelectronics*, (I.S.H.M.), 3(1), 1–7, 1980.
34. H. Fricke; "The Theory of Electrolytic Polarization," *Philosophical Magazine*, Series 7, Volume 14, Issue 90, 310–318, 1932.
35. P. Debye; Polar Molecules, *Chemical Catalogue Company,* New York, p. 94, 1929.
36. H. P. Schwan; "Alternating Current Spectroscopy of Biological Substances," *Proceedings of the IRE*, 47, 1841–1859, 1959.
37. R. J. Friauf; "Polarization Effects in Ionic Conductivity of Silver Bromide," Journal of Chemical Physics, 22(8), 1329–1338, 1954.
38. J. H. Beaumont and P. W. M. Jacobs; "Polarization in Potassium Chloride Crystals," *Journal of Physics and Chemistry of Solids*, 28, 657–667, 1967.
39. J. R. Macdonald; "Electrical Response of Materials Containing Space Charge with Discharge at the Electrodes, *Journal of Chemical Physics*, 54, 2026–2050, 1971.
40. K. J. Vetter; "*Electrochemical Kinetics: Theoretical and Experimental Aspects*," Academic Press, New York, 1967.
41. A. G. Milnes; "*Deep Impurities in Semiconductors*," John Wiley & Sons, Inc., New York, 1973.
42. H. G. Grimmeiss; "Deep Level Impurities in Semiconductors," *Annual Review of Materials Science,* 7, 341–376, 1977.
43. G. L. Miller, D. V. Lang and L. C. Kimerling; "Capacitance Transient Spectroscopy," *Annual Review of Materials Science,* 7, 377–448, 1977.
44. F. A. Kroger; "Defect Chemistry in Crystalline Solids," *Annual Review of Materials Science*, 7, 449–475, 1977.
45. C. Wagner; "Point Defects and Their Interaction," *Annual Review of Materials Science*, 7, 1–22, 1977.
46. D. M. Smyth; "*The Defect Chemistry of Metal Oxides*," Oxford University Press, New York, 2000.
47. E. A. Irene; *Electronic Materials Science*, John Wiley & Sons, Inc. Hoboken, New Jersey, 2005.

48. S. M. Sze and K. K. Ng; *Physics of Semiconductor Devices*, 3rd Edition, Wiley Interscience Publication, Hoboken, N.J., 2007.
49. S. M. Sze; *Physics of Semiconductor Devices,* 2nd Edition, Wiley Interscience Publication, New York, 1981.
50. S. M. Sze; *Semiconductor Devices, Physics, and Technology*, Wiley, New York, 1985.
51. W. Shockley and W. T. Read; "Statistics of the Recombinations of Holes and Electrons," *Physical Review,* 87(5), 835–842, 1952.
52. R. N. Hall; "Electron-Hole Recombination in Germanium," Physical Review, 87, 387, 1952.
53. W. Shockley and J. T. Last; "Statistics of the Charge Distribution for a Localized Flaw in a Semiconductor," *Physical Review*, 107(2), 392–396, 1957.
54. L. C. Kimerling; "Influence of Deep Traps on the Measurement of Free-Carrier Distributions in Semiconductors by Junction Capacitance Techniques," *Journal of Applied Physics,* 45(4), 183–1845, 1974.
55. D. V. Lang; "Fast Capacitance Transient Apparatus: Application to ZnO and O Centers in GaP P-N Junctions," *Journal of Applied Physics*, 45(7), 3014–3022, 1974.
56. D. V. Lang; "Deep-Level Transient Spectroscopy: A New Method to Characterize Traps in Semiconductors," *Journal of Applied Physics*, 45(7), 3023–3032, 1974.
57. D. L. Lossee; "Admittance Spectroscopy of Impurity Levels in Schottky Barriers," *Journal of Applied Physics,* 46_5, 2204–2214, 1975.
58. D. L. Lossee; "Admittance Spectroscopy of Deep Impurity Levels: ZnTe Schottky Barriers," *Applied Physics Letters*, 21, 54–56, 1972.
59. G. Vincent, A. Chantre, and D. Bois; "Electric Field on the Thermal Emission of Traps in Semiconductor Junctions," *Journal of Applied Physics,* 50(8), 5484–5487, 1979.
60. G. Vincent, D. Bois, and P. Pinard; "Conductance and Capacitance Studies in GaP Schottky Barriers," *Journal of Applied Physics,* 46(12), 5173–5178, 1975.
61. M. Lax; "Giant Traps," *Journal of Physics and Chemistry of Solids*, 8, 66–73, 1959.
62. M. Lax; "Cascade Capture of Electrons in Solids," *Physical Review,* 119(5), 1502–1523, 1960.
63. F. A. Grant; "Use of Complex Conductivity in the Representation of Dielectric Phenomena," *Journal of Applied Physics*, 29(1), 76–80, 1958.
64. E. S. Yang; *Fundamentals of Semiconductor Devices,* McGraw-Hill Book Co., New York, 1978.
65. R. S. Muller and T. I. Kamins; *Device Electronics for Integrated Circuits*, John Wiley & Sons, Inc., New York, 1977.
66. D. A. Neamen; *An Introduction to Semiconductor Devices,* McGraw-Hill Book Co., Boston, 2006.

67. D. R. Askeland, P. P. Fulay, and W. J. Wright; *The Science and Engineering of Materials*, 6th edition, Cengage Learning, Stamford, Connecticut, 2011.
68. B. S. Mitchell; *An Introduction to Materials Engineering and Science*, John Wiley & Sons, Inc., Hoboken, New Jersey, 2004.
69. E. J. Mittemeijer; *Fundamentals of Materials Science*, Springer-Verlag, Berlin, Heidelberg, 2010.
70. L. H. V. Vlack; *Elements of Materials Science and Engineering*, 4th edition, Addison-Wesley Publishing Company, Reading, Massachusetts, 1980.
71. A. G. Guy; *Introduction to Materials Science*, McGraw-Hill Book Company, New York, 1972.
72. C. A. Wert and R. M. Thomson; *Physics of Solids*, 2nd edition, McGraw-Hill Book Company, New York, 1970.
73. M. A. Alim, S. Bhattacharjee, S. Khanam, S. Runa, and N. Muna; "The Behavior of the ZnO-Bi_2O_3-Based Varistor Capacitance," *Journal of Active and Passive Electronic Devices*, 9(2), 157–184, 2014.
74. M. A. Alim; "Influence of Intrinsic Trapping on the Performance Characteristics of ZnO-Bi_2O_3 Based Varistors," *Active and Passive Electronic Components*, 17, 99–118, 1994.
75. M. A. Alim; "Influence of Multiple Trapping Phenomena on the Applications Criteria of ZnO-Bi_2O_3 Based Varistors," *Active and Passive Electronic Components*, 17, 57–65, 1994.
76. M. A. Alim and M. A. Seitz; "Singular Nature of Preferential Conducting Paths at High Electric Fields in ZnO-Based Varistors," *Journal of the American Ceramic Society*, 71, C246–C249, 1988.
77. M. A. Alim, M. A. Seitz and R. W. Hirthe; "High-Temperature/Field Alternating Current Behavior of ZnO-Based Varistors," *Journal of the American Ceramic Society*, 71, C52–C55, 1988.
78. M. A. Alim; "High-Frequency Terminal Resonance in ZnO-Bi_2O_3-Based Varistors," *Journal of Applied Physics*, 74(9), 5850–5853, 1993.
79. M. A. Alim; "An Analysis of the Mott-Schottky Behavior in ZnO-Bi_2O_3 Based Varistors," *Journal of Applied Physics*, 78(7), 4776–4779, 1995.
80. H. E. Brown; *Zinc Oxide – Properties and Applications*, International Lead and Zinc Research Organization, Inc., N.Y., 1976.
81. H. E. Brown; *Zinc Oxide Rediscovered*, The New Jersey Zinc Company, Horse Head Products, N.Y., 1957.
82. G. Heiland, E. Mollwo, and F. Stockmann; *Solid State Physics* – Volume 8, F. Seitz and D. Turnbull edited, Academic Press, New York, 191–323, 1959.

4

Ideal Equivalent Circuits and Models

4.1 Concept of Equivalent Circuit

An interesting concept of developing the equivalent circuit arises from linear circuit theory [1]. Thus, the equivalent circuit originates from the basics of the linear circuits using applications of Ohm's law, Kirchhoff's laws, and superposition principle. No matter how complicated the network, the scenario between the two terminals across the network behaves as if it consists only of a source and associated impedance. From this basic argument any linear circuit can be solved.

A linear circuit is also known as the network comprising of sources and linear elements. Two forms of an equivalent circuit for a network using a power source and a resistor are known as the Thévenin equivalent circuit and the Norton equivalent circuit [1–3], respectively. The concept of these two equivalent circuits is displayed for a linear circuit network in Figure 4.1. The load resistor is selected arbitrarily in the linear circuit network. For a Thévenin circuit the load resistor between terminals A and B is in series with the equivalent resistance (R_{eq}) of the network and the voltage source (V_{OC}), whereas for a Norton circuit the load resistor between terminals A and B is in parallel with the parallel combination of the equivalent resistance (R_{eq}) of the network and the current source (I_{SC}). For practical cases R_{eq} is the lumped quantity, which is subject to generalization for the equivalent impedance Z_{eq} along with the load impedance.

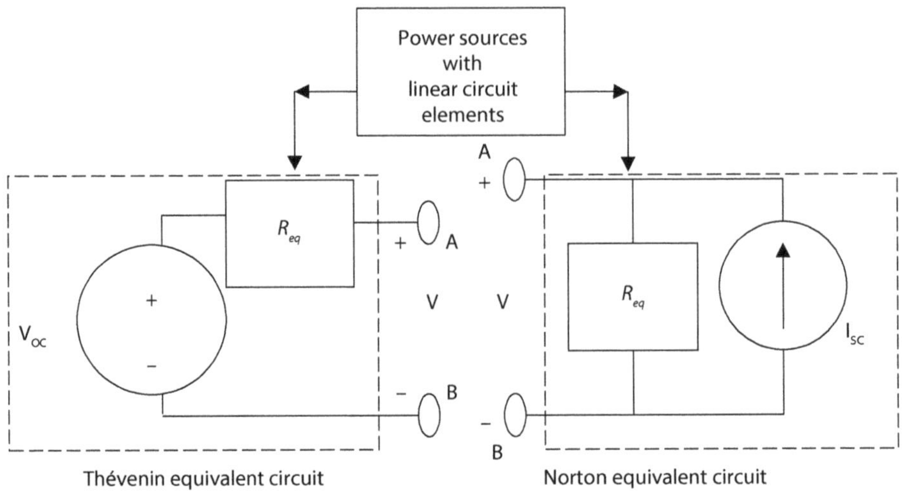

Figure 4.1 Thévenin's equivalent circuit with voltage source V_{OC} is shown at the left and Norton's (current source) equivalent circuit with the current source I_{SC} at the right. The impedance R_{eq} is the same in both cases and the source values are related to each other by $V_{OC} = R_{eq}\, I_{SC}$.

Both Thévenin and Norton circuits are equivalent to each other and to the linear circuit network as well. The Thévenin circuit provides voltage dividing feature with a voltage source whereas the Norton circuit provides current dividing feature with a current source. These two circuits give rise to successive source transformation to solve linear circuit problems [1]. Therefore, an equivalent circuit provides a meaningful concept or model for a complicated network that allows understanding the complexity. This concept can be extended to an unknown material system or the DUT for which the underlying operative mechanisms can be derived via each lumped element of the equivalent circuit. In this way an equivalent circuit of an unknown DUT can be used in developing any product or process or controlling the quality.

4.2 Simple and Basic Circuits in Complex Planes: R, C, R-C Series, and R-C Parallel

For a pure resistance (i.e. corresponding conductance) $Y' = G_p$ and $\delta = 90°$. In this case $\tan \delta$ becomes ∞ for the real current approaching a finite value and the imaginary current turning to zero. Thus, a pure resistance possessing conductance is represented by a point on the x-axis for any frequency in the Y^*-plane plot with angular frequency ω increasing in the upward direction. The same resistance will be represented by a point on the x-axis for any frequency in the Z*-plane plot with angular frequency ω increasing in the downward direction. Again, the same resistance will appear as a point for each frequency along the y-axis in the C*-plane plot with angular frequency ω increasing in the downward direction. Further the same resistance will appear as a point for each frequency along the y-axis in the M*-plane plot with angular frequency increasing in the upward direction. These plots in four complex planes are depicted in Figure 4.2. By the same token a pure capacitance is displayed in the same four complex planes as shown in Figure 4.3 where an arrow indicates the direction of the angular frequency ω. By using the necessary condition for

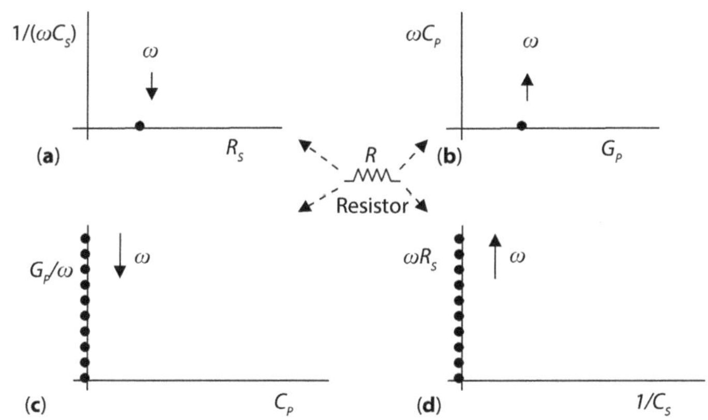

Figure 4.2 Representation of the resistance R or conductance G in four complex planes: Z*, Y*, C*, and M* where an arrow indicates the direction of the increasing measurement angular frequency ω.

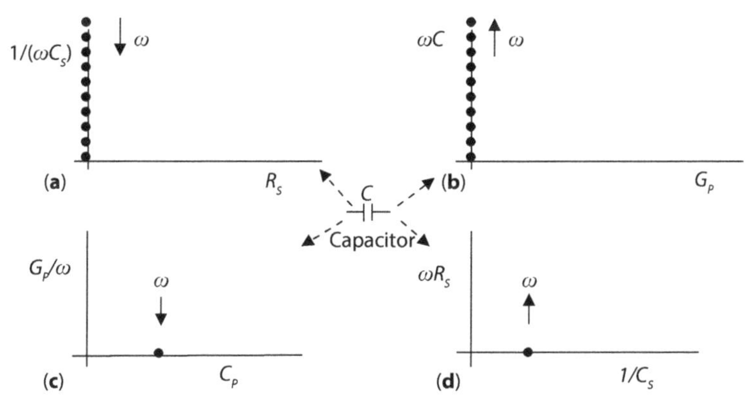

Figure 4.3 Representation of a capacitor C in four complex planes: Z*, Y*, C*, and M* where an arrow indicates direction of the increasing measurement angular frequency ω.

the pure inductance similar representation can be constructed in the same four complex planes.

Knowing the behavior of pure resistance and pure capacitance via Figures 4.2 and 4.3 it is possible to comprehend several simplistic scenarios for the R-C series and R-C parallel circuits in four complex planes. The simplistic representation of the R-C series circuit in four complex planes is shown in Figure 4.4 where τ_S is the time constant and an arrow indicates the direction of the angular frequency ω. In general, a resistor allows conduction under both ac and dc conditions while capacitor allows conduction only under the ac condition. It blocks the current flow under the dc condition. Thus, the R-C series combination represents a blocking circuit only under the dc condition. Under the matching condition of impedance or admittance, single visible semicircular relaxation may be obtained in the Y*- and C*-planes. The arc of the semicircle may appear as a straight line segment if the conductance is much larger than the susceptance in the Y*-plane. The arc of a semicircle may appear as a straight line segment for a gigantic semicircle. Likewise if the real capacitance is extremely larger than the imaginary capacitance then the straight line segment appears in the C*-plane. The response in the Z*- and M*-planes may be obtained as a vertical line with intercepts R_S and $1/C_S$, respectively. The vertical line represents dc blocking circuit in both Z*- and M*-planes whereas semicircular relaxation represents ac non-blocking circuit in both Y*- and C*-planes for the R-C series combination.

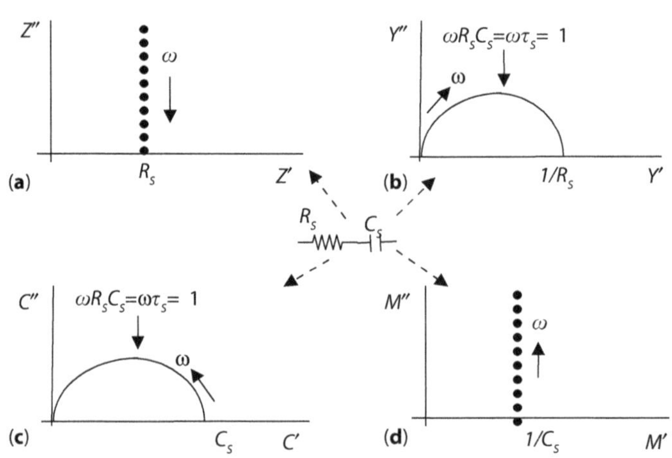

Figure 4.4 Simplistic representation of one R-C series circuit in four complex planes: Z*, Y*, C*, and M* where τ_S is the time constant and an arrow indicates the direction of the increasing measurement angular frequency ω.

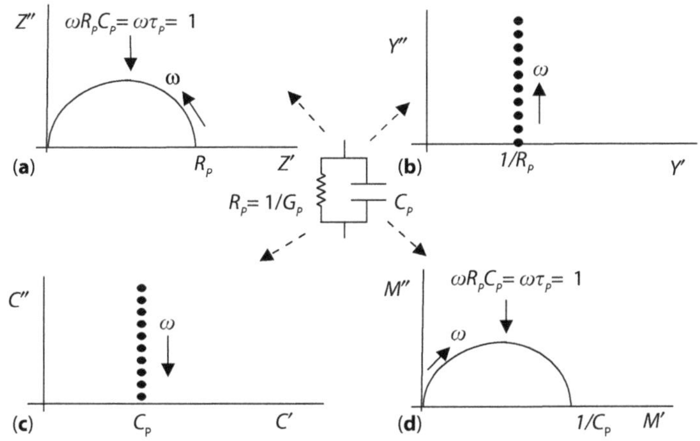

Figure 4.5 Simplistic representation of one R-C parallel circuit in four complex planes: Z*, Y*, C*, and M* where τ_p is the time constant and an arrow indicates the direction of the increasing measurement angular frequency ω.

The simplistic representation of the R-C parallel circuit in four complex planes depicted in Figure 4.5 where an arrow indicates the direction of the angular frequency ω and τ_p is the time constant. For a single semicircular relaxation in the Z*- and M*-planes yield this equivalent circuit as shown in Figure 4.5(a) and 4.5(d). It is a conductive (non-blocking) circuit under both ac and dc conditions. Under the matching condition of impedance or admittance, single visible semicircular relaxation may be obtained in the Z*- and M*-planes. The arc of the semicircle may appear as the straight line segment in the Z*-plane if the resistance is much larger than the reactance. Once again, the arc of a semicircle may appear as the straight line segment for a gigantic semicircle. Likewise if the real elastance is extremely larger than the imaginary elastance then the straight line segment may appear in the M*-plane. The response in the Y*- and C*-planes may be obtained as a vertical line as shown in Figure 4.5(b) and 4.5(c) with intercepts G_p (= $1/R_p$) and C_p, respectively. Here, the vertical line represents dc blocking for the capacitance in the Y*-plane and the ac conduction for the resistance in the C*-plane whereas semicircular relaxation represents non-blocking ac circuit in both Z*- and M*-planes for the R-C parallel combination.

4.3 Debye Circuits: Single Relaxation

A simplistic series R-C or parallel R-C equivalent circuit corresponds to the single Debye semicircular relaxation for each complex plane. For each relaxation there is a finite intercept on the left-side of the semicircle whose center lies on the x-axis.

The admittance of an R-C (i.e. R_P – C_P) parallel circuit shown in Figure 4.6(a) is given by:

$$Y^* = G_P + j(\omega C_P). \tag{4.1}$$

The corresponding impedance Z* of the same circuit is given by:

$$Z^* = \frac{1}{Y^*} = \frac{1}{G_P + j\omega C_P} \tag{4.2}$$

An easy manipulation of equation (4.2) taking G_P as a factor in the denominator gives:

$$Z^* = \left[\frac{(1/G_P)}{1 + j\omega\tau_P}\right] = \frac{R_P}{1 + j\omega\tau_P} \tag{4.3}$$

where $\tau_P = (C_P/G_P)$. Equation (4.3) is the simplistic form of the single relaxation referred to as the Debye equation for the R-C parallel circuit having τ_P as the relaxation time. Using the complex conjugate in equation (4.3), the Debye equation further separates real and imaginary parts:

$$Z^* = \frac{R_P(1 - j\omega\tau_P)}{1 + \omega^2\tau_P^2} = \frac{R_P}{1 + \omega^2\tau_P^2} - j\left[\frac{R_P\omega\tau_P}{1 + \omega^2\tau_P^2}\right] = Z' - jZ'' \tag{4.4}$$

where the magnitude of R_P is defined as $(R_O - R_\infty)$ with R_O corresponds to extremely low frequency f_O (or angular frequency ω_O) and R_∞ corresponds to extremely high frequency f_∞ (or angular frequency ω_∞). Thus, real and imaginary parts of Z* are given by:

$$Z' = \frac{R_P}{1 + \omega^2\tau_P^2} = \frac{R_o - R_\infty}{1 + \omega^2\tau_P^2} \tag{4.5}$$

and

$$Z'' = \frac{R_P\omega\tau_P}{1 + \omega^2\tau_P^2} = \frac{(R_o - R_\infty)\omega\tau}{1 + \omega^2\tau_P^2}. \tag{4.6}$$

At this conjecture if a plot of Z′ (on x-axis) versus Z″ (on y-axis) is prepared then a semicircle is obtained. The center of this semicircle will lie on the x-axis having no intercept on the x-axis. Now using (R_O - R_∞) for R_P equation 4.3 becomes:

$$Z^* = \frac{R_P}{1+j\omega\tau_P} = \frac{R_o - R_\infty}{1+j\omega\tau_P} \tag{4.7}$$

which requires a finite intercept of magnitude R_∞ corresponding to extremely high frequency upon generalization of equations 4.5 through 4.7. Thus, equation 4.7 becomes:

$$Z^* = R_\infty + \frac{R_P}{1+j\omega\tau_P} = R_\infty + \frac{R_o - R_\infty}{1+j\omega\tau_P}. \tag{4.8}$$

Since the resistance is on the real (x) axis and reactance is on the imaginary (y) axis of the Z*-plot, once again R_∞ will lie on the x-axis. Rewriting generalized real and imaginary parts of Z*, and then upon rearrangement of these equations give an equation of a circle [4]:

$$\left(Z' - \frac{R_o + R_\infty}{2}\right)^2 + (Z'')^2 = \left(\frac{R_o - R_\infty}{2}\right)^2. \tag{4.9}$$

That is why impedance data provide semicircular relaxation in the Z*-plot where the measurement frequency runs anti-clockwise direction along the arc of the semicircle. Likewise complex capacitance (C*) or permittivity plot (ε*) equation provides semicircular representation. Often Z*-plot or C*-plot are referred to as Cole-Cole plot [5] which was originally presented as the ε*-plot [5]where the direction of the measurement frequency runs anti-clockwise direction along the arc of the semicircle. Nevertheless, semicircular response in the Y*-plot or M*-plot is identical to the Z*- or C*-plot that provides equation of a circle where the measurement frequency runs clockwise along the arc.

The time constant or relaxation time τ_P is obtained from the peak of the semicircle when the real part of the immittance is plotted on the x-axis (abscissa) and the imaginary part of the immittance is plotted on the y-axis (ordinate). For the Z*-plot at two ends of the terminal frequency limits f_∞ implying high frequency and f_{dc} or f_o implying low frequency corresponding to R_∞ and R_{dc} or R_o, respectively, can be measured so that

$R_P = (R_o - R_\infty) = (R_{dc} - R_\infty)$. Using the value of R_P and τ_P in $\omega\tau_P = \omega R_P C_P = 1$, the value of C_P is obtained.

Invariably this kind of R-C parallel circuit will exhibit semicircular relaxation either in the Z*-plane or in the M*-plane depending on the situation based on the matching values between the real and imaginary parts. This is ideal or Debye relaxation for which the center of the semicircle lies on the x-axis. Unless there is a perfect resistor or a perfect capacitor it is absolutely impossible to obtain the center of the semicircular relaxation on the x-axis. For any high precision fitting procedure the presence of minute or ultra-small depression angle becomes obvious.

An equivalent circuit can be complicated based on the nature of the DUT. When multiple operative mechanisms are simultaneously present then based on the complicacy of the relaxations, the equivalent circuit can be derived. Multiple relaxations in a SCP can be satisfied with more than one equivalent circuit. Definitely more than one meaningful circuit cannot be extracted from the multiple relaxations in the SCP. The arrangement of the circuit elements must match with the underlying operative mechanisms. The single relaxation of the Z*-plane or M*-plane of Figure 4.5 suggests a simplistic R-C parallel circuit. Thus, for three successive semicircular relaxations in the same Z*-plane or M*-plane can obviously yield three R-C parallel combination in series known as the voltage dividing Voigt circuit. To validate such a circuit three consecutive events must be determined between the electrodes.

Single semicircle response often refers to clear material system such as depletion region of the single-crystal based p-n junction. Depending on the value of the resistance and capacitance of the bulk with respect to the resistance and capacitance of the p-n junction region (depletion layer) visible intercept or another semicircle on the left-side of the low-frequency semicircle can appear. For small magnitude of the bulk parameters this behavior may be suppressed or become non-visible in the Z*- and M*-planes with respect to the capacitance and resistance of the p-n junction. While fitting the semicircle a tiny value of the intercept is more than likely to be visible which eliminates the capacitance of the bulk related to semiconductor's permittivity.

The Voigt circuit is purely voltage dividing in nature. Each relaxation segment experiences certain voltage across it. Thus, each R-C parallel combination is associated with a voltage drop for the multiple relaxations in the Z*- or M*-plane. Again each R-C parallel segment follows series or successive event within the DUT. Maxwell circuit represents current dividing feature for each branch or leg or segment found in the multiple relaxations in the Y*- or C*-plane. The Ladder circuit reflects both voltage and current dividing segments for the multiple relaxations which is somewhat complicated but requires explanation for such arrangement of the elements.

To understand voltage dividing and current dividing equivalent circuits in conjunction with the dc blocking and non-blocking (i.e., conductive) features as a function of applied frequency f, it is necessary to consider the type of passive elements obtained via complex plane plot. Figures 4.6,

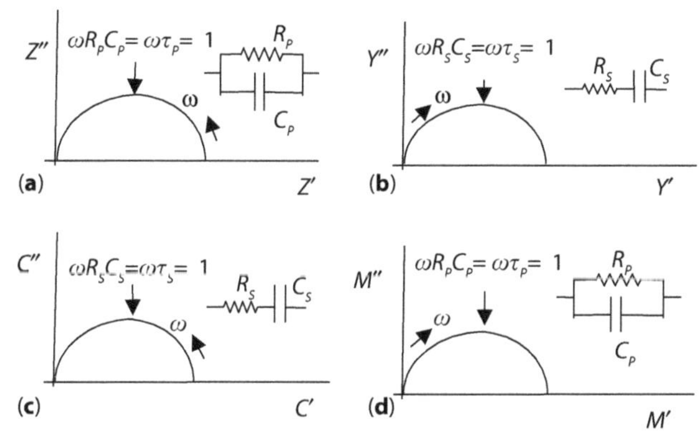

Figure 4.6 Simplistic equivalent circuit for single relaxation without the intercept in four complex planes where τ is the time constant in the respective complex plane and an arrow indicates the direction of the increasing measurement angular frequency ω.

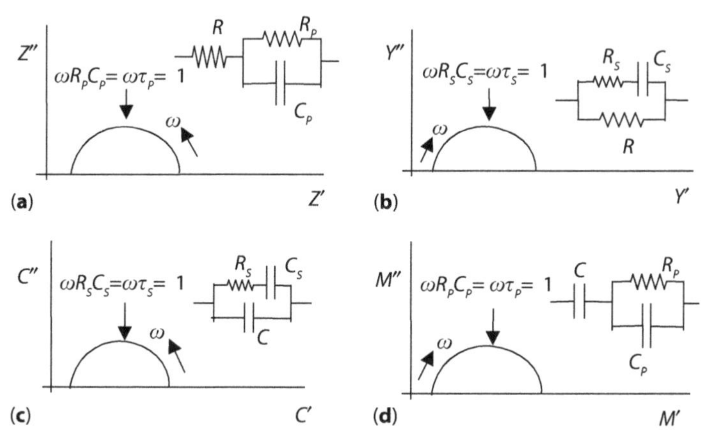

Figure 4.7 Simplistic equivalent circuit for single relaxation with intercept in each of the four complex planes where τ is the time constant in the respective complex plane and an arrow indicates the direction of the increasing measurement angular frequency ω.

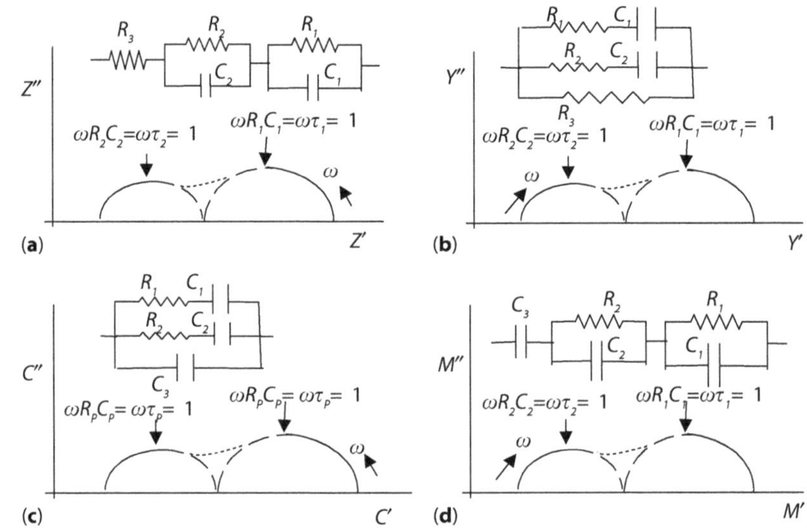

Figure 4.8 Simplistic equivalent circuit for two relaxations with intercept in each of the four complex planes where τ is the time constant in the respective complex plane and an arrow indicates the direction of the increasing measurement angular frequency ω.

4.7, and 4.8 are presented with single semicircle without intercept, single semicircle with intercept, and two semicircles with intercept for each of the complex planes. For single relaxation without intercept for the Figure 4.6 gives an R-C parallel circuit for the Z*- and M*-planes whereas an R-C series circuit in the Y*- and C*-planes. The intercept of Figure 4.7 provides the series element with an R-C parallel circuit for the Z*- and M*-planes, and the parallel element with R-C series circuit in the Y*- and C*-planes. Careful inspection of the Z*- and M*-planes indicates R replaced by C and C replaced by R in the equivalent circuit whereas Y*- and C*-planes indicate C replaced by R and R replaced by C in the equivalent circuit. The same observation is the case for the relaxation responses and corresponding equivalent circuit model for each complex plane shown in Figure 4.8.

Upon inspecting Figure 4.5 it is clear that the equivalent circuit obtained in the Z*-plane is always conductive (or non-blocking) in nature regardless of the number of relaxations. Therefore, any combination of the circuit elements must correspond to the conductive or non-blocking feature. Multiple relaxations simply give series segments via Voigt circuit indicating voltage dividing feature. The intercept on the left-side of the semicircular relaxation is the part of the voltage division for the entire circuit. Since the x-axis represents real part of the impedance Z*, the intercept is always resistance portraying conductive aspect of the equivalent circuit.

This behavior is realized in Figures 4.6(a), 4.7(a), and 4.8(a). There are two segments for the Voigt circuit beside the intercept in Figure 4.8(a). No segment is considered as shorted at any measurement frequency but getting switched to the next segment leaving behind the finite resistance in the frequency domain.

Figure 4.6 indicates that the single semicircular relaxation in both Z*- and M*-planes give conductive or non-blocking circuit whereas both Y*- and C*-planes give dc blocking circuit. When an intercept exists on the left-side of the semicircle, as shown in Figure 4.7, then both Z*- and Y*-planes give conductive or non-blocking circuit whereas both C*- and M*-planes give dc blocking circuit. In any situation Z*-plane always gives conductive or non-blocking circuit, and C*-plane always gives dc blocking circuit. Thus, features in Z*-plane is often important for any non-insulating DUT, and in C*-plane refers to the insulating type DUT. It is important to know at the outset whether the DUT is conductive type or insulating type.

In general, the equivalent circuit obtained in the Y*-plane for single relaxation without intercept shown in Figure 4.6(b) is blocking in nature. Existence of intercept in the Y*-plane portrays non-blocking or conductive feature of the ultimate current dividing equivalent circuit provided in Figures 4.7(b) and 4.8(b). Referring to Figure 4.8(b) the conductive path initiates at low frequencies yielding the intercept R_3 at the outset of the measurement, and then gradually exhausts the next low-frequency semicircular capacitance C_2 yielding the associated series resistance R_2 constituting the time constant τ_2. This path never shorts out the electrical conduction but gets diverted to the next higher frequency semicircle. Upon exhausting the capacitance C_1 at high frequencies the associated series resistance R_1 constitutes time constant τ_1. Since the x-axis represents real part of the admittance Y*, the intercept is always conductance portraying conductive aspect of the equivalent circuit. This behavior is realized in Figures 4.7(b) and 4.8(b) As a current dividing circuit, no branch or leg or segment is shorted at any measurement frequency.

Each equivalent circuit shown in each complex plane representation of Figures 4.6, 4.7, and 4.8 refers to usual straightforward model. Whenever multiple relaxations appear, modeling of the equivalent circuit may become tedious. It is important to comprehend that for a single relaxation it is possible to visualize two potential circuits though the ac small-signal electrical data may not exhibit more than one relaxation in more than one complex plane.

The equivalent circuit obtained in the C*-plane for single relaxation without intercept shown in Figure 4.6(c) is dc blocking in nature. Existence of intercept in the C*-plane portrays blocking feature of the ultimate current dividing equivalent circuit provided in Figures 4.7(c) and 4.8(c). In the circuit of Figure 4.8(c), the low-frequency semicircular capacitance C_1 is gradually exhausted to yield the associated series resistance R_1 that constitutes the time constant τ_1. This path never shorts out the electrical conduction but gets diverted to the next higher frequency semicircular path. Upon exhausting the capacitance C_2 at high frequencies the associated series resistance R_2 reveals time constant τ_2. Finally, the dc blocking element C_3 emerges as intercept at extremely high frequency domain. Since the x-axis represents the real part of the complex capacitance C*, the intercept is always capacitance portraying dc blocking aspect of the equivalent circuit that corresponds to the high frequency response. The dc blocking behavior is realized in Figures 4.6(c), 4.7(c), and 4.8(c). As a current dividing circuit no branch or leg or segment is shorted at any measurement frequency.

Upon inspecting Figure 4.5(d) or Figure 4.6(d), it is clear that the equivalent circuit obtained in the M*-plane is always conductive (or non-blocking) for single relaxation without intercept. The intercept on the left-side of the semicircular relaxation is the part of the voltage division for the entire circuit but it blocks the dc conduction. Since the x-axis represents real part of the modulus M*, the intercept is always elastance implying reciprocal of the capacitance portraying dc blocking aspect of the equivalent circuit at the outset of the measurement at low frequencies. The next gradual branch that gets exhausted is the low-frequency semicircular capacitance C_2 yielding the associated parallel resistance R_2 constituting the time constant τ_2. This path never shorts out the electrical conduction but gets diverted to the next higher frequency semicircle. Upon exhausting the capacitance C_1 at high frequencies the associated parallel resistance R_1 constitutes time constant τ_1. This behavior is realized in Figures 4.6(d), 4.7(d), and 4.8(d). Two segments of the Voigt circuit beside the intercept in Figure 4.8(d) is considered as shorted at any measurement frequency but getting switched to the next segment leaving behind the finite parallel resistance. It is learnt from the foregoing features of the complex planes that capacitance does not get shorted as frequency increases but allows switching the segment for the exhaustion to the next level of capacitance at higher frequencies. Further it is also noticed that the capacitance in the current dividing circuit obtained in the Y*- and C*-planes adds together because of the parallel nature for them.

The appearance of the equivalent circuit may look identical for the non-Debye relaxations with the ideal Debye relaxations. The conversion of these elements requires the non-Debye factor to obtain the Debye equivalent. Debye and non-Debye inter-mixing conversion is very common among users.

4.4 Duality of the Equivalent Circuits: Multiple Circuits for a Single Plane

Derivation of the equivalent circuit model from the multiple relaxations in a single complex plane is often tedious if multiple relaxations are observed for the DUT which is complicated. For a simplistic scenario the Voigt circuit is straightforward. However, complexity can arise with the consideration of Maxwell or Ladder type circuits for the Z*-plane. Figure 4.9 shows a situation where three possible circuits may arise with two relaxations having an intercept in the Z*-plane. The investigator needs to understand about the complexity of the DUT thoroughly to select the appropriate circuit satisfying the relaxations. This complexity of the equivalent circuit is applicable for every complex plane that exhibits multiple relaxations with an intercept.

For the Y*-plane, Maxwell circuit is expected but consideration of Voigt or Ladder circuit is crucial. Similarly for the C*-plane, the Maxwell circuit

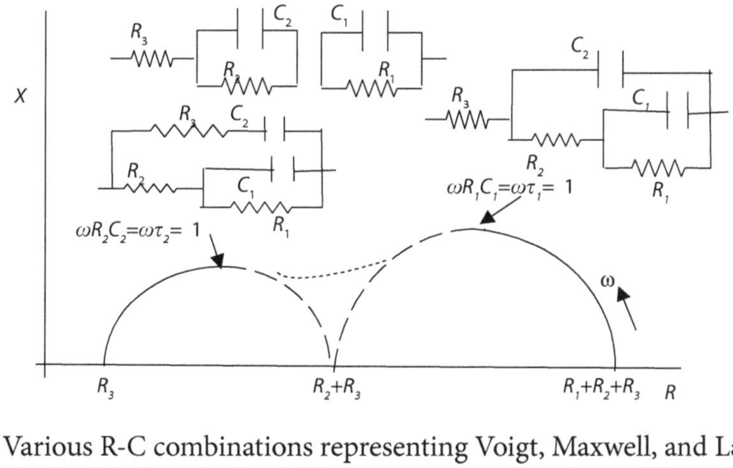

Figure 4.9 Various R-C combinations representing Voigt, Maxwell, and Ladder type circuits in the Z*-plane (= R + j X) showing two relaxations with an intercept where τ is the time constant and an arrow indicates the direction of the increasing measurement angular frequency ω.

is expected but consideration of a Voigt or Ladder circuit requires critical examination. In the same way for the M*-plane, the Voigt circuit is expected but consideration of Maxwell or Ladder requires validity. In fact, the Ladder type circuit has no room to accommodate in any of the complex plane relaxations. Therefore, consideration of the Ladder circuit can arise as an offshoot model for the DUT. For the solid state systems Ladder circuit is rarely accommodated for interpretation regardless of the complex plane representation of the relaxations. Discussion on Voigt, Maxwell, and Ladder circuits are found elsewhere [5, 6].

4.5 Duality of Equivalent Circuits between Z*- and M*-Planes for Relaxations without Intercept

Matching resistance and reactance can provide one relaxation in the Z*-plane as well as in the M*-plane for the same data. Extracted resistance, capacitance, and relaxation time in each complex plane may coincide each of these parameters closely for the two complex planes. For two relaxations in Z*-plane and one relaxation in the M*-plane the scenario is different for the same set of data. Respective relaxations are displayed in Figure 4.10. From the simplistic equivalent circuits:

$$R_M = R_1 + R_2 \tag{4.10}$$

where R_M is extracted from the elastance and time constant of the M*-plane. Both R_1 and R_2 appear to be on the same order of magnitude regardless of $R_1 > R_2$ or $R_1 < R_2$ as they are obtained from the finite chord of each semicircle in the Z*-plane. Of course the size of the semicircle determines the value of the chord. This means that if R_1 is large then the

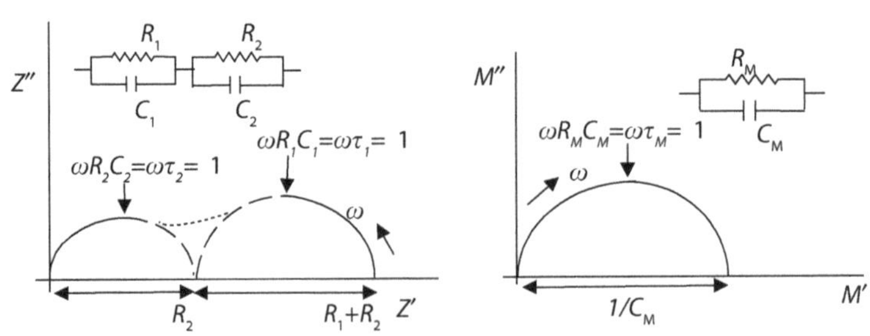

Figure 4.10 Duality between two relaxations in Z*-plane and one relaxation in M*-plane without intercept where τ is the time constant and an arrow indicates the direction of the increasing measurement angular frequency ω.

semicircle is large which obviously extends to wide frequency range for the data. Likewise if R_2 is large then wide frequency range is expected for this relaxation. The feature of capacitance in the Z*-plane clearly indicates that there are two distinct time constants visibly differing that originates from the magnitude of the capacitances. For the M*-plane, the elastance obtained from the chord constitutes single capacitance where the smaller valued capacitance dominates in the lumping response. It is expected that the difference between the two capacitances is at least an order of magnitude to distinguish the two time constants in the Z*-plane. The M*-plane representation does not clearly distinguish the presence of two distinct time constants whereas the Z*-plane does.

Figure 4.11 displays two relaxations in M*-plane and one relaxation in the Z*-plane for the same set of data. From the simplistic equivalent circuits:

$$R_Z = R_1 + R_2 \tag{4.11}$$

where R_Z is extracted from the chord of the Z*-plane. Both C_1 and C_2 appear to be on the same order of magnitude regardless of $C_1 > C_2$ or $C_1 < C_2$ as they are obtained from the chord representing elastance of each semicircle in the M*-plane. The size of the semicircle determines the value of the chord. This means that if C_1 is large then the semicircle is large which obviously extends to wide frequency range for the data. Likewise if C_2 is large then wide frequency range is expected for this relaxation. The feature of resistance in the M*-plane clearly indicates that there are two distinct time constants visibly differing that originates from the magnitude of the resistances. For the Z*-plane, the chord constitutes single

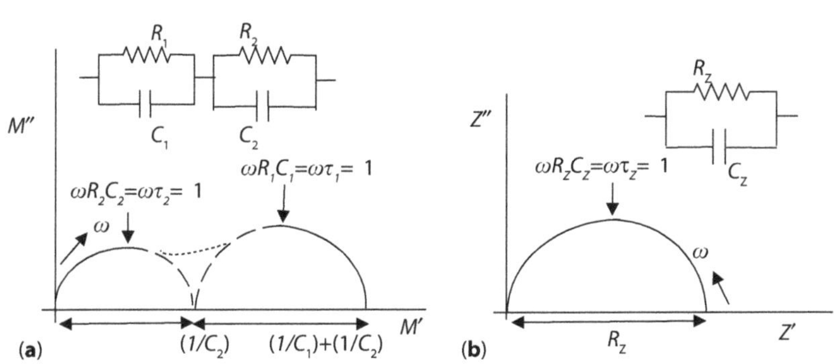

Figure 4.11 Duality between two relaxations in M*-plane and one relaxation in Z*-plane without intercept where τ is the time constant and an arrow indicates the direction of the increasing measurement angular frequency ω.

resistance where the smaller valued resistance dominates in the lumping response. It is expected that the difference between the two resistances is at least an order of magnitude to distinguish the two time constants in the M*-plane. The Z*-plane representation does not clearly distinguish the presence of two distinct time constants whereas the M*-plane does [7].

4.6 Duality of Equivalent Circuits between Y*- and C*-Planes for Relaxations without Intercept

Figure 4.12 displays two relaxations in Y*-plane and one relaxation in the C*-plane for the same set of data. From the simplistic equivalent circuits:

$$C_C = C_1 + C_2 \tag{4.12}$$

where C_C is extracted from the chord of the C*-plane. C_C is dominated by the larger value either for C_1 or for C_2. Both R_1 and R_2 appear to be on the same order of magnitude regardless of $R_1 > R_2$ or $R_1 < R_2$ as they are obtained from the chord of each semicircle in the Y*-plane. The size of the semicircle determines the value of the chord. This means that if the conductance G_1 is large then R_1 is small, and if the conductance G_2 is small then R_2 is large. Thus, for the large semicircle resistance is small extending to short frequency range for the data. Likewise for the small semicircle wide frequency range is expected for this relaxation. The feature of resistance in the Y*-plane clearly indicates that there are two distinct time constants visibly differing that originates from the magnitude of the capacitances. For the C*-plane, the chord constitutes single capacitance C_C where the smaller-valued capacitance dominates in the lumping response.

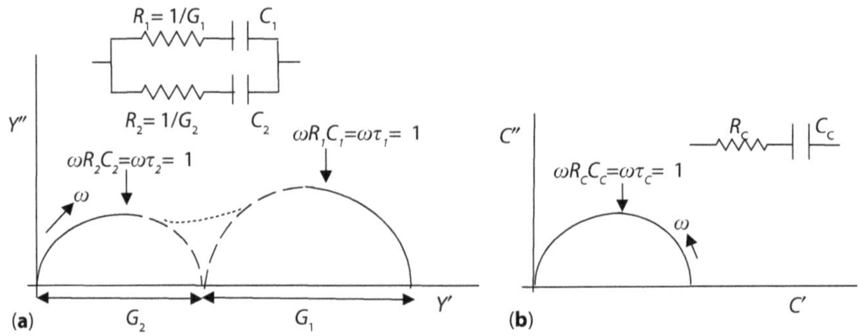

Figure 4.12 Duality between two relaxations in Y*-plane and one relaxation in C*-plane without intercept where τ is the time constant and an arrow indicates the direction of the increasing measurement angular frequency ω.

It is expected that the difference between the two resistances is at least an order of magnitude to distinguish the two time constants in the Y*-plane. The C*-plane representation does not clearly distinguish the presence of two distinct time constants whereas the Y*-plane does.

Figure 4.13 displays two relaxations in C*-plane and one relaxation in the Y*-plane for the same set of data. From the simplistic equivalent circuits:

$$C_Y = C_1 + C_2 \tag{4.13}$$

where C_Y is extracted from the time constant of the Y*-plane. C_Y is the summation of C_1 and C_2. Both C_1 and C_2 appear to be on the same order of magnitude regardless of $C_1 > C_2$ or $C_1 < C_2$ as they are obtained from the chord of each semicircle in the C*-plane. The size of the semicircle determines the value of the chord. This means that if C_1 is large then extending data exists in the semicircular arc. If C_2 is small then short range data exists in the semicircular arc. The feature of resistance in the C*-plane clearly indicates that there are two distinct time constants visibly differing that originates from the magnitude of the resistances. For the Y*-plane, the chord constitutes single resistance R_Y where the smaller valued resistance dominates in the lumping response. It is expected that the difference between the two resistances is at least an order of magnitude to distinguish the two time constants in the C*-plane. The Y*-plane representation does not clearly distinguish the presence of two distinct time constants whereas the C*-plane does.

Based on duality presented in Figures 4.10, 4.11, 4.12, and 4.13 it is possible to draw a straightforward conclusion for the conditions on relative magnitude of the elements for which four complex plane plots provide best separation of multiple time constant systems or DUTs.[17, 18] This is summarized in Figure 4.14.

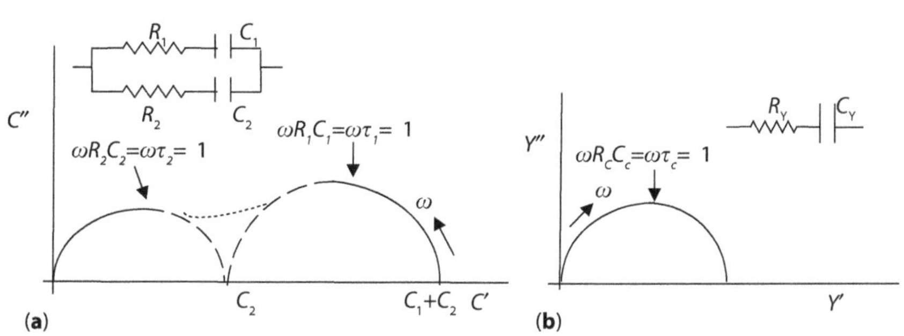

Figure 4.13 Duality of equivalent circuits between two relaxations in C*-plane and one relaxation in Y*-plane without intercept where τ is the time constant and an arrow indicates the direction of the increasing measurement angular frequency ω.

Conditions on magnitude of resistance (R's) and capacitances (C's)	Complex plane plots	
R's equal, C's varied	Z*	Y*
R's varied, C's equal	M*	C*

Figure 4.14 Conditions on magnitude of resistance and capacitance to obtain duality with another complex plane providing best separation of multiple time constant systems or DUTs.

4.7 Duality of Equivalent Circuits for Simultaneous Z*-, Y*-, C*-, and M*-Planes' Relaxations

Two situations can be considered for the data giving one relaxation in all four complex planes simultaneously having intercept and no intercept on the left side of the semicircle. The first situation without intercept is depicted in Figure 4.6, and the second situation with intercept is depicted in Figure 4.7. Circuits in Figure 4.6(a) and 4.6(d) coincide each other satisfying ac/dc non-blocking condition, whereas circuits in Figure 4.6(b) and 4.6(c) coincide each other having dc blocking condition. Since dc blocking situation requires explanation, it is absolutely necessary to comprehend as how the DUT turned to near perfect insulator and what has caused this to happen. Overall from the simplistic scenario of the circuit stand point, proper interpretation of the selection requires determination.

The equivalent circuits of Figure 4.7(a) and 4.7(d) coincide each other as it is found in Figure 4.7(b) and 4.7(c). As seen from Figure 4.7(c) and 4.7(d) the circuits are the dc blocking type. Also these two circuits are difficult to convert each other's form, and then justify the underlying operative mechanisms for the DUT. Each circuit requires valid logic for the dc blocking behavior. Figure 4.7(a) and 4.7(b) are giving non-blocking circuits but conversion of one to the other is a tedious task although it is needed to verify the value of each element. It is essential to find out why this duality exists, and related interpretation for the elements.

So far the ZnO-based varistor exhibited some form of relaxation simultaneously in Z*- and Y*-planes, and multiple relaxations in the C*-plane, and were found meaningful [8–11] in a series of papers in 1988 that continued thereafter [12–15]. The fourth complex plane provided dc blocking behavior and was excluded in reporting during that time as the varistor does not support the dc blocking behavior due to the valid leakage conduction or industrially concerned watts loss. However, the same M*-plane was presented [16] in 1997 where no intercept was indicated which is ridiculous to trust for the same plane. Only inadequate data in the low frequency, particularly lacking in the sub-Hz domain, can mislead this way.

4.8 Proposition of Equivalent Circuit: Polycrystalline Grains and Grain Boundaries

Polycrystalline materials are often complicated due to the presence of multiple phases and unlimited types of defect states. Grain-to-grain contact is never found to have uniform pattern. An ideal model of grain arrangement having grain-to-grain contact is shown in Figure 4.15(a). This is similar to the ideal orientation of grains having a brick-wall type model as if all bricks are uniformly placed at a certain interval. For such a model, the equivalent circuit comprises of an alternate R-C parallel combination in series comprising of grain and grain boundary between the electrodes. This is because the grain and grain boundary are forming two distinct relaxations because of the difference in resistance as shown in Figure 4.15(b). The same uniform grains as in Figure 4.15(a), having identical orientation in succession along with the grain boundary as shown in Figure 4.15(c), constitute a single relaxation time when grains and grain boundaries are lumped together. This is a strong lumped behavior of grain and grain boundary together as shown in Figure 4.15(d). Such a lumped behavior is possible when capacitance and resistance of grain are identical with that of the grain boundary. In polycrystalline material grain orientations are random or different from each other as shown in idealized manner in Figure 4.15(e). For each grain and grain boundary is shown in succession as $R_{1\text{-}G}$-$C_{1\text{-}G}$ and $R_{1\text{-}GB}$ –$C_{1\text{-}GB}$ parallel combination as each pair with dotted line for N grains. This is shown in Figure 4.15(f). In reality such a concept ultimately forms a lumped response. Since the circuit is, overall, a parallel R-C combination it is likely that such a trap-less idealized polycrystalline material might exhibit Z*-plane relaxation.

For the systematic progression of idealized layered grains of Figure 4.16(a) on the development of the equivalent circuit is shown in Figure 4.16(b). Overall, this idealized layered arrangement may constitute

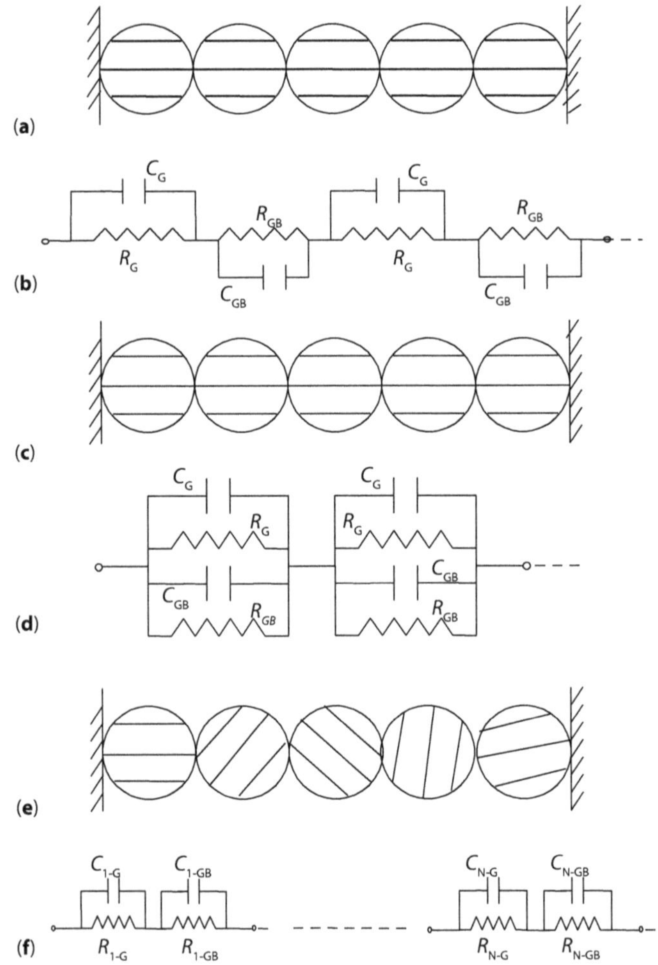

Figure 4.15 Idealized polycrystals in (a) show the equivalent circuit in (b) where grains and grain boundaries are distinguishable, while idealized polycrystals in (c) show the equivalent circuit in (d) where grains and grain boundaries are not distinguishable, and for a random orientation of grains in (e) shows the equivalent circuit in (f).

an R-C parallel combination in series for the distinguishable lumped grains and the lumped interfaces designated as the grain-boundaries due to the difference in time constant. Each layered system possesses an R-C parallel combination that exercises as the lumped response in the eventual modeling.

Two-layered polycrystalline material system possessing two distinct time constants is shown in Figure 4.17(a). Such a system can exhibit two R-C parallel combinations in series as the equivalent circuit. If two R's are identical then two relaxations can be obtained in the Z^*-plane whereas if

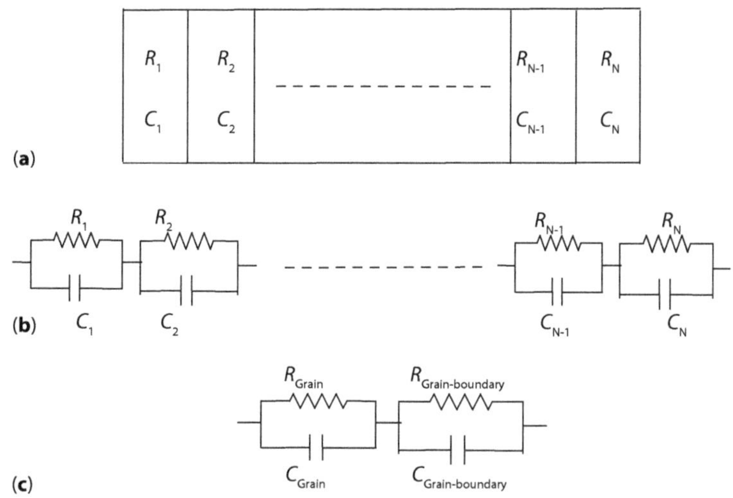

Figure 4.16 Systematic progression of idealized layered grains shown in (a) is modeled in (b) while the eventual circuit is displayed in (c).

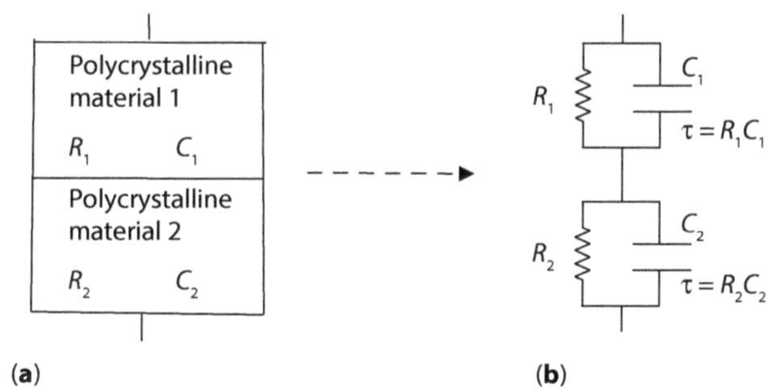

Figure 4.17 Two-layered polycrystalline material system of (a) possessing two distinct time constants shown in (b).

two C's are identical then two relaxations can be obtained as non-blocking in the M*-plane.

Problems

1. What equivalent circuit can be expected for a metal-semiconductor junction? What equivalent circuit is expected for a p-n junction? If an insulator is used as the third material to form a junction for these two devices then what kind of equivalent circuit can be expected? Is ac small-signal measurement

possible for the third device? If the measurement is limited, then explain the reasons.

2. What is the major difference between a voltage dividing circuit and a current dividing circuit? How does the Ladder circuit differ from these two circuits? What mechanism or mechanisms would be necessary for each of these circuits? Explain the answer.
3. What is the difference between a dc blocking circuit and a conductive circuit? Can one parallel R-C circuit be replaced by one series R-C circuit in a network or vice versa?
4. In an all-purpose network in a segment or a path, if a parallel R-C circuit has 10 kΩ with 10 nF combination coinciding with that of a p-n diode having the same values measured at any frequency, then could the p-n diode be used in lieu of the parallel R-C circuit in that segment or path? Explain the answer.
5. In an all-purpose network in a segment or a path, if a parallel R-C circuit has 10 kΩ with 10 nF combination coinciding with that of a p-n diode having the same values only at 1 MHz, then could the p-n diode be used in lieu of the parallel R-C circuit in that segment or path? Explain the answer.

References

1. J. D. Irwin and R. M. Nelms; *Basic Engineering Circuit Analysis,* 10th Edition, John Wiley & Sons, Inc., New York, 2011.
2. D. H. Johnson; "*Origins of the Equivalent Circuit Concept: The Voltage-Source Equivalent,*" Proceedings of the IEEE, 91, 636–640, 2003.
3. D. H. Johnson; "*Origins of the Equivalent Circuit Concept: The Current-Source Equivalent,*" Proceedings of the IEEE, 91, 817–821, 2003.
4. C. P. Smyth; "*Dielectric Behavior and Structure,*" McGraw-Hill Book Co., N.Y., 1955.
5. E. Barsoukov and J. R. MacDonald; *Impedance Spectroscopy: Theory, Experiment, and Applications,* Edited, John Wiley & Sons, Inc., Hoboken, New Jersey, 2005.
6. J. R. MacDonald; *Impedance Spectroscopy: Emphasizing Solid materials and Systems,* Edited, John. Wiley & Sons, Inc., Hoboken, New Jersey, 1987.
7. J. J. Ackmann and M. A. Seitz; 'Methods of Complex Impedance Measurements in Biologic Tissue," *CRC Critical Reviews in Biomedical Engineering,* 11(4), 281–311, 1984.

8. M. A. Alim, M. A. Seitz and R. W. Hirthe; "Complex Plane Analysis of Trapping Phenomena in Zinc Oxide Varistors," *Journal of Applied Physics*, 63(7), 2337–2345, 1988.
9. M. A. Alim and M. A. Seitz; "Singular Nature of Preferential Conducting Paths at High Electric Fields in ZnO-Based Varistors," *Journal of the American Ceramic Society*, 71, C246–C249, 1988.
10. M. A. Alim, M. A. Seitz and R. W. Hirthe; "High-Temperature/Field Alternating Current Behavior of ZnO-Based Varistors," *Journal of the American Ceramic Society*, 71, C52–C55, 1988.
11. L. C. Sletson, M. E. Potter and M. A. Alim; "Influence of Sintering Temperature on Intrinsic Trapping in Zinc Oxide-Based Varistors," *Journal of the American Ceramic Society*, 71, 909–913, 1988.
12. M. A. Alim; "Admittance-Frequency Response in Zinc Oxide Varistor Ceramics," *Journal of the American Ceramic Society*, 72(1), 28–32, 1989.
13. M. A. Alim; "High-Frequency Terminal Resonance in ZnO-Bi_2O_3-Based Varistors," *Journal of Applied Physics*, 74(9), 5850–5853, 1993.
14. M. A. Alim; "Influence of Intrinsic Trapping on the Performance Characteristics of ZnO-Bi_2O_3 Based Varistors," *Active and Passive Electronic Components*, 17(2), 99–118, 1994.
15. M. A. Alim; "Influence of Multiple Trapping Phenomena on the Applications Criteria of ZnO-Bi_2O_3 Based Varistors," *Active and passive Electronic Components*, 17(1), 57–65, 1994.
16. M. Andres-Verges and A. R. West; "Impedance and Modulus Spectroscopy of ZnO Varistors," *Journal of Electroceramics*, 1–2, 125–132, 1997.
17. J. J. Ackmann and M. A. Seitz; 'Methods of Complex Impedance Measurements in Biologic Tissue," *CRC Critical Reviews in Biomedical Engineering*, 11(4), 281–311, 1984.
18. M. A. Seitz; "Study of Heterogeneous Composite Materials via Lumped Parameter/Complex Plane Analysis," *International Journal of Hybrid Microelectronics* (I.S.H.M.), 3(1), 1–7, 1980.

5

Debye and Non-Debye Relaxations

5.1 Ideal Systems

The ideal or Debye relaxation process from the measured ac small-signal electrical data is very rarely observed for the engineering material systems and devices [1–4]. The Debye behavior represented in the form of the semicircular loci in the complex plane formalism [2], where the center essentially lies on the real axis (x-axis), is described in its general form as:

$$I^{*}(\omega) = I_{\infty} + \frac{I_{dc} - I_{\infty}}{1 + j\omega\tau} \tag{5.1}$$

where I portrays immittance parameter. The subscript indicates values corresponding to f → 0 Hz (dc) and f → ∞ (high frequency). Essentially equation (5.1) refers to the impedance (Z^*) or capacitance (C^*) plot where measurement frequency increases anti-clockwise direction along the semicircular loci. Invariably the generalized equations for admittance (Y^*) and modulus (M^*) are a little different than equation (5.1) because the direction of the measurement frequency increases in the clockwise direction along the semicircular loci.

For Debye relaxation conversion of one immittance function to another gives ideal parameters. This means resistance R and/or capacitance C in the equivalent circuit are ideal elements implying a resistor made of metal film conductor and capacitor constitutes parallel plate configuration.[5] Thus, a series R-C or parallel R-C combination provides ideal Debye relaxation depending on the specific complex plane representation. Essentially engineering material systems and devices behaving like a metal film resistor and a parallel plate capacitor for their equivalent circuit model can be referred to as ideal systems.

In order to derive Debye relaxation, the measured form of the DUT as a function of measurement frequency considered in the equivalent parallel R_p-C_p mode shown in Figure 5.1 yields admittance [6] expression as:

$$Y^* = G_p + j\,(\omega C_p) \tag{5.2}$$

where relaxation time τ (= R_p C_p), G_p = $1/R_p$, and subscript P denotes the mode of measurement. The impedance Z^* for the same R_p-C_p parallel combination becomes

$$Z^* = \frac{1}{Y^*} = \frac{1}{G_p + j\omega C_p} = \frac{1}{G_p\left(1 + j\omega\left[\dfrac{C_p}{G_p}\right]\right)} \tag{5.3a}$$

or

$$Z^* = \frac{R_p}{1 + j\omega R_p C_p} = \frac{R_p}{1 + j\omega\tau}. \tag{5.3b}$$

The generalization of equation (5.3) requires a series resistance R with the R_p-C_p parallel combination per Figure 5.1(b). This series resistor R simply adds to the impedance in the real part while the relaxation portion remains unchanged. Thus, equation (5.3) becomes:

$$Z^* = R + \frac{R_p}{1 + j\omega R_p C_p} = R + \frac{R_p}{1 + j\omega\tau}. \tag{5.4}$$

The second term in equation (5.3) or (5.4) represents Debye relaxation via a semicircular loci in the complex plane desired as impedance (Z^*) plot to yield R-C parallel equivalent circuit per Figure 1(a). The parameter R_p represents the value of the equivalent resistance which comes from the difference between the two extreme ends of the terminal frequencies designated as low frequency (LF) where $f \to 0$ Hz (dc) and high frequency (HF) where $f \to \infty$. Thus,

$$R_p = R_{dc} - R_\infty \tag{5.5}$$

can be used in equation (5.4) to obtain the idealized Debye form:

$$Z^* = R + \frac{R_{dc} - R_\infty}{1 + j\,\omega\,R_p\,C_p} = R + \frac{R_{dc} - R_\infty}{1 + j\,\omega\,\tau} = Z' - j\,Z''. \tag{5.6}$$

Once again for no left-side intercept equation (5.3) satisfies equivalent circuit of Figure 5.1(a), and the presence of the left-side intercept equation (5.6) satisfies Figure 5.1(b). From the equivalent circuit viewpoint, R is obtained as intercept at high frequency $f \rightarrow \infty$ on the left side of the semicircular relaxation in the Z*-plane depicted in Figure 5.2. The real part of the impedance is Z′ and the imaginary part of the impedance is Z″ gives ideal Debye semicircle having the center on the x-axis means the parameter R_p (= R_{dc} - R_∞) is the diameter. The generic Z*-plot of Figure 5.2 must possess an equal unit magnitude on both abscissa (x-axis) and ordinate (y-axis) so that the same value can be ascribed for the distance known as the radius or diameter of the semicircle.

If the equivalent circuit corresponds to Figure 5.1(a) then the data can be converted to modulus using:

$$M^* = (j\,\omega)\,Z^* = M' + j\,M'', \tag{5.7}$$

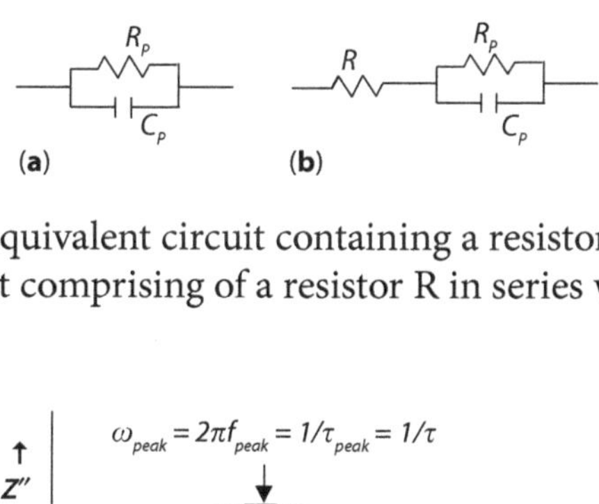

Figure 5.1 (a) Parallel equivalent circuit containing a resistor R_p and a capacitor C_p. (b) Equivalent circuit comprising of a resistor R in series with the parallel R_p-C_p combination.

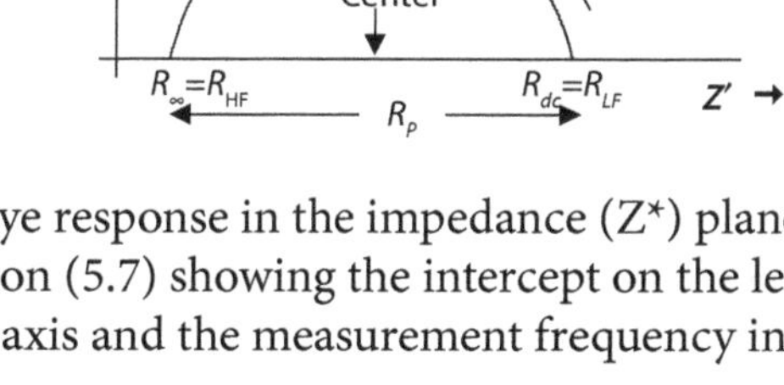

Figure 5.2 Ideal Debye response in the impedance (Z*) plane via the semicircular response using equation (5.7) showing the intercept on the left side of it where the center is on the x-axis and the measurement frequency in the anti-clockwise direction.

and can be applied to the relaxation process when plotted M′ (real axis) versus M″ (imaginary axis). This is because the data corresponding to equation (5.4) are used in the form of equation (5.7). In this case there is a high possibility of obtaining semicircular relaxation in the M*-plane unless extremely non-compatible or non-matching values exist for both resistance and reactance of the DUT. The circuit of Figure 5.1(a) can be demonstrated as the Debye response in both Z*- and M*-planes concerning the analysis of the data where no intercept [7] was present. For the data corresponding to equation (5.4) or (5.6) the conversion to M* using equation (5.7) may yield the equivalent circuit with heavy distortion in the semicircular relaxation at high frequency showing vertical line as M″ (= ωR_S) grows upward with increasing measurement frequency. This vertical line is due to the presence of R as the series resistor of Figure 5.1(b).

There may be two situations for the role of the series resistor R concerning Figure 5.1(b) with the M*-plane relaxation depicted in Figure 5.3. If R is a low frequency visible parameter then Figure 5.3(a) is feasible. In the Z*-plane R is the high frequency visible parameter as it is derived from the left-side intercept of the semicircular relaxation on the x-axis. This R is expected to be the high frequency parameter in the M*-plane as well. Thus, the semicircular relaxation must exhibit distortion toward the high frequency (f → ∞) end. For this situation Figure 5.3(a) is not the plausible appearance but Figure 5.3(b) is a more reasonable appearance of the entire process. If the response resembles Figure 5.3(a), then the investigator must comprehend as to how R can be legitimate as the low frequency visible parameter in the M*-plane.

The entire discussion on Z*- and M*-planes is applicable for the equivalent circuits shown in Figure 5.3(a) and Figure 5.3(b). In the same way discussion on Y*- and C*-planes is applicable for the equivalent circuits shown in Figure 5.4(a) and Figure 5.4(b). The semicircular relaxation of

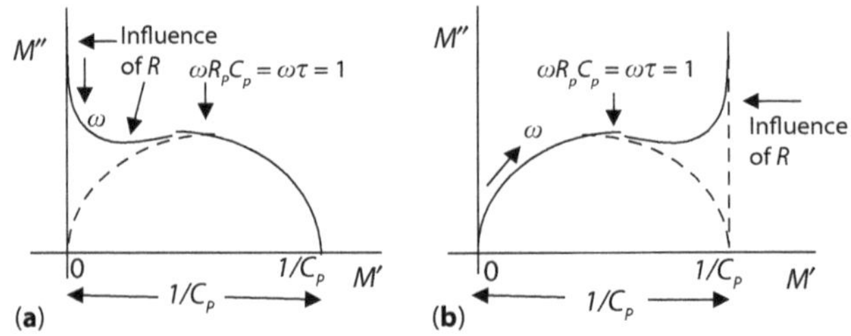

Figure 5.3 Two possible relaxation situations in the M*-plane representation of the parallel R_p-C_p combination in series with R where τ is the time constant and an arrow indicates the direction of the increasing measurement angular frequency ω.

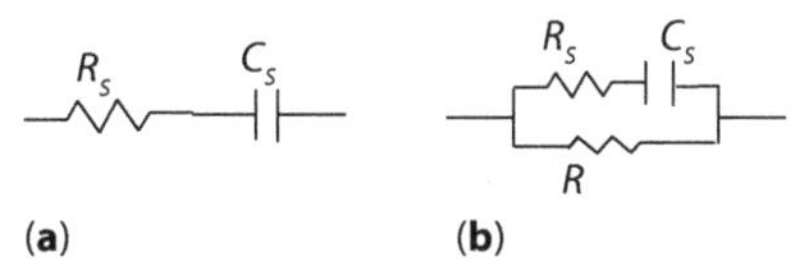

Figure 5.4 (a) Series equivalent circuit containing a resistor R_S and a capacitor C_S. (b) Equivalent circuit comprising of a resistor R in parallel with the series R_S-C_S combination.

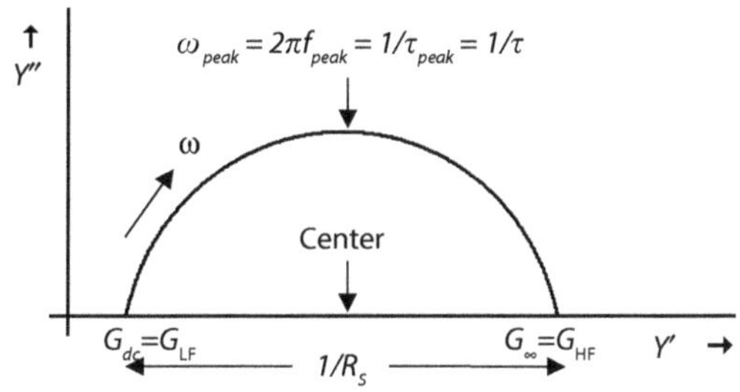

Figure 5.5 Ideal Debye response in the Y*-plane via the semicircular response using equation (5.2) showing the intercept on the left side of it where the center is on the x-axis and the measurement frequency in the anti-clockwise direction.

Figure 5.4(a) refers to the series R_S-C_S combination. When the intercept is achieved on the left side there must be a resistor in parallel with the series R_P-C_P combination as in Figure 5.4(b). From the equivalent circuit viewpoint, R is obtained as the intercept at low frequency f → dc (0 Hz) on the left side of the semicircular relaxation in the Y*-plane depicted in Figure 5.5.

If the equivalent circuit corresponds to Figure 5.4(a) then the data can be converted to complex capacitance using:

$$C^* = Y^*/(j\,\omega) = C' + jC'', \tag{5.8}$$

and can be applied to the relaxation process when plotted C′ (real axis) versus C″ (imaginary axis). In this case there is a high possibility of obtaining semicircular relaxation in the C*-plane unless extremely non-compatible or non-matching values exist for both resistance and reactance of the DUT. Two circuits of Figure 5.4 can be demonstrated as the Debye response in both Y*- and C*-planes.

There may be two situations for the role of the parallel resistor R of Figure 5.4(b) with respect to the C*-plane relaxation depicted in Figure 5.6. If R is a high frequency visible parameter then Figure 5.6(a) is feasible. In the Y*-plane R is the low frequency visible parameter as it is derived from

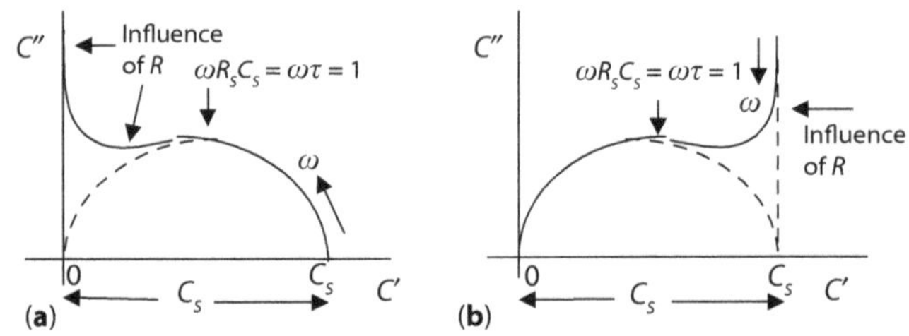

Figure 5.6 Two possible relaxation situations in the C*-plane representation of the series R_S-C_S combination in parallel with R where τ is the time constant and an arrow indicates the direction of the increasing measurement angular frequency ω.

the left-side intercept of the semicircular relaxation on the x-axis. This R is expected to be the low frequency parameter in the C*-plane as well. Thus, the semicircular relaxation must exhibit distortion toward the low frequency (f → dc) end. For this situation Figure 5.6(a) is not the plausible appearance but Figure 5.6(b) is a more reasonable appearance of the entire process. If the response resembles Figure 5.6(a) then the investigator has to comprehend as how R can be legitimate as the high frequency visible parameter in the C*-plane.

At this conjecture the classic illustration of Grant [8] provided in Figure 5.7 depicts duality of the semicircles between two complex planes. Four equivalent circuits were demonstrated in both complex conductivity (σ^*) and complex permittivity (ε^*) planes. These two complex planes are directly derived from the non-normalized admittance (Y^*) plane for the complex conductivity (σ^*), and non-normalized complex capacitance (C^*) plane for the complex permittivity (ε^*). Grant demonstrated the duality between these two complex planes for a single circuit, for which underlying viewpoints are illustrated by Daniel [9] and later by Bunget and Popescu [10]. Despite the existence of this kind of pioneering illustration of using dual complex planes in a simultaneous fashion, widespread examination or evaluation of the data did not get popularity although it was rarely noted [11–15].

The presence of steady conductivity σ (dc or low frequency resistance) distorts the permittivity (or C*-plane) plot at low frequency while the presence of ε_∞ (high frequency capacitance) distorts the conductivity (or Y*-plane) plot. However, elimination of ε_∞ can give uniform or complete semicircular response in the σ^*-plot. This can be achieved via subtracting the identified value of ε_∞ or the capacitance at high frequency. Such a process must not be practiced as subtraction route eliminates the capacitance presumably related to geometric contribution within the DUT implying

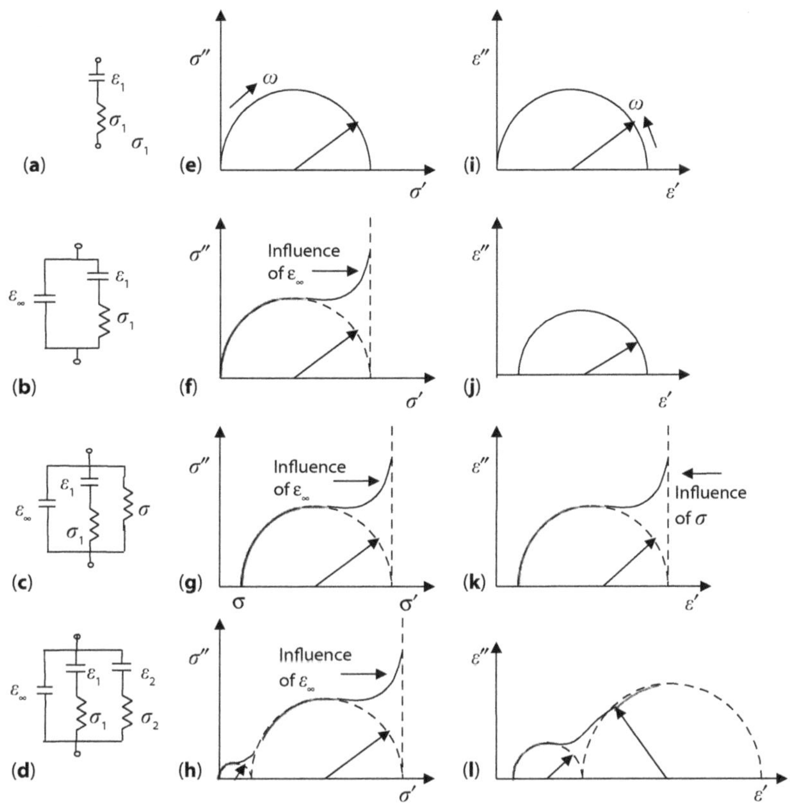

Figure 5.7 Reproduction of complex conductivity (σ' versus σ'') plot and complex permittivity (ε' versus ε'') plot corresponding to four equivalent circuits from the classic paper of Grant [7].

depletion region or barrier layer beside bulk behavior. Thus, two contributions based on the geometric configuration for the DUT belongs to ε_∞ or high frequency capacitance. Since ε_∞ can be delineated via the left-side intercept in the ε^*-plane (i.e., C^*-plane) relaxation, it is absolutely reasonable to establish two geometrical capacitances existing in parallel within the DUT. It may be noted that the depletion region and bulk behavior operate in a parallel fashion between the two opposite electrodes for the DUT.

The low frequency conductivity σ implying $f \rightarrow 0$ Hz (dc) causes distortion in the permittivity plot. Thus, the relaxation curve varies with the level of initial value of the dc conductivity of the DUT. The influence of dc conductivity on the semicircular relaxation loci is illustrated by Coelho [16] which is presented in Figure 5.8. As the level of dc conductivity increases, the loci start to distort gradually. For such a behavior the equivalent circuit stands the same as in Figure 5.7(c). The inset of Figure 5.8 illustrates

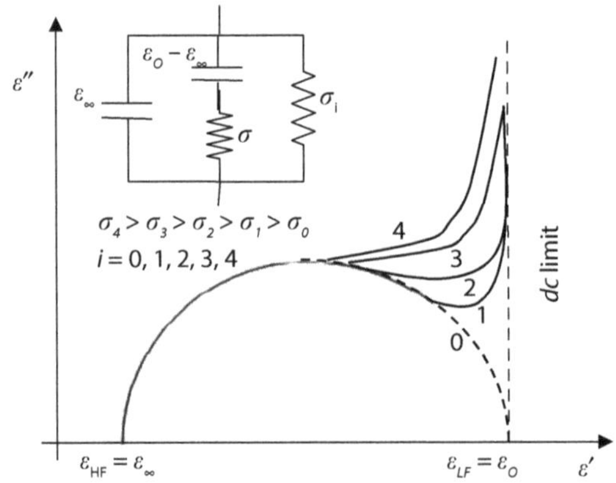

Figure 5.8 Influence of the dc conductivity σ on the dielectric relaxation presented in the permittivity plane plot for the equivalent circuit shown in the inset where curves 1 through 4 represent different dc conductivity values and curve 0 represents perfect insulator having $\sigma_1 = 0$.

the same equivalent circuit for the curves σ_1, σ_2, σ_3, and σ_4. These curves represent a gradual increase in the dc conductivity while σ_0 represents perfect insulator-like behavior depicting dotted line for the segment of the semicircular relaxation. This conductivity is often used in defining the leakage current of the DUT. It means higher the leakage current is, then higher the conductivity is. Since the ZnO-based varistor is used in surge protection, leakage current stabilization and/or reduction is important for the surge arrester. Varistors exhibit variation in the dc conductivity that influences the distortion in the relaxation loci at low frequencies observed via the C*-plane plot [17]. Reduction in the dc conductivity is a viable developmental effort for the varistors for surge arrester applications.

5.2 Non-Ideal Systems

A non-ideal system can be recognized via the display of the measured ac small-signal electrical data in any of the complex plane formalisms. The same data can be converted to other complex plane formalisms employing mathematical expression or equation. Often these mathematical forms are good for the Debye relaxation and corresponding equivalent circuit model. This is because these convertible equations are idealized satisfying only Debye type behavior.

The equivalent circuit for the non-ideal system extracted from a single complex plane relaxation is, usually, complicated. This is because the equivalent circuit elements are directly not legitimate to convert to another form. An example can clarify this complication. An R-C parallel circuit can be converted to its R-C series form using straightforward mathematical expressions. The values obtained for the series form can also be used to reversely obtain the values for the parallel combination. These back-and-forth values of the R-C parallel and R-C series are now established as valid or correct. This is only possible when the R-C parallel combination represents the Debye form of admittance or the R-C series form represents the Debye form of impedance.

If the R-C parallel circuit is associated with the non-ideal system then the corresponding admittance or impedance will certainly be non-ideal or non-Debye. The conversion method of this R-C parallel circuit to R-C series form employs mathematical expression which is idealized but never contains a non-ideal corrective factor. Thus, R-C parallel combination and its converted R-C series form are not equal or equivalent or identical. The number or value of each element reflects validity but the real world scenario is far from this situation. This is because the relaxation process of the R-C parallel form has never had Debye type as its origin because the semicircular relaxation does not possess the center on the x-axis. Certainly back-and forth conversion of the extracted circuit traces the origin. If the semicircular relaxation for the R-C parallel combination has a depression parameter h or a depression angle θ as shown in Figure 5.9, then the converted R-C series parameters require consideration of perturbation in the pertinent equation. As a regular practice, investigators use a conversion style as if each element is Debye without hesitation or prior attention. Basically the equivalent circuit of Figure 5.1(b) satisfies Figure 5.2 as "Debye origin" whereas Figure 5.9 does not satisfy as "non-Debye origin" for all purposes. Originally Cole-Cole [3] considered necessary perturbation for the associated element for the ε^*-plane plot. For simplicity and for the time being, if the presumption is the same value for each circuit element from these two illustrations then the attitude is naïve which is not free from erroneous concept.

Now consider a simulation of the equivalent circuit to rejuvenate the plot. If the simulation is done correctly using usual admittance equation then the plot of Figure 5.2 can be achieved perfectly. Yet, Figure 5.9 can never be achieved. This is because the simulating equation used the Debye values obtained from Figure 5.2. Again if the values of Figure 5.9 are used to simulate then Figure 5.2 will be reproduced, not Figure 5.9. This is due to the use of Debye equation each time for the simulation. Nevertheless, the simulating equation requires incorporation of the non-Debye parameters obtained as depression parameter h or depression angle θ as the

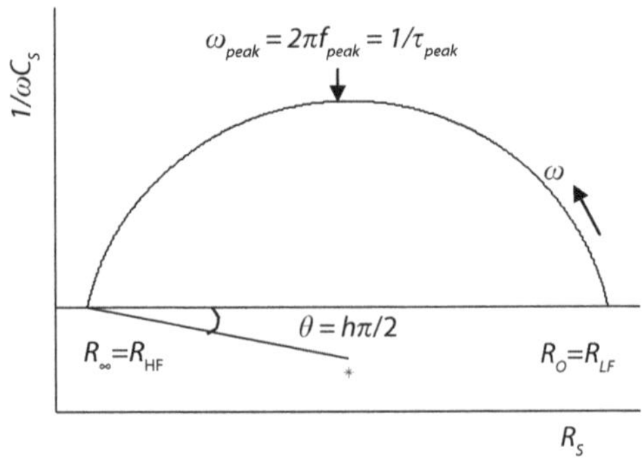

Figure 5.9 Schematic illustration of a depressed semicircular relaxation in the Z*-plane showing how the electrical parameters are extracted using the complex plane analysis where θ is the depression angle, *h* is the depression parameter, and U_i and V_I are the chords of the semicircular loci.

correction factor. Getting back to equation (5.4), the semicircular relaxation function F*(ω) is given by:

$$F^*(\omega) = \frac{1}{1 + j\omega\tau} \tag{5.9}$$

where essentially

$$\tau = \frac{1}{\omega_{peak}} = \frac{1}{2\pi f_{peak}} \tag{5.10}$$

defined in equation (5.2) satisfying

$$\omega_{peak}\,\tau = \omega_{peak}\,\tau_{peak} = 1 \tag{5.11}$$

when represented in the complex plane reveals a depression angle equal to zero. The ideal semicircular relaxation containing an intercept on the left side of the semicircle is presumed to be either zero or not included in F*(ω) in equation (5.9). In representing a semicircular behavior the ordinate (y-axis) and abscissa (x-axis) must have the same or equal plotting grid scale so that the relaxation in the complex plane can clearly be delineated. This means that each unit grid or graphical segment in length on the ordinate must be equal to the abscissa. Often many investigators display simplistic x- versus y- plot without considering equal plotting grid. This must be avoided and curve fitting procedure must be installed.

A semicircle whose center lies on the x-axis is defined as Debye or Debye-like relaxation, whereas for a non-Debye relaxation the center of the semicircle lies below the x-axis giving a finite depression angle (θ) measured from the point at the left-intercept to the center of the semicircle below the x-axis. The depression angle gives rise to the amount of the depression demonstrated originally by Cole-Cole [3] as the ratio of θ and π/2(= 90°) [4]. Thus, the Cole-Cole (C-C) type depression parameter *h* is given by [16]:

$$h = \frac{\theta}{\pi/2} = \frac{2\theta}{\pi}, \tag{5.12}$$

satisfying the relaxation function

$$F^*(\omega) = \frac{1}{1+(j\omega\tau)^{1-h}} = \frac{1}{1+(j\omega\tau)^{\alpha}}. \tag{5.13}$$

The parameter h ranges between 0 and 1 depending on the idealized Debye and/or extreme non-Debye responses for the semicircular relaxation process. This means $h \to 0$ describes idealized Debye and $h \to 1$ describes extreme non-Debye. The parameter α is related to the depression parameter h showing dependence on each other by:

$$1 - h = \alpha. \tag{5.14}$$

The value of α ranges between 0 and 1. Equation (5.13) takes Debye form when $h \to 0$ for which $\alpha \to 1$, and becomes non-Debye when $h \to 1$ for which $\alpha \to 0$. Then the equation (5.14) leads to a point on the x-axis when $h \to 1$ implying $\alpha = 0$. The extreme non-Debye response is a point on the x-axis for $\alpha = 0$. The ideal Debye or Debye-like behavior can be achieved as $\theta \to 0°$ implying $\alpha \to 1$, and an extreme simplistic non-Debye response can be visualized when $\theta \to 90°$ implying $\alpha \to 0$. That is why α may be termed as the "inverse depression parameter" in the present context.

Any deviation from the ideal Debye response reveals non-ideal or non-Debye behavior usually via the presence of the depression parameter for the semicircular relaxation and/or skewed nature of the semicircular loci. The Cole-Cole [3] type non-Debye relaxation does not have skewed loci whereas Davidson-Cole [18, 19] and Havriliak-Negami [20, 21] relaxations possess skewed behavior of the semicircular loci. These non-Debye dielectric responses are often described as Cole-Cole (C-C),

Davidson-Cole (D-C), and Havriliak-Negami (H-N) relaxations in the complex plane formalism via complex permittivity (ε^*) plot which is the normalized presentation of the complex capacitance (C^*) plot where the measurement frequency increases in the anti-clockwise direction along the relaxation curve.

The complex plane representation of the ac small-signal electrical data possessing this feature is usually referred to as the Cole-Cole [3] plot. The depression parameter for each of the non-Debye responses derived from the semicircular loci represents complexity in the overall relaxation processes. The curve fitting procedure [4, 7] of the data points in the complex plane formalism often provides extraction of the parameters associated with the semicircular relaxations. Some effort has been issued to find out the nature of the skewed behavior [7, 22] apart from the presence of the depression behavior. Overall, two non-Debye relaxation functions given by

$$F^*(\omega) = \frac{1}{[1+(j\omega\tau)]^{\beta}}, \tag{5.15}$$

and

$$F^*(\omega) = \frac{1}{[1+(j\omega\tau)^{\alpha}]^{\beta}} \tag{5.16}$$

where α and β are the depression parameters that portray the skewed behavior in the relaxation loci. The scenario of α and β are much different than the parameter α described in equation (5.14). In no way α and β can be ascertained that their origin is only tied to the depression angle alone as in the Cole-Cole representation given by equation (5.13). Thus, equations (5.9), (5.13), and (5.15) can be described by imposing generalized conditions on α and β in equation (5.16) that range between 0 and 1. These generic Debye and non-Debye relaxations are presented in Figure 5.10.

The non-skewed C-C [3] relaxation is given by

$$I^*_{C-C}(\omega) = I_{C-C\infty} + \frac{(I_{C-CO} - I_{C-C\infty})}{1+(j\,\omega\,\tau_{C-C})^{\alpha}} \tag{5.17}$$

whereas the skewed behavior described as the Davidson-Cole [17, 18] relaxation is given by

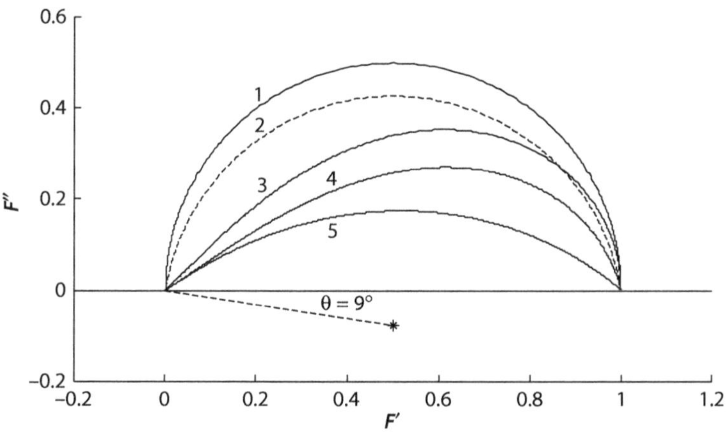

Figure 5.10 Generic representation of the relaxation curves in complex plane for the relaxation function $F^*(\omega)$ given in equation (5.16): (1) Debye formalism with depression angle $\theta = 0°$ or $\alpha = 1$ and $\beta = 1$, (2) non-Debye Cole-Cole formalism with depression angle $\theta = 9°$ or $\alpha = 0.9$ and $\beta = 1$, (3) Davidson-Cole formalism with $\alpha = 1$ and $\beta = 0.5$, (4) Havriliak-Negami formalism with $\alpha = 0.85$ and $\beta = 0.45$; and (5) Havriliak-Negami formalism with $\alpha = 0.45$ and $\beta = 0.85$.

$$I^*_{D-C}(\omega) = I_{D-C\infty} + \frac{(I_{D-CO} - I_{D-C\infty})}{(1 + j\,\omega\,\tau_{D-C})^{\beta}}. \tag{5.18}$$

and the Havriliak-Negami [19, 20] relaxation is given by

$$I^*_{H-N}(\omega) = I_{H-N\infty} + \frac{(I_{H-NO} - I_{H-N\infty})}{\left[1 + (j\,\omega\,\tau_{H-N})^{\alpha}\right]^{\beta}}. \tag{5.19}$$

Originally equations (5.17), (5.18), and (5.19) were represented in the ε*-plane (or C*-plane) where the measurement frequency increases in the anti-clockwise direction along the relaxation loci. The subscript C-C, D-C, and H-N represent Cole-Cole, Davidson-Cole, and Havriliak-Negami relaxations, respectively. The subscript 0 denotes the dc (or static condition) measurement corresponding to $f \rightarrow 0$ Hz, subscript ∞ indicates high frequency end corresponding to $f \rightarrow \infty$, and τ is the relaxation time for respective situation. Both α and β are the depression parameters that are used in describing ideal Debye, non-skewed non-Debye (C-C), and skewed non-Debye (D-C and H-N) relaxation loci in the ε*- or C*-plane.

5.3 Non-Ideal Systems Implying Distributed Time Constants

The presence of a finite depression angle in the semicircular relaxation obtained in any of the complex planes represents non-ideal or non-Debye behavior for the DUT. In general, this kind of DUT is more than likely to be a MPCHPH based system where no electrode effect is considered or encountered as each contact is considered to be ohmic. Of course certain liquid systems are different in that they exhibit depression angle in the semicircular relaxation. Surely this system is far from the SCSJ system concerning the homogeneity of the microstructural fabric. Usually a MPCHPH system contains discontinuous species as opposed to the continuous homogeneity of the SCSJ system [4]. Thus, homogeneous system allows uniform current flow at any given cross-section between the electrodes emphasizing ideal or Debye conduction processes. For inhomogeneous systems variation in impedance takes place at any given cross-section between the two opposite electrodes.

When measurement frequency f is applied for the ac small-signal stress then there are certain domains within the DUT where this stress is distributed. These domains are referred to as the electrical field falling regions. More precisely ac small-signal stress is distributed across each grain boundary containing two adjacent or successive grains where depletion region or space charge layer exist. Thus, depletion region or space charge layer is the electrical field falling region [4, 23–28]. Besides this situation, a small amount of the ac small-signal stress may be experienced across each grain interior without the depletion length for each grain which is sometimes referred to as the bulk. For a better practical assumption, there are two physical regions (grain interior and depletion region) where the ac small-signal stress is dropped (or experienced) and can easily be referred to the electrical field falling regions [4, 23–28].

The presence of the depression angle indicates very complicated situation as the ordinate (y-axis) parameter is depressed. For each complex plane ordinate parameter is different but associated with angular frequency ω as the multiplying factor or the dividing factor for the resistance (or conductance) or the capacitance. This makes the ordinate parameter completely frequency-dependent. Another aspect of the semicircle is that the chord obtained as the straight line segment on the x-axis corresponding to two endpoints (i.e., intercepts on left- and right-side) is independent of frequency. However, finding the root cause and meaning of the depression angle for the DUT depends on the particular complex plane and the investigator's comprehensive view point about the DUT [4, 17, 23–28].

Cole-Cole [3]gave a clear concept of the equivalent circuit for the single depressed semicircular relaxation observed in the C*-plane provided as the generic plot in Figure 5.11(a). This is the classic C-C plot where the ac small-signal electrical data are normalized to obtain complex permittivity ε^* (= ε' – j ε''). For the semicircular relaxation the parameter R is associated with the depression parameter originally defined as α in the C-C empirical relation:

$$\varepsilon^* = \varepsilon_\infty + \frac{(\varepsilon_o - \varepsilon_\infty)}{1+(j\omega\tau)^{1-\alpha}} \tag{5.20}$$

where symbols have their usual meanings, i.e., ε_∞ (ε_{HF}) is the intercept on the left side of the semicircle at high frequency (HF), ε_O (ε_{LF}) is the

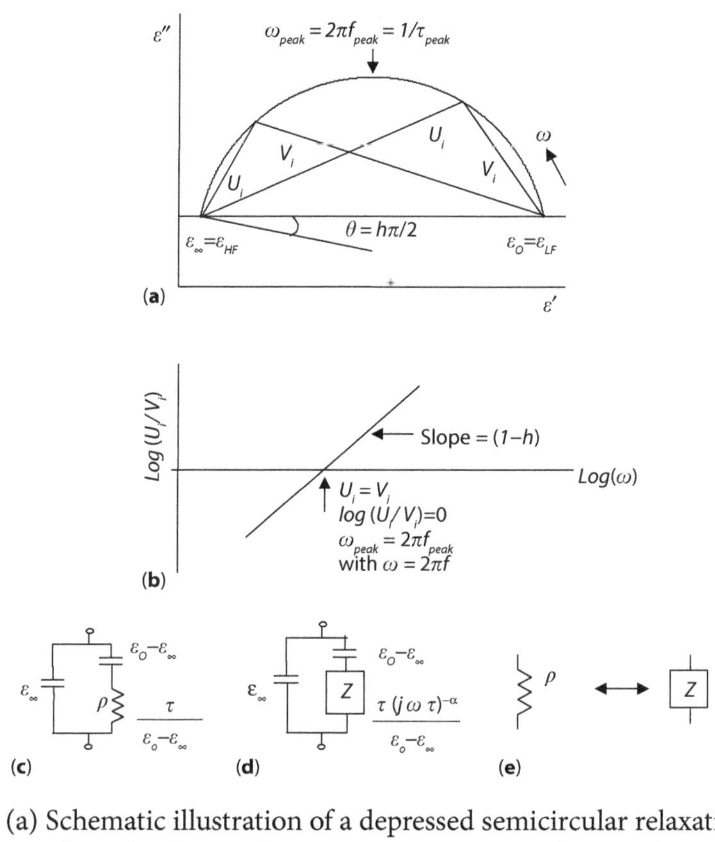

Figure 5.11 (a) Schematic illustration of a depressed semicircular relaxation in the ε^*-plane showing how the electrical parameters are extracted using the complex plane analysis where θ is the depression angle, h is the depression parameter, and U_i and V_I are the chords of the semicircular loci; (b) chord fitting verification via Log(U_i / V_i) versus Log(ω) plot; (c) ideal Debye circuit presuming Debye relaxation in (a); (d) Cole-Cole proposed circuit termed as non-Debye circuit for the relaxation in (a); (e) Debye resistor (resistivity) of (c) replaced by Cole-Cole (non-Debye) frequency-independent impedance.

intercept on the right side of the semicircle at low frequency (LF), and α is the depression parameter defined as the ratio of the depression angle θ and $\pi/2$ illustrated in Figure 5.11(a). The parameters U_i and V_i are the chords measured from any point on the semicircular loci corresponding to the measurement frequency f making $U_i = V_i$ at the peak of the semicircle. When log (U_i/V_i) is plotted against log f a straight line is obtained with slope $(1 - \alpha)$ as shown in Figure 5.11(b). For a Debye semicircle in the same complex plane one would obtain simple R-C series circuit as in Figure 5.7(b) corresponding to Figure 5.7(j) with a parallel C originating from the left-intercept ε_∞ (= C) of the semicircle. Thus, for Debye relaxation in the same complex plane the circuit displayed in Figure 5.7(a) corresponds to Figure 5.7(i).

The depression parameter α ranges between 0 and 1 implying depression angle θ ranging between 0° and 90°. For $\alpha = 0$, equation (5.21) yields Debye form whereas for the other extreme case, $\alpha = 1$, it is difficult to comprehend about the relaxation process or the dispersion of the DUT. For a finite value of α Cole-Cole [3] determined the associated impedance of the element per Figure 5.11(a) as:

$$Z = \frac{\tau(j\omega_{peak}\,\tau_{peak})^{-\alpha}}{\varepsilon_o - \varepsilon_\infty} \tag{5.21}$$

with a constant phase angle $\alpha\pi/2$ which is independent of frequency. Cole-Cole regarded the element in the form of impedance instead of discovering the term constant phase element. As of today, many investigators consistently use this impedance as a constant phase element or the acronym CPE. The CPE is generally believed to originate from a distribution in the current density along the electrode surface as a result of surface and bulk inhomogeneity. Use of the term or nomenclature CPE is simply an extraneous misnomer to identify the Cole-Cole impedance element. Thus, the depression parameter α or the depression angle $(\alpha\pi)/2$ is a straightforward terminology for the situation that Cole-Cole [3] observed as the impedance in the equivalent circuit expressed in equation (5.21). Consistent effort did not yield a meaningful or satisfactory interpretation for the depression parameter α or the depression angle $\alpha\pi/2$, though there is no shortage of concepts and ideas about it since then.

The parameter noted as Z in Figure 5.11(d) is the "complex impedance" of the element shown as the box which is independent of frequency. If the single element of Figure 5.11(d) possesses the "complex impedance" then this element is not singular in nature. The term "complex impedance" was originally emphasized by Cole-Cole [3] presumably because of the expression in equation (5.21). It contains the imaginary quantity "j" which refers to Z as the "complex quantity" for the element in Figure 5.11(d); otherwise "impedance" itself is a complex number. So there is no redundancy in the term "complex impedance" as used by Cole-Cole [3].

In general, the presence of the depression angle reflects overall nonsingular nature of the observed relaxation. The non-zero (finite) depression angle θ of a semicircular relaxation is presumably related to the non-uniformity in the ordinate parameter. Therefore, the interpretation of the presence of the depression angle is somehow related to the terminal frequencies. The meaning of singular and nonsingular nature of the observed relaxation requires explanation as the depression angle is likely to be present because of the degree or level of uniformity and non-uniformity of the conduction processes or paths between the electrodes of the DUT [4, 17, 22–30]. The time constant in equation (5.21) is complicated and presumed to be a distributed parameter for the DUT. Though the chord of the semicircle and the intercept on the left side of the semicircle constitute the real quantity as capacitance, the resistance turned out to be "Cole-Cole impedance" per equation (5.21). The complicated nature of the Cole-Cole impedance is surely associated with the y-axis parameter. Therefore, the depressed parameter is on the ordinate. During the conversion of the Cole-Cole impedance it is essential that equation (5.21) must be used in conjunction as a perturbing factor for the converting Debye equation.

There are several reasons for the presence of the depression angle for a DUT but it is never the case that any single reason alone is responsible for the presence of the depression angle. The tentative reasons listed are not limited to the following events [4, 17, 22–30] for the DUT as the:

1. modulation of the Fermi level at the intersection (crossover) with the defect energy varies about the equilibrium position from junction to junction (i.e., grain boundary to grain boundary) or at the electrical field falling region (basically depletion region or space charge layer) corresponding to the ac small-signal voltage;
2. variation in the capture cross-section and capture/emission rate of the traps or defect states from junction to junction (i.e., grain boundary to grain boundary) or at the electrical field falling region;

3. variation in the energy depth of the traps or defect states from junction to junction (i.e., grain boundary to grain boundary) or at the electrical field falling region;
4. variation in the relaxation time or time constant of the traps or defect states from junction to junction (i.e., grain boundary to grain boundary) or at the electrical field falling region;
5. variation in the process of holding charges and releasing charges at a particular site or defect state from junction to junction (i.e., grain boundary to grain boundary) or at the electrical field falling region;
6. variation in the defect states comprising of the combination of the donor-like traps and the acceptor-like traps from junction to junction (i.e., grain boundary to grain boundary) or at the electrical field falling region; and
7. variation in the density of each trap or defect state type from junction to junction (i.e., grain boundary to grain boundary) or at the electrical field falling region.

For the liquid system there are variations in the conduction path between the opposite electrodes. Therefore, the system needs to be defined critically where species may be involved that are causing depression angle. The conduction path may need evaluation deeply where charge transport, trapping, and de-trapping may be possible mechanisms that are causing variation in Fermi level (chemical potential) equivalent. If the fluid system is just similar to the solid then viewpoint requires some incorporation of the solid-state views for the liquid system.

The bulk capacitance of a semiconductor single crystal is not supposed to be thermally activated whereas the associated resistance can be thermally activated because of the thermal carrier generation possibly causing by the ionization of the constituting element(s). This is because the capacitance of a single crystal determines the dielectric constant ε_r of that single crystal which is usually invariant with the external conditions like temperature, pressure, etc. Incorporation of very high amount of contaminants or very large amount of impurities to the single crystal may cause some finite change in the terminal capacitance or corresponding dielectric constant ε_r. In that case terminal capacitance may be thermally activated beside the thermal activation of the resistance. Skewing from the continuity of the lattice structure to possess more than one phase in the so-called single-crystal configuration can cause deviation in the thermal behavior. In that case this skewing behavior is causing the onset of non-ideal or non-Debye response via the presence of the depression angle in the semicircular relaxation. Invariably this type of observation is often too complex to determine physical location within the DUT unless tedious single junction study is conducted. Certain behavior of the ZnO-Bi_2O_3-based multi-component polycrystalline varistor in this aspect is strong, and still lacking thorough understanding [4, 17, 27–31] though in many cases there are controlled fixed performance characteristics for the surge protection applications.

For almost all commercial varistors it is observed that in the close vicinity of current density 1 mA/cm² corresponding to the electric field E_{1mA/cm^2} the device enters a homogeneous (ideal or Debye) conduction process from inhomogeneous (non-Debye) conduction process [24, 30, 31]. This electric field or current density is the transition or demarcation between the non-Debye conduction processes at low voltages and Debye conduction processes at high voltages. This distinction is sharply noted at E_{1mA/cm^2} obtained via the depression angle of the semicircular behavior in the impedance (Z*) plane [30] for the ac small-signal electrical data where $\theta_{im} \rightarrow 0°$ as shown in Figure 5.12. It is then a fact that the tunneling is more than likely to be an involved mechanism that caused Debye conduction processes. Nevertheless, the current density within the device is operative in a homogeneous manner at this conduction process [30, 32].

Almost all existing models of the varistors are based on simple Debye or singular Debye-like relaxation concept of the conduction processes

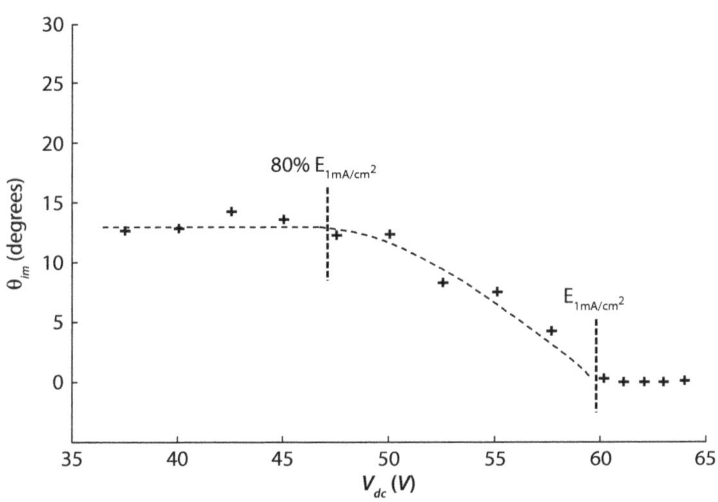

Figure 5.12 The dc voltage dependence of the depression angle θ_{im} obtained in the impedance (Z*) plane semicircular relaxation for the well-formed commercial ZnO-Bi_2O_3-based varistor.

across the grain boundaries which is indeed never observed in the devices whether prepared in the laboratory or in the manufacturing set-up. Also these varistors do not attribute to the dependence of the recipe specific or processing conditions. As a definite rule, ZnO varistors always exhibit non-Debye [17, 30] conduction processes for electric field $E < E_{1mA/cm^2}$. For electric field $E \geq E_{1mA/cm^2}$ the device makes a sharp transition from non-Debye to Debye conduction process, and attributed to the conduction "path channeling" and brought the device to a singular Debye-like relaxation. At a later date the same response was termed as the current localization [32]. Often generalized Debye model [33] is considered in determining grain-boundary admittance which is too simplistic scenario for the multi-component ZnO varistors. Experimental observations for these varistors do not represent the assumed situation as in a bicrystal [34]. The terminal capacitance of varistors under the dc condition is the total contribution of the traps [17, 29] between the two terminals. Nevertheless these traps are non-Debye and, thus, the capture cross-section and/or capture rate turns out to be a distributed parameter. Quantification of this distribution is realized but it poses a tedious task due to the distributed relaxation. Trapping is a series event for holding charges and releasing mobile carriers at a site or defect state causing R-C series combination as the equivalent circuit [17]. Holding charges at a site gives rise to the storage as capacitance while releasing charges from the same site gives mobile carriers causing conduction in the form of resistance or conductance.

At this conjecture Figure 5.13 depicts a curve fitting style for the non-Debye relaxation using chord ratio for complex capacitance (C^*) and impedance (Z^*) planes where the measurement frequency increases in the anti-clockwise direction. Also Figure 5.14 depicts curve fitting for the non-Debye relaxation using chord ratio for admittance (Y^*) and modulus (M^*) planes where the measurement frequency increases in the clockwise direction.

5.4 D-C Representation, Depression Parameter, and Equivalent Circuit: Conventional Domain

The Davidson-Cole empirical equation representing the skewed relaxation curve (arc) is reproduced from equation (5.18) in the ε^*-plane as:

$$\varepsilon^*(\omega) = \frac{(\varepsilon_s - \varepsilon_\infty)}{(1 + j\,\omega\,\tau_0)^\beta}. \tag{5.22}$$

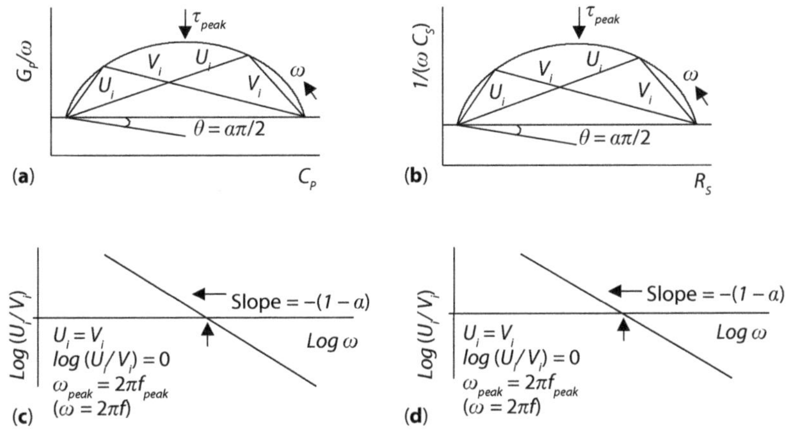

Figure 5.13 Fitting procedure of the semicircle in the C*-plane in (a) and Z*-plane in (b) where angular frequency ω increases in the anti-clockwise direction showing chord ratio U_i/V_i versus ω in the log-log scale in (c) and in (d), respectively, yielding depression parameter α, depression angle θ (= $\alpha\,\pi/2$), peak angular frequency ω_{peak} (= $2\,\pi\,f_{peak}$), and time constant τ_{peak}.

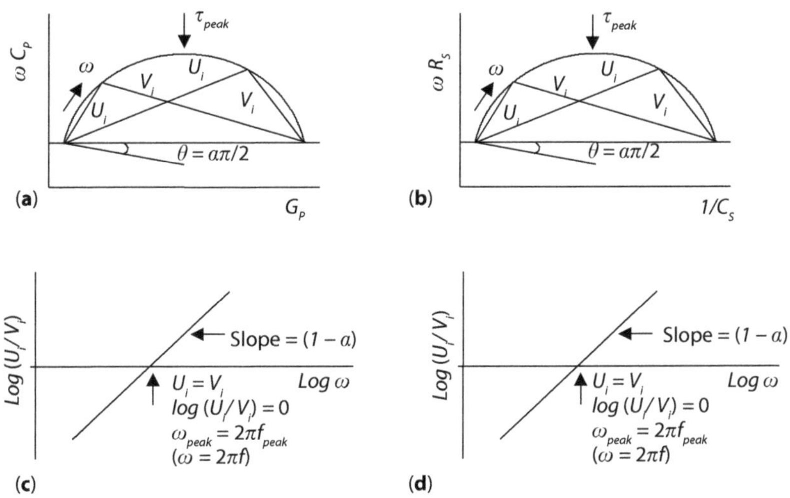

Figure 5.14 Fitting procedure of the semicircle in the Y*-plane in (a) and M*-plane in (b) where angular frequency ω increases in the clockwise direction showing chord ratio U_i/V_i versus ω in the log-log scale in (c) and in (d), respectively, yielding depression parameter α, depression angle θ (=$\alpha\pi/2$), peak angular frequency ω_{peak} (= $2\,\pi\,f_{peak}$), and time constant τ_{peak}.

Separating real and imaginary terms for equation (5.20) [22]:

$$\varepsilon'(\omega) = (\varepsilon_s - \varepsilon_\infty)\cos\phi^{\beta}\cos\beta\phi, \tag{5.23a}$$

and

$$\varepsilon''(\omega) = (\varepsilon_s - \varepsilon_\infty) \cos\phi^\beta \sin\beta\phi, \tag{5.23b}$$

where

$$\varphi = \text{arc tangent}(\omega\, \tau) = \tan^{-1}(\omega\, \tau), \tag{5.24}$$

and

$$1 + j\,\omega\,\tau_o = e^{j\phi}\sqrt{1+\omega^2\,\tau^2} = e^{j\phi}/\cos\phi, \tag{5.25}$$

where τ_o is the relaxation time involving the product of the R-C (resistance-capacitance) parameters corresponding to the equivalent circuit, and τ is the running parameter as a function of frequency. The Davidson-Cole empirical relation can be represented by the impedance quantities since the frequency-dependent data goes in the same anti-clockwise direction as does for the permittivity data. The entire feature remains the same if this simple transformation is done from complex permittivity (ε^*) to impedance (Z^*). Eliminating the geometrical parameters from the complex permittivity, ε^*, it becomes simply complex capacitance (C^*). The complex resistivity (ρ^*) becomes impedance, Z^*, by eliminating the geometrical parameters [4].

Rewriting equations (5.22) through (5.25) in the form of impedance terms [22]:

$$Z^*(\omega) = \frac{(Z_s - Z_\infty)}{(1 + j\,\omega\,\tau_o)^\beta} = Z' - j\,Z''. \tag{5.26}$$

Separating real and imaginary terms for equation (5.25) [22]:

$$Z'(\omega) = (Z_s - Z_\infty) \cos\phi^\beta \cos\beta\phi, \tag{5.27a}$$

and

$$Z''(\omega) = (Z_s - Z_\infty) \cos\phi^\beta \sin\beta\phi, \tag{5.27b}$$

with

$$\varphi = \text{arc tangent}\,(\omega\, \tau) = \tan^{-1}(\omega\, \tau), \tag{5.28}$$

and

$$1 + j\,\omega\,\tau_o = e^{j\phi}\sqrt{1+\omega^2\,\tau^2} = e^{j\phi}/\cos\phi. \tag{5.29}$$

The Davidson-Cole response can be achieved in the impedance plot where the measurement frequency moves anti-clockwise for each measurement point on the curve representing the total data. The intercept on the left side of the arc has a magnitude R_1 while the relaxation portion on the real axis (x-axis) gives $(Z_s - Z_\infty) = R_2$. The generic Davidson-Cole representation is provided in Figure 5.15. The real world situation exists for the Ni-Cd battery [22] for which the equivalent circuit may be considered per Figure 5.1(b). This circuit is widely familiar for general interpretation with the concept of time constant $\omega\tau = 1$ at the peak of the skewed semicircular curve. Due to the skewed nature of the Davidson-Cole curve the time constant is not necessarily corresponding to the conventional $\omega\tau = 1$ at the peak. Thus, this concept is likely to be dubious while deriving the simplified equivalent circuit like this one.

Rearranging real and imaginary parts of the D-C empirical relation [22]:

$$\begin{aligned} Z'(\omega) = \Delta Z \cos\varphi^\beta \cos(\beta\,\varphi) &= (Z_s - Z_\infty)\cos\varphi^\beta \cos(\beta\,\varphi) \\ &= (Z_{dc} - Z_\infty)\cos\varphi^\beta \cos(\beta\,\varphi), \end{aligned} \tag{5.30}$$

and

$$\begin{aligned} Z''(\omega) = \Delta Z \cos\varphi^\beta \sin(\beta\,\varphi) &= (Z_s - Z_\infty)\cos\varphi^\beta \sin(\beta\,\varphi) \\ &= (Z_{dc} - Z_\infty)\cos\varphi^\beta \sin(\beta\,\varphi). \end{aligned} \tag{5.31}$$

Setting $\Delta Z = (Z_s - Z_\infty) = R_2$ indicates the resistance via an equivalent circuit parameter constituting the relaxation time. The intercept magnitude on the left side of the skewed semicircular loci in Figure 5.15 is represented by the equivalent circuit parameter R_1. Taking the ratio of imaginary and real parts:

$$\tan(\beta\phi) = \frac{Z''}{Z'} = \frac{R_2\cos\phi^\beta\,\sin(\beta\phi)}{R_2\cos\phi^\beta\,\cos(\beta\phi)}. \tag{5.32}$$

Differentiating imaginary part Z" in equation (5.31) with respect to φ:

$$\frac{dZ^*}{d\phi} = -R_2.\cos\phi^\beta.\,\beta.\left[\frac{\sin\phi}{\cos\phi}\right]\sin(\beta\phi) + R_2.\cos\phi^\beta.\,\beta.\cos(\beta\phi). \tag{5.33}$$

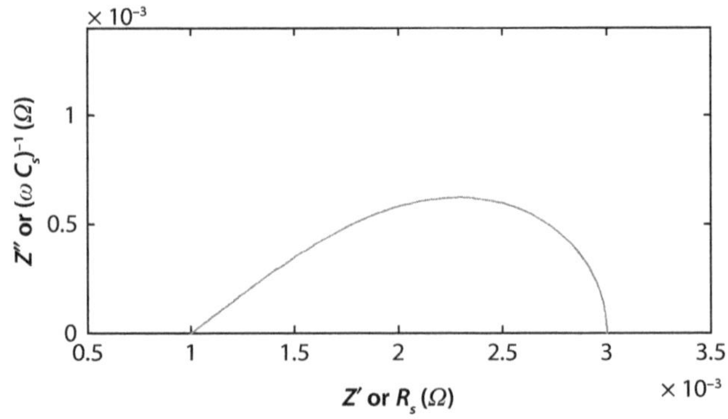

Figure 5.15 The generic impedance representation obtained from the Davidson-Cole relation given in equation (5.26) for the depression parameter $\beta = 0.4$. [Imaginary part shows positive.]

At the maximum magnitude of the y-axis parameter [$Z'' = 1/(\omega\ C_s)$] on the loci of the skewed semicircular behavior in Figure 5.15 frequency corresponds to the relaxation time. Terming this frequency as the "peak frequency," equation (5.23) may be set to zero. Thus,

$$\frac{dZ''_{pk}}{d\phi} = 0 = -R_2 .\cos\phi^{\beta}.\beta.\left[\frac{\sin\phi}{\cos\phi}\right].\sin(\beta\phi) + R_2 .\cos\phi^{\beta}.\beta.\cos(\beta\phi). \tag{5.34}$$

Then equation (5.31) becomes

$$\frac{\sin\phi_{pk}}{\cos\phi_{pk}} = \frac{\cos(\beta\phi_{pk})}{\sin(\beta\phi_{pk})}, \tag{5.35}$$

where φ is distinguished as φ_{pk}. Subsequent rearranging of equation (5.35) gives

$$\tan\phi_{pk} = \frac{1}{\tan(\beta\phi_{pk})}. \tag{5.36}$$

Using equations (5.32) and (5.36) φ_{pk} is obtained as

$$\phi_{pk} = \tan^{-1}\left(\frac{1}{\tan(\beta\phi_{pk})}\right) = \tan^{-1}\left(\frac{Z'_{pk}}{Z''_{pk}}\right) = \arctan\left(\frac{Z'_{pk}}{Z''_{pk}}\right) \tag{5.37}$$

and then

$$\tan(\phi_{pk}) = \frac{1}{\tan(\beta\phi_{pk})} = \frac{Z'_{pk}}{Z''_{pk}}. \tag{5.38}$$

Solving for β using equations (5.32) and (5.37):

$$\tan(\beta\phi_{pk}) = \frac{Z''_{pk}}{Z'_{pk}} \tag{5.39}$$

and then

$$\beta\phi_{pk} = \tan^{-1}\left(\frac{Z''_{pk}}{Z'_{pk}}\right) = \arctan\left(\frac{Z''_{pk}}{Z'_{pk}}\right), \tag{5.40}$$

yielding

$$\beta = \left(\frac{1}{\phi_{pk}}\right).\tan^{-1}\left(\frac{Z''_{pk}}{Z'_{pk}}\right) = \frac{\pi}{2\phi_{pk}} - 1. \tag{5.41}$$

The expression for β in equation (5.41) is a real quantity while the equivalent circuit parameter R_2 is a real quantity as input to the calculation. The parameter β is a highly complicated reflecting perturbation in the relaxation time in equation (5.25).

In order to extract R_2, substituting φ_{pk} in equations (5.26) and (5.27):

$$R_2 = \frac{Z'_{pk}}{\cos\phi_{pk}^{\beta}\cos(\beta\phi_{pk})} = \frac{Z''_{pk}}{\cos\phi_{pk}^{\beta}\sin(\beta\phi_{pk})}. \tag{5.42}$$

From equation (5.28),

$$\phi_{pk} = \tan^{-1}(\omega_{pk}\tau_o)\arctan(\omega_{pk}\tau_o), \tag{5.43}$$

where $\tau = \tau_o = R_2 C_2$, and ω_{pk} is the angular frequency at the peak of the relaxation arc corresponding to the maximum value of Z". Using equations (5.38) and (5.43) upon re-arrangement gives

$$\omega_{pk}\ \tau_0 = \tan\phi_{pk} = \frac{1}{\tan(\beta\phi_{pk})} = \frac{Z'_{pk}}{Z''_{pk}}. \tag{5.44}$$

Note that $\omega_{pk}\ \tau_o = 1$ only in the case of an ideal or Debye relaxation where $\beta = 1$ and $\varphi_{pk} = \pi/4$. Substituting τ_o using equivalent circuit elements:

$$R_2C_2 = \tau_o = \frac{1}{\omega_{pk}} \cdot \left(\frac{Z'_{pk}}{Z''_{pk}}\right) \tag{5.45}$$

in which ω_{pk} is the angular frequency at the maximum of the imaginary axis (y-axis). This gives

$$C_2 = \frac{1}{R_2\omega_{pk}} \cdot \left(\frac{Z'_{pk}}{Z''_{pk}}\right), \tag{5.46}$$

where Z_{pk}'/Z_{pk}'' is a scale factor or correction for the ideal or Debye relaxation. The depression parameter β and the equivalent circuit elements R_2 and C_2 obtained in equations (5.41), (5.42), and (5.46) are the real quantities. This is the usual expectation for the Davidson-Cole representation in the conventional real world domain.

5.5 Depression Parameter Based on $\omega\tau_{peak} = 1$: Complex Domain

The Davidson-Cole response in the impedance form may be rearranged using proper symbols for the measurement frequencies as:

$$\frac{Z - Z_{HF}}{Z_{LF} - Z_{HF}} = \frac{1}{\left[1+(j\omega\tau)\right]^{\beta}}, \tag{5.47}$$

where $Z = Z^* = Z(\omega)$, $\tau = \tau_o = R_2\ C_2$ = relaxation time, $\omega = 2\pi f$ = angular frequency with f is the measurement frequency. The adoption of $\tau = \tau_o = R_2\ C_2$ leads to the concept of $\omega\tau_{peak} = 1$. The parameter β is known as the depression parameter, representing both compressed and skewed behavior of the semicircular-like data representation when compared to the ideal Debye-like or C-C like semicircular behavior. The ideal Debye-like semicircular behavior has the center on the x-axis or C-C like semicircular behavior has the center below the x-axis while D-C response does not portray any specific center of the skewed behavior.

With $\alpha = 1$ in the Havriliak-Negami response of equation (5.19) gives a Davidson-Cole response as in equation (5.18) or in (5.20) or in (5.47). Taking log on both sides of equation (5.47) and then rearranging it as:

$$\ln\left(Z - Z_{HF}\right) = -\beta \ln(1 + j\omega\tau) + \ln(Z_{LF} - Z_{HF}). \tag{5.48}$$

Upon further re-arrangement of equation (5.48) the straight line equation of the form

$$y = ax + b \tag{5.49}$$

is obtained with

$$y = \ln\left(Z - Z_{HF}\right), \tag{5.50}$$

$$a = -\beta, \tag{5.51}$$

$$x = \ln\left(1 + j\omega\tau\right), \tag{5.52}$$

$$b = \ln\left(Z_{LF} - Z_{HF}\right) = \ln(R_2), \tag{5.53}$$

and $R_2 = Z_{LF} - Z_{HF}$ with R_1 as the intercept on the left side of the skewed semicircular response of the Ni-Cd battery data [22] published in the D-C plane. The parameters β and R_2 can be obtained through the linear least square fitting of the data [35, 36]. In equation (5.49):

$$a = \frac{n\sum x_i y_i - \left(\sum x_i\right)\left(\sum y_i\right)}{n\sum x_i^2 - \left(\sum x_i\right)\left(\sum x_i\right)} \tag{5.54}$$

and

$$b = \frac{\left(\sum y_i\right)\left(x_i^2\right) - \left(\sum x_i\right)\left(\sum x_i y_i\right)}{n\sum x_i^2 - \left(\sum x_i\right)\left(\sum x_i\right)}, \tag{5.55}$$

where x and y are the regular coordinates representing the parameters in Figure 5.15 with *i* as the datum point and *n* is the number of observations or the data points corresponding to each measurement frequency.

Equations (5.54) and (5.55) are extended further, using equations (5.50) and (5.52) as substitutions for x and y. It is then obtained:

$$a = \frac{n\sum ln\left(1+j\omega_i\tau\right)\times ln\left(Z_i - Z_{HF}\right) - \sum ln\left(1+j\omega_i\tau\right)\times \sum ln\left(Z_i - Z_{HF}\right)}{n\sum ln\left(1+j\omega_i\tau\right)\times ln\left(1+j\omega_i\tau\right) - \sum ln\left(1+j\omega_i\tau\right)\times \sum ln\left(1+j\omega_i\tau\right)} \tag{5.56}$$

and

$$b = \frac{\begin{array}{c}\sum \ln\left(Z_i - Z_{HF}\right)\times \sum \ln\left(1+j\omega_i\tau\right)\times \ln\left(1+j\omega_i\tau\right) \\ -\sum \ln\left(1+j\omega_i\tau\right)\times \sum \ln\left(1+j\omega_i\tau\right)\times \ln\left(Z_i - Z_{HF}\right)\end{array}}{n\sum \ln\left(1+j\omega_i\tau\right)\times \ln\left(1+j\omega_i\tau\right) - \sum \ln\left(1+j\omega_i\tau\right)\times \sum \ln\left(1+j\omega_i\tau\right)}. \tag{5.57}$$

From equations (5.52) and (5.53) two parameters β and R_2 can be calculated as

$$\beta = -a, \tag{5.58}$$

and using equation (5.53), it is obtained:

$$R_2 = Z_{LF} - Z_{HF} = e^{\,b}. \tag{5.59}$$

The following example elaborates a and b numerically for a set of experimental data obtained from the published data [22] for the Ni-Cd battery with

$$a = \frac{\begin{array}{c}(23)\left(-281.53 - j\,240.95\right) \\ -\left(33.063 + j\,25.049\right)\left(-192.7 - j\,14.605\right)\end{array}}{(23)\left(48.83 + j\,93.73\right) - \left(33.063 + j\,25.04\right)^2} \tag{5.60}$$

or

$$a = -\,0.6232 + j\,(0.1206) \tag{5.61}$$

and

$$b = \frac{\begin{array}{c}\left(-192.7 - j\,14.605\right)\left(48.83 + j\,93.73\right) \\ -\left(-281.53 - j\,240.95\right)\left(33.063 + j\,25.04\right)\end{array}}{(23)\left(48.83 + j\,93.73\right) - \left(33.063 + j\,25.04\right)^2} \tag{5.62}$$

or

$$b = -7.35 - j(0.129). \tag{5.63}$$

From equations (5.60) through (5.63) and using equations (5.51) and (5.59), it is obtained:

$$\beta = -a = 0.6232 - j(0.1206), \tag{5.64}$$

and

$$R_2 = Z_{LF} - Z_{HF} = e^b = 6.36 \times 10^{-4} - j(8.29 \times 10^{-5}). \tag{5.65}$$

The complex nature of β and R_2 containing the imaginary part is not clear in terms of a physical interpretation concerning the Ni-Cd battery. Further it is not ascertained via the electrical circuit elements when the equivalent circuit is considered for the D-C response. Based on the complex nature of β, it is presumed to be more complicated than the simple non-Debye or C-C situation where a straightforward depressed semicircular relaxation is observed. Nevertheless, the complex appearance possibly originates from the distortion of the usual semicircular type relaxation behavior. When the distortion is excluded via the simulated data for β then ideal D-C data can be generated. From this ideal data β can be reproduced as the real quantity using equations (5.48) through (5.64). This indicates that the data must possess ideal D-C like feature giving only the real face of β.

The extraction of the capacitance C_2 corresponding to R_2 that form the relaxation time ($\tau = \tau_{pk} = R_2C_2$) using equation (5.65) obviously becomes a complex quantity. Reiterating, the relaxation time that corresponds to the frequency making the maximum magnitude of the imaginary part Z" obviously yields a complex quantity. In this way β, R_2, and C_2 are complex quantities.

5.6 Optimization of Z_{HF}: Complex Domain

In equation (5.48) there are three unknowns: β, R_2 (= Z_{LF} - Z_{HF}), and Z_{HF}. However, equations (5.54) and (5.55) provide solutions to β, and R_2. Further the unknown Z_{HF} can be resolved by applying the condition that the experimental values and theoretical (or calculated) values of $Z(\omega)$ must match for every measured data, including the peak corresponding to the maximum of the imaginary part of impedance on the y-axis. To match the data in the peak region or elsewhere both real and imaginary parts of the impedance $Z(\omega)$ must be at their minimum distance from the experimental data. The minimum error at the peak used as the criterion for determining Z_{HF} is given by [22]:

$$Z_{RMS} = \sqrt{\left(Z'^{T}_{pk} - Z'^{E}_{pk}\right)^2 + \left(Z''^{T}_{pk} - Z''^{E}_{pk}\right)}, \tag{5.66}$$

where Z_{pk} is the peak value of $Z''(\omega)$ representing the parameter on the y-axis. However, corresponding value of Z' is used as a pair of observations constituting Z^* at the peak frequency yielding the relaxation time, τ. More precisely Z'_{pk} corresponds to the real part of Z^* where Z''_{pk} is maximum reflecting the magnitude on the y-axis. The superscripts T and E indicate calculated (theoretical) value and experimental value, respectively, with Z' (= R_s) as the real part known as the series resistance and $Z'' = (\omega\ C_s)^{-1}$ is the imaginary part known as the reactance.

Out of three unknown parameters (β, R_2, and Z_{HF}) in equation (5.47), two (β and R_2) of them are resolved via complex domain analysis. The third unknown parameter Z_{HF} may be solved via trial and error. To obtain the optimized value for Z_{HF} the error associated with the maximum value corresponding to the imaginary part Z'' must be minimized. This means that the optimum value of Z_{HF} corresponding to the peak value of the imaginary part Z'' must result in the minimum error. In order to accomplish reasonable or acceptable meaning of the optimization approach of Z_{HF}, selection of the left-intercept of the skewed semicircular-like D-C relaxation behavior on the real axis (Z'-axis) is critical. Thus, an interval on the real axis is selected carefully to sweep a specific range for Z_{HF} to determine the minimum error value of the maximum of the imaginary axis (Z''-axis) corresponding to the peak-frequency (ω_{pk} or f_{pk}) associated in that region or segment of the skewed arc.

The error in the D-C representation of the data as a function of Z_{HF} and the parameters β, R_2, and C_2 derived in the conventional and complex domains are provided in the published paper [22]. The parameter C_2 is extracted from the time constant corresponding to $\omega\tau_{peak} = 1$. The real situation with the Ni-Cd battery is dealt [22] where the data represented in the impedance plane satisfied the D-C behavior. The electrical model displayed in Figure 5.1(b) can be represented by the impedance

$$Z^*(\omega) = R_1 + \frac{R_2}{1 + j\omega\tau}, \tag{5.67}$$

where

$$\tau = R_2 C_2, \tag{5.68}$$

and should display a semicircular pattern when the data are plotted in the Z*-plane. The peak value has the condition where $\omega = \frac{1}{\tau}$ having R_2 is the diameter of the semicircle and R_1 is the distance that the curve is translated rightward from the origin. In reality the actual response is not the ideal response of equation (5.67) but rather a depressed and skewed semicircle with a more linear slope evident at higher frequencies. This tendency can be better represented by the D-C [18, 19] expression

$$Z^*(\omega) = R_1 + \frac{R_2}{(1 + j\omega\tau)^{\beta}} \tag{5.69}$$

where

$$\tau = R_2 C_2. \tag{5.70}$$

When $\beta = 1$ the condition of equation (5.69) represents ideal Debye response, and then equations (5.67) and (5.69) become equivalent.

5.7 Depression Parameter β Based on $\omega\tau_{peak} = 1$

The meaning of the D-C depression parameter β in equation (5.18) or in equation (5.69) is highly complicated in terms of interpreting the equivalent circuit model. It is not quite straightforward as in the case of the Debye or C-C (simple non-Debye) relaxation representation. Thus, the equivalent circuit for the D-C relaxation is not as simple as it is extracted from a regular semicircular relaxation in the impedance plane or C-C plane. In this way β is not a simple distributed parameter as often justified for α in the C-C response in equation (5.17).

In order to extract the meaning of β in the D-C relaxation it is essential to comprehend the meaning of α in the C-C relaxation. Invariably the prime difference between the C-C and D-C relaxations lies with the skewed nature apart from the depressed feature of the relaxation curve. Thus, the concept of the semicircular-like behavior extending over D-C response is not useful. The complicated nature of this relaxation in conjunction with the depression-like response does not give straightforward meaning of β as often distinct depression angle θ does for the C-C depression parameter α in equation (5.17). For the depressed semicircle α

becomes a distributed parameter realizing that the relaxation time τ is the average time constant. The presence of the depression angle θ indicates considerable degree of non-uniformity in the conductance term (y-axis parameter) of the C-C plane. The presence of the depression angle (θ) or the depression parameter α (or h) further signifies the variation in the inhomogeneity of the root source of the relaxation process causing the conductance to vary. This is likely to happen from one relaxation site to the other relaxation site within the system.

In the C-C relaxation process the peak of the semicircle represents the maximum value of the y-axis parameter. The relaxation time τ corresponds to the frequency of the peak and is given by $\omega\tau_{peak} = 1$. Thus, this frequency is designated as the peak-frequency (f_{peak}) having the angular frequency ω_{peak}. Corresponding to ω_{peak} the relaxation time becomes

$$\tau_{peak} = \frac{1}{2\pi f_{peak}} \tag{5.71}$$

satisfying

$$\omega_{peak}\,\tau_{peak} = 1. \tag{5.72}$$

The analysis and subsequent interpretation of the relaxation process in the D-C plane refers to the conventional visualization of $\omega_{peak}\,\tau_{peak} = 1$ at the peak of the y-axis parameter of the relaxation curve. The validity of this concept for the D-C plane needs thorough examination using the relaxation function

$$F^*(\omega) = \frac{1}{(1 + j\omega\tau_{peak})^{\beta}}. \tag{5.73}$$

In this way the conventional concept of $\omega\tau_{peak} = 1$ at the peak of the y-axis parameter of the relaxation curve is a simplistic scenario for the underlying relaxation process. The interpretation of β, therefore, deserves two-fold descriptions. For the Z*-plane response the C-C type relaxation becomes

$$Z^*(\omega) = Z_\infty + \frac{(Z_s - Z_\infty)}{1 + (j\omega\tau_0)^{\alpha}}, \tag{5.74}$$

where α is not associated with the conductance like the complex permittivity plane (i.e., C-C) plot that originally Cole-Cole [3] referred to. In the case of Z*-plane the reactance or capacitance in the y-axis is depressed implying the capacitance as a distributed parameter yielding average time constant.

Two situations noted for the complex permittivity (C-C) and the impedance planes yield identical logistics concerning the interpretation of the depression angle or the depression parameter (h or α) but the underlying meaning is different. Concurrently the derivation of the equivalent circuit from each of these two planes is not the same. The interpretation of the contributing circuit elements as well as the meaning or origin of each element is different. Thus, the equations (5.17) and (5.74) have identical appearance but the extracted information from each of the complex planes is not the same or is much different.

The relaxation process in the semiconductor type materials or junction devices represented by the equations (5.17) and (5.74) is observed depending on the type of species. In most cases trapping associated behavior is recognized via the ε^*- or C^*-plane [4, 17, 25, 26, 29] and σ^*- or Y^*-plane with or without the presence of a depression angle. In any of these complex planes the relaxation time τ is the distributed or average parameter yielding R-C series combination for the depressed semicircular behavior [29]. In the impedance plane an identical semicircular response having a depression angle yields the R-C parallel combination with relaxation time τ. Therefore, the origin of the relaxation becomes completely different regardless of the system investigated. For the polycrystalline semiconductors the depressed semicircle portrays a more complicated situation both in the permittivity (C-C) plane as well as in the Z^*-plane than those found in the case of the single-crystal based junction devices.

A back-to-back energy band diagram is considered in Figure 5.16 to illustrate the meaning of the relaxation time τ obtained from the depressed semicircle either from the C-C plane or from the Z^*-plane. In general, the back-to-back energy band diagram refers to the polycrystalline systems where grain boundary is depicted having two successive grains. The donor states and the trapping states are the potential states available below the conduction band. The trapping states are not shown in Figure 5.16 but are presumed to have a range in the band gap below the conduction band, causing the finite depression angle θ. For zero depression angle (i.e., $\theta = 0°$) the range of the trapping states becomes singular or simply a single state below the conduction band.

Upon the application of the ac small-signal voltage, these states will sweep through the amplitude of the same signal. For the polycrystalline device, these sweeping states are unlikely to be singular in nature for the defect states. Therefore, the time constant τ comprises of the charge

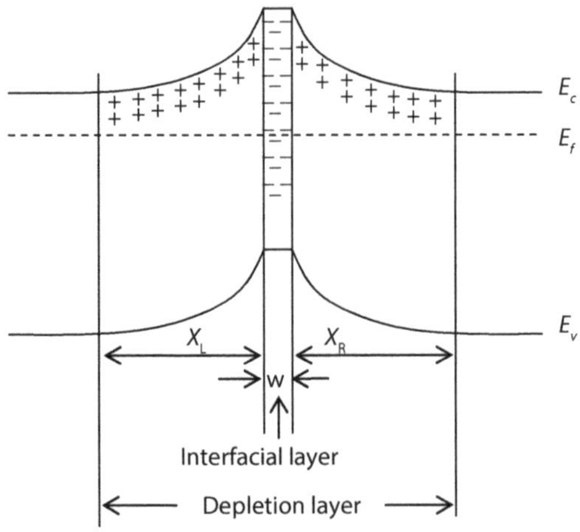

Figure 5.16 Energy band diagram for the back-to-back Schottky barrier device in the polycrystalline semiconductors where grain-to-grain junction provides variation in the relaxation processes associated with the defect states in conjunction with the barrier layer relaxation.

mobility represented by the conductance and the charge storage represented by the capacitance for the defect states. For the non-identical defect states, the conductance may be identical due to the uniform mobility of the mobile carriers while the capacitance may not be the same from location to location for the same defect states. This may attribute to the charge state of the defects signifying positive or negative cluster, capture cross-section, complexity of the ionized defects, etc. Thus, the trapping contribution in the series event for the single charge arises via storing for out-of-phase and subsequently releasing for in-phase conduction. Upon releasing, the charge becomes a mobile carrier until trapped by the next site. Such an event essentially gives the concept of the R-C series combination having relaxation time τ as a distributed parameter, which in turn can be extracted from the complex capacitance or permittivity (i.e., C-C) plane. Thus, the presence of a depressed semicircle in the C-C plane is a plausible concept. In reality this happens for many polycrystalline systems [4, 17, 25, 26].

For the R-C parallel combination form of the relaxation time τ, the space charge capacitance may not be uniform because of the variation in the gradient of the doping distribution instead of the concept of the abrupt region. The conductance of the same region remains very much static, giving rise to a singular value. Thus, the relaxation time τ may be a distributed parameter attributing to the variation in the space charge capacitance. Essentially the space charge region experiences a drop of most of the electric field apart from the grains when ac small-signal voltage is applied. This scenario may provide discontinuous or non-uniform space charge width causing the geometric capacitance to vary. This variation gives rise to the concept of the finite depression angle θ. The space charge region is shown in Figure 5.16 as the depletion layer. For zero depression angle (i.e., $\theta = 0°$) the space charge region provides a singular value regardless of averaging, considering multiple junctions within the electrical field falling regions across the grain boundaries. The aforementioned two conceptual contributions involving (1) trapping and (2) space charge region within the semiconducting device, regardless of the single-crystal or polycrystal based technology, may display the ac electrical data either in the C-C plane or in the Z^*-plane. Accordingly the meaning of the depression angle θ or the depression parameter h (or α) can be extracted.

Having discussed the situation with the depression parameter α for the C-C and the Z^*- planes it is obviously a tedious task to find the meaning of β for the D-C relaxation. The meaning of β in the C-C plane is different from that in the Z^*-plane. This is because the parameter in the y-axis is the dictating quantity. However, one may find an analogy of identical consideration for the concept of the equivalent circuit whether applicable for the C-C plane or for the Z^*-plane. Since the data [22] belong to the practical Ni-Cd battery and display nicely in the Z^*-plane the discussion is restricted to only within this plane. Therefore, the meaning of β in the C-C plane is not extended from that of the D-C [18, 19] work.

Considering the practical aspect of the investigated systems, the possible explanation for the skewed or distorted curve results from the electrochemical activation energy which is not homogeneous across the electrode surface as for the Ni-Cd battery, and this energy is directly related to the time constant τ. The Davidson-Cole response is not symmetrical about the peak of the curve per Figure 5.15. The Z^*-plane representation of the ac electrical data is noted herein as the D-C response. In the D-C response, there is no specific center of the skewed semicircle. Further, it is asymmetrical about the so-called relaxation time τ. The peak of the curve corresponds to a specific frequency for the maximum value of the imaginary part Z''. This specific frequency for the peak of the curve can be termed as the peak-frequency. These facts do not reveal a straightforward equivalent circuit for the system. Due to the asymmetrical nature of the D-C curve, it is presumed that the relaxation time τ corresponds to a complicated nature of the equivalent circuit comprising of the R-C parallel combination that attributes to the semicircular relaxation in the Z^*-plane per Figure 5.15.

Invariably the concept of the R-C parallel combination is a naïve approach that comes from the standard Z^*-plane plotting having a single Debye-like semicircular relaxation. A perturbation of the Debye-like relaxation appears in the C-C plane where a finite depression angle is observed in the impedance plane. The presence of the depression angle means that the center of the semicircular response lies below the real (x) axis. Indeed this is not the case for the D-C response.

Based on the concept of the C-C response, the relaxation time τ is frequently described as the distributed parameter [4, 17, 25, 26, 29, 30, 37–40].Often this τ is described as the average time constant or the average relaxation time in the C-C scenario [4, 17, 29, 30, 37–40].Thus, the depression parameter α in the C-C relation given in equation (5.17) is a distributed parameter. In this way the concept of the distributed parameter α leads to the average time constant or average relaxation time.

In the C-C representation the y-axis parameter is depressed. Thus, the capacitive contribution via reactance of the terminal impedance is depressed [4, 17, 25, 26, 29, 30, 37–40].This implies that the capacitance is distributed, causing τ to be an average time constant, while the resistance is obtained from the fixed magnitude of the chord of the semicircle on the x-axis. The magnitude of the depression angle of the C-C semicircle in the Z^*-plane [27] gives a feel for the nature of the distribution of τ. This explanation does not apply to the D-C representation as there is no semicircular behavior to emphasize as Debye of the C-C relaxation.

In Figure 5.15 the intercept on the left side of the skewed semicircle may be termed as the left-intercept. The ordinate (y-axis) and abscissa (x-axis) must have the same plotting scale so that the semicircular relaxation in the impedance plane can clearly be delineated. This means that each unit grid or graphical segment in length on the ordinate must be equal to the abscissa. A semicircle whose center lies on the x-axis is defined as a Debye or Debye-like relaxation. For a non-Debye relaxation the center of the semicircle lies below the x-axis, giving rise to a depression angle (θ) measured from the point at the left-intercept to the center of the semicircle below x-axis. The value of α ranges between 0 and 1 depending on the idealized Debye and/or extreme non-Debye response of a semicircular relaxation. Thus, Debye or Debye-like behavior can be achieved as $\theta \rightarrow 0°$ implying $\alpha \rightarrow 1$, and an extreme non-Debye response can be visualized with $\theta \rightarrow 90°$ implying $\alpha \rightarrow 0$.

Taking the approach of the C-C interpretation of τ in conjunction with α, the meaning of β in the D-C response is further complicated. Consideration of the R-C parallel equivalent circuit may be logical and reasonable, but the relaxation time τ for such an equivalent circuit is essentially too tedious to visualize. It may not be as simple as described by $\omega\tau = 1$. A few equivalent circuits may be reviewed considering the skewed behavior of the D-C representation. Invariably the D-C arc is not symmetric about the maximum of the y-axis parameter. Therefore, for a relaxation process the product $\omega\tau$ may not be equal to 1 corresponding to the maximum of the y-axis parameter. This argument has to be viewed with respect to the ideal Debye or Cole-Cole relaxation viewpoint where essentially $\omega\tau = 1$.

5.8 Feature of the Depression Parameter β Based on $\omega\tau \neq 1$

The conventional analytical procedure implies identical extraction of the relaxation information for both the idealized C-C semicircular curve ($\omega\tau = 1$) and the D-C skewed relaxation curve. The meaning of β becomes too simplistic for the idealized concept of $\omega\tau = 1$ extracted from the maximum value of the y-axis parameter. It is worthy to examine the D-C relaxation function given in equation (5.73) and verify the peak of the nature of the relaxation time using the concept of the product of $\omega \times \tau$. Alternately the product $\omega \times \tau$ for the D-C skewed curve may be examined for a wide range of values as a function of the imaginary parameter of the impedance such that $m \leq \omega\tau \leq n$ where m is a fraction below 1 (i.e., $m \leq 1$) and n is either integer or fraction above 1 (i.e., $n \leq 1$). Since the imaginary part of the impedance provides the maximum value corresponding to the ω_{peak} or τ_{peak} for the C-C semicircular relaxation curve then it is expected that the same would appear for the D-C skewed relaxation curve. This implies that $\omega\, \tau_{peak} \equiv 1$.

By setting various values of the product $\omega \times \tau$ in the relaxation function F(ω) in equation (5.73) it is possible to obtain the impedance function $Z^*(\omega)$ or permittivity function $\varepsilon^*(\omega)$ given in equations (5.5), (5.7), (5.17), (5.18), (5.69), and (5.74). This function can be used in plotting the product $\omega \times \tau$ on the x-axis versus imaginary part of $Z^*(\omega)$ on the y-axis for a specific value of β. This is displayed in Figure 5.17 [22]. The variation of the D-C depression parameter β as a function of the product $\omega \times \tau$ is displayed in Figure 5.18 [22]. This product corresponds to the maximum of β obtained in Figure 5.17. Two insets show the variation of F'' as a function of the product $\omega \times \tau$ and the relation between β and F'', respectively.

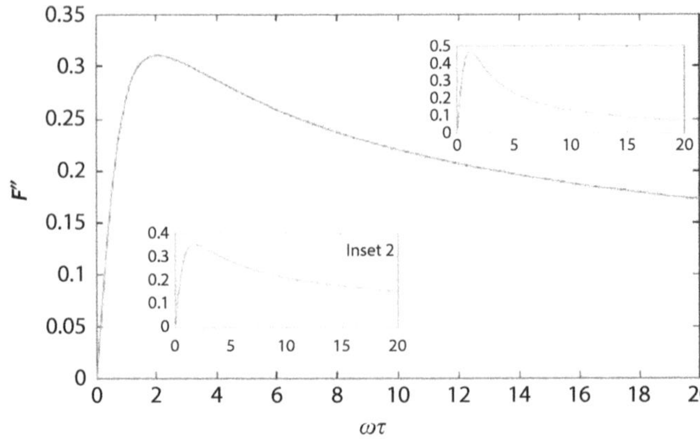

Figure 5.17 The product ω Í τ is displayed as a function of the imaginary part of the impedance for $\beta = 0.40$ showing $\omega\tau_{peak} = 208$ and $Z'' = \text{Im}Z^* = 310$. Inset 1 shows $\beta = 0.85$ with $\omega\tau_{peak} = 1.14$ and $Z'' = \text{Im}Z^* = 0.464$; and inset 2 shows $\beta = 0.5$ with $\omega\tau_{peak} = 1.74$ and $Z'' = \text{Im}Z^* = 0.353$.

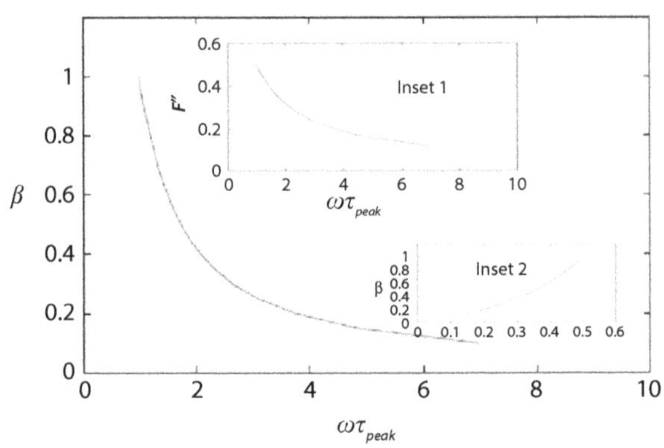

Figure 5.18 The variation of the D-C depression parameter β as a function of the product $\omega \times \tau$. Inset 1 shows variation of F'' as a function of the product $\omega \times \tau$; and inset 2 shows the relation between β and F''.

5.9 Analysis of the Havriliak-Negami Representation

The H-N relaxation function can be ascribed from equation (5.19) representing the skewed relaxation curve (arc) given by [41]

$$F^*(\omega) = \frac{1}{[1 + (j\omega\tau)^a]^\beta} \tag{5.75a}$$

$$= \frac{1}{\left[1 + (\omega\tau)^a \cos\left(\frac{a\pi}{2}\right) + j(\omega\tau)^a \sin\left(\frac{a\pi}{2}\right)\right]^\beta}. \tag{5.75b}$$

Both depression and skewed nature of the H-N relaxation curve are represented by the parameters α and β. In order to separate real and imaginary parts of equation (5.75a) $F^*(\omega)$ is expressed in the form:

$$F^*(\omega)=\frac{1}{(A+jB)^{\beta}} \tag{5.76}$$

with

$$A=1+(\omega\tau)^{a}\cos\left(\frac{a\pi}{2}\right) \tag{5.77}$$

and

$$B=(\omega\tau)^{a}\sin\left(\frac{a\pi}{2}\right). \tag{5.78}$$

Then

$$F^*(\omega)=\frac{1}{(A+jB)^{\beta}}=\frac{1}{[X\exp(j\gamma)]^{\beta}} \tag{5.79a}$$

$$=X^{-\beta}\exp(-j\beta\gamma)=X^{-\beta}\left[\cos(\beta\gamma)-j\sin(\beta\gamma)\right] \tag{5.79b}$$

where

$$X^2=A^2+B^2=\left[1+(\omega\tau)^{a}\cos\left(\frac{a\pi}{2}\right)\right]^2+(\omega\tau)^{2a}\sin^2\left(\frac{a\pi}{2}\right) \tag{5.80a}$$

$$=1+(\omega\tau)^{2a}+2(\omega\tau)^{a}\cos\left(\frac{a\pi}{2}\right) \tag{5.80b}$$

and

$$\tan(\gamma)=\frac{B}{A}=\frac{(\omega\tau)^{a}\sin\left(\frac{a\pi}{2}\right)}{1+(\omega\tau)^{a}\cos\left(\frac{a\pi}{2}\right)}. \tag{5.81}$$

Thus, real and imaginary parts become:

$$F'=X^{-\beta}\cos(\beta\gamma) \tag{5.82a}$$

and

$$F'' = X^{-\beta} \sin(\beta\gamma) \tag{5.82b}$$

with

$$\frac{F''}{F'} = \tan(\beta\gamma). \tag{5.83}$$

The H-N response can be achieved either in the impedance plot or in the complex capacitance plot where the measurement frequency moves anti-clockwise for each measurement point on the curve representing the total data. The presence of the intercept on the left side of the relaxation arc has a finite magnitude while the relaxation portion can be expressed as in empirical equation (5.76). Like the D-C representation the skewed nature of the generic H-N curve is shown in Figure 5.10. The time constant is not necessarily corresponding to the conventional $\omega\tau = 1$ at the peak of the skewed arc. Thus, $\omega\tau = 1$ is likely to be dubious while deriving the simplified equivalent R-C (resistance-capacitance) circuit for the relaxation function $F^*(\omega)$ shown in Figure 5.1(a).

Differentiating the imaginary part $F''(\omega)$ in equation (5.82) with respect to $(\omega\tau)^{\alpha}$:

$$\frac{dF''}{d(\omega\tau)^{a}} = -\beta X^{-(\beta+1)} \sin(\beta\gamma) \frac{dX}{d(\omega\tau)^{a}} + \beta X^{-\beta} \cos(\beta\gamma) \frac{d\gamma}{d(\omega\tau)^{a}}, \tag{5.84}$$

where

$$\frac{dX}{d(\omega\tau)^{a}} = \frac{1}{2}\left[1+(\omega\tau)^{2a} + 2(\omega\tau)^{a} \cos\left(\frac{a\pi}{2}\right)\right]^{-\frac{1}{2}} 2\left[(\omega\tau)^{a} + \cos\left(\frac{a\pi}{2}\right)\right] \tag{5.85a}$$

$$= \left[1+(\omega\tau)^{2a} + 2(\omega\tau)^{a} \cos\left(\frac{a\pi}{2}\right)\right]^{-\frac{1}{2}} \left[(\omega\tau)^{a} + \cos\left(\frac{a\pi}{2}\right)\right] \tag{5.85b}$$

with

$$\gamma = \tan^{-1}\left[\frac{(\omega\tau)^{a} \sin\left(\frac{a\pi}{2}\right)}{1+(\omega\tau)^{a} \cos\left(\frac{a\pi}{2}\right)}\right] = \tan^{-1}\varphi. \tag{5.86}$$

Differentiating γ in equation (5.86) with respect to $(\omega\tau)^{\alpha}$:

$$\frac{d\gamma}{d(\omega\tau)^{a}} = \frac{1}{1+\varphi^{2}} \frac{d\varphi}{d(\omega\tau)^{a}} \tag{5.87}$$

where

$$\begin{aligned} 1+\varphi^{2} &= 1+\frac{(\omega\tau)^{2a}\sin^{2}\left(\frac{a\pi}{2}\right)}{\left[1+(\omega\tau)^{a}\cos\left(\frac{a\pi}{2}\right)\right]^{2}} \\ &= \frac{1+(\omega\tau)^{2a}+2(\omega\tau)^{a}\cos\left(\frac{a\pi}{2}\right)}{\left[1+(\omega\tau)^{a}\cos\left(\frac{a\pi}{2}\right)\right]^{2}} \end{aligned} \tag{5.88}$$

and

$$\frac{d\varphi}{d(\omega\tau)^{a}} = \frac{d}{d(\omega\tau)^{a}}\left[\frac{(\omega\tau)^{a}\sin\left(\frac{a\pi}{2}\right)}{1+(\omega\tau)^{a}\cos\left(\frac{a\pi}{2}\right)}\right] = \frac{\sin\left(\frac{a\pi}{2}\right)}{\left[1+(\omega\tau)^{a}\cos\left(\frac{a\pi}{2}\right)\right]^{2}}. \tag{5.89}$$

Rearranging equation (5.87), it is obtained:

$$\frac{d\gamma}{d(\omega\tau)^{a}} = \frac{\sin\left(\frac{a\pi}{2}\right)}{1+(\omega\tau)^{2a}+2(\omega\tau)^{a}\cos\left(\frac{a\pi}{2}\right)}. \tag{5.90}$$

In order to obtain the maximum of $F''(\omega)$, setting equation (5.84) as [41]:

$$\frac{dF''}{d(\omega\tau)^{a}} = 0 \tag{5.91}$$

Thus,

$$\frac{dF''}{d(\omega\tau)^a} = -\beta X^{-(\beta+1)} \sin(\beta\gamma) \frac{dX}{d(\omega\tau)^a} + X^{-\beta}\beta\cos(\beta\gamma)\frac{d\gamma}{d(\omega\tau)^a} = 0 \tag{5.92a}$$

$$= -X^{-1}\sin(\beta\gamma)\frac{dX}{d(\omega\tau)^a} + \cos(\beta\gamma)\frac{d\gamma}{d(\omega\tau)^a} = 0 \tag{5.92b}$$

$$= -\left[1+(\omega\tau)^{2a}+2(\omega\tau)^a\cos\left(\frac{a\pi}{2}\right)\right]^{-\frac{1}{2}}\sin(\beta\gamma)\left[1+(\omega\tau)^{2a}+2(\omega\tau)^a\cos\left(\frac{a\pi}{2}\right)\right]^{-\frac{1}{2}}$$

$$\times\left[(\omega\tau)^a+\cos\left(\frac{a\pi}{2}\right)\right] + \frac{\cos(\beta\gamma)\sin\left(\frac{a\pi}{2}\right)}{1+(\omega\tau)^{2a}+2(\omega\tau)^a\cos\left(\frac{a\pi}{2}\right)} = 0 \tag{5.92c}$$

Switching the quantities of both sides of the equal sign, equation (5.92c) becomes [41]:

$$\left[1+(\omega\tau)^{2a}+2(\omega\tau)^a\cos\left(\frac{a\pi}{2}\right)\right]^{-1}\sin(\beta\gamma)\left[(\omega\tau)^a+\cos\left(\frac{a\pi}{2}\right)\right]$$

$$= \frac{\cos(\beta\gamma)\sin\left(\frac{a\pi}{2}\right)}{1+(\omega\tau)^{2a}+2(\omega\tau)^a\cos\left(\frac{a\pi}{2}\right)}. \tag{5.93}$$

Rearranging equation (5.93):

$$\sin(\beta\gamma)\left[(\omega\tau)^a+\cos\left(\frac{a\pi}{2}\right)\right] = \sin\left(\frac{a\pi}{2}\right)\cos(\beta\gamma). \tag{5.94}$$

Thus, at the maximum value of F″, $\beta\gamma = (\beta\gamma)_{peak}$, and $\omega\tau = (\omega\tau)_{peak}$, equation (5.94) becomes:

$$\tan(\beta\gamma)_{peak} = \frac{\sin\left(\frac{a\pi}{2}\right)}{(\omega\tau)^a_{peak}+\cos\left(\frac{a\pi}{2}\right)}. \tag{5.95}$$

Equation (5.85) represents the condition when $F'' = F''_{max}$. When $\alpha \to 1$, equation (5.95) then leads to:

$$(\omega\tau)_{peak} \tan(\beta\gamma)_{peak} = 1 \tag{5.96}$$

and equation (5.81) yields

$$\tan(\gamma_{peak}) = (\omega\tau)_{peak}. \tag{5.97}$$

Then equation (5.96) becomes D-C response [20–22]:

$$\tan(\gamma_{peak}) \tan(\beta\gamma)_{peak} = 1. \tag{5.98}$$

Expression of β appears as real quantity for α being real using equations (5.95):

$$\beta = \left(\frac{1}{\gamma_{peak}}\right) \arctan\left[\frac{\sin\left(\frac{\alpha\pi}{2}\right)}{(\omega\tau)^{\alpha}_{peak} + \cos\left(\frac{\alpha\pi}{2}\right)}\right] \tag{5.99}$$

where, γ and $\alpha\pi$ are in radians; and $(\omega\tau)_{peak}$ corresponds to the maximum value of $F''(\omega)$.

5.10 Geometrical Interpretation of H-N Relaxation at the Limiting Case

Figure 5.19(a) displays a generic representation of the H-N relaxation formalism [41]. This feature has a similarity with the D-C feature [18, 19] with $\alpha = 1$. Equation (5.83) gives the ratio between the imaginary and real part of the H-N relaxation function yielding $\tan(\beta\gamma)$. From equation (5.81) it is seen that $\tan(\gamma)$ increases as ω increases anti-clockwise, i.e., the angle $\beta\gamma$ increases. Further examination of the imaginary and real parts of the H-N relaxation function and imposing limiting case, equation (5.83) gives [20, 21, 41]:

$$\lim_{\omega \to \infty} \frac{F''}{F'} = \lim_{\omega \to \infty} \tan(\beta\gamma) = \tan(\beta\gamma_{max}) \tag{5.100}$$

$$\tan \gamma_{\omega \to \infty} = \lim_{\omega \to \infty} \left(\frac{\sin\left(\frac{a\pi}{2}\right)}{(\omega\tau)^{-a} + \cos\left(\frac{a\pi}{2}\right)} \right) = \tan\left(\frac{a\pi}{2}\right) = \tan\left(\gamma_{\max}\right) \tag{5.101}$$

$$\gamma_{\omega \to \infty} = \gamma_{\max} = \frac{a\pi}{2}. \tag{5.102}$$

Combining equations (5.100) and (5.102), it is obtained:

$$\lim_{\omega \to \infty} \frac{F''}{F'} = \tan\left(\beta\gamma_{\omega \to \infty}\right) = \tan\left(\beta\gamma_{\max}\right) = \tan\left(\frac{a\beta\pi}{2}\right) \tag{5.103}$$

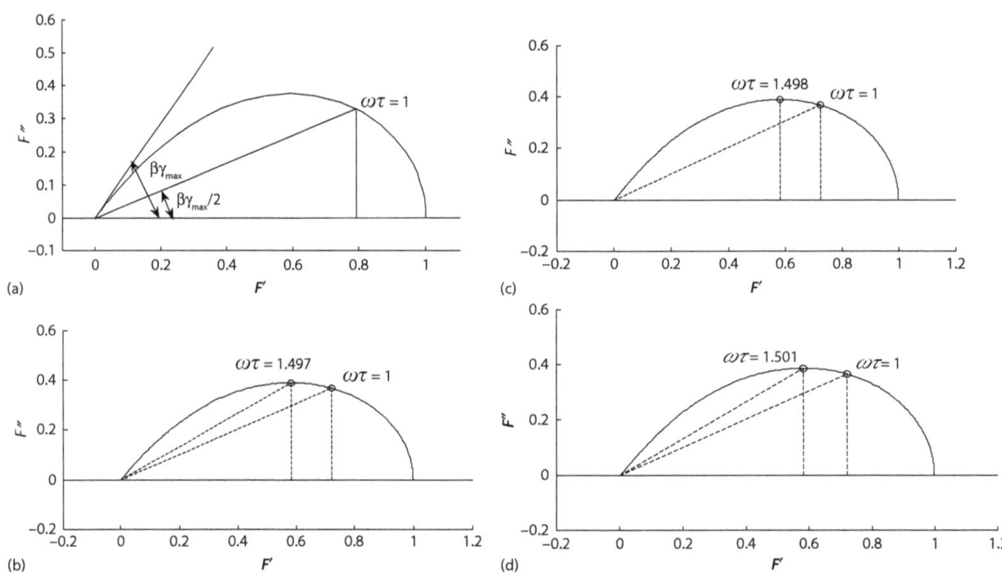

Figure 5.19 (a) Havriliak-Negami representation of the relaxation function $F^*(\omega)$ per equation (5.75a) where measurement frequency runs from right to the left. (b) Havriliak-Negami representation of the relaxation function $F^*(\omega)$ with $\alpha = 1$, $\beta = 0.6$ that corresponds to $F'_{peak} = F'_{\omega\tau=1.497} = 0.5843$ and $F''_{peak} = F''_{\omega\tau=1.497} = 0.3905$ along with $F'_{\omega\tau=1} = 0.7237$, $F''_{\omega\tau=1} = 0.3888$. (c) Havriliak-Negami representation of the relaxation function $F^*(\omega)$ with $\alpha = 0.999$, $\beta = 0.6$ that corresponds to $F'_{peak} = F'_{\omega\tau=1.498} = 0.5842$ and $F''_{peak} = F''_{\omega\tau=1.498} = 0.3899$ along with $F'_{\omega\tau=1} = 0.7236$, $F''_{\omega\tau=1} = 0.3682$. (d) Havriliak-Negami representation of the relaxation function $F^*(\omega)$ with $\alpha = 0.995$, $\beta = 0.6$ that corresponds to $F'_{peak} = F'_{\omega\tau=1.501} = 0.5841$ and $F''_{peak} = F''_{\omega\tau=1.501} = 0.3877$ along with $F'_{\omega\tau=1} = 0.7229$, $F''_{\omega\tau=1} = 0.3662$.

where

$$\beta\gamma_{max} = (\beta\gamma)_{max} = \frac{\alpha\beta\pi}{2}. \tag{5.104}$$

The product $\alpha \times \beta$ has a proven connection with the Kohlrausch-Williams-Watts (KWW) function [41–43]. Equation (5.101) indicates the ratio between imaginary and real part of the H-N relaxation function attains its maximum value when $\omega \to \infty$. This situation is shown in Figure 5.19(a) where the bisector of the angle $\beta\gamma_{max}$ intersects the relaxation curve [41] at the point $\omega\tau = 1$. The angle created by this bisector with the x-axis is given by:

$$\frac{\beta\gamma_{max}}{2} = \frac{\alpha\beta\pi}{4} \tag{5.105}$$

Now, when $\omega\tau = 1$, equation (5.81) yields:

$$\tan\gamma|_{\omega\tau=1} = \frac{\sin\left(\frac{\alpha\pi}{2}\right)}{1+\cos\left(\frac{\alpha\pi}{2}\right)} = \tan\left(\frac{\alpha\pi}{4}\right). \tag{5.106}$$

Therefore,

$$\gamma|_{\omega\tau=1} = \frac{\alpha\pi}{4} = \frac{\gamma_{max}}{2}, \tag{5.107}$$

and

$$\tan\left(\frac{\beta\gamma_{max}}{2}\right) = \tan\left(\frac{\beta\alpha\pi}{4}\right) = \tan\left(\beta\gamma|_{\omega\tau=1}\right) = \left.\frac{F''}{F'}\right|_{\omega\tau=1}. \tag{5.108}$$

Equation (5.108) proves that the bisector of the angle $\beta\gamma_{max}$ intersects the H-N relaxation curve at a point where $\omega\tau = 1$. This important information can be used to determine the value of τ.

Gerischer impedance [44] representation in the impedance plane lies for $\alpha = 1$ and $\beta = 0.5$ which is a special case. This impedance gives $\omega\tau = 1$ in the low-frequency domain whereas the peak of the loci gives $\omega\tau = \sqrt{3} \approx 1.732$ and the corresponding coordinate gives $F'_{peak} = F'_{\omega\tau=1.732} = \sqrt{3/8} \approx 0.6124$ and $F''_{peak} = F''_{\omega\tau=1.732} = 1/\sqrt{8} \approx 0.3536$. Since the Ni-Cd battery data indicated $\beta = 0.6$ with $\alpha < 1$, the D-C relaxation formalism is not emphasized. Figure 5.19(b) refers to the D-C formalism (for $\alpha = 1$) while Figures 5.19(c) and 5.19(d) represent the H-N formalism [41]. Thus, $\alpha = 0.999$ and $\beta = 0.6$ as well as $\alpha = 0.995$ and $\beta = 0.6$ indicate H-N representation of the experimental data.

5.11 Extraction of the Relaxation Time τ and the H-N Depression Parameters α and β

The H-N impedance representation given by [41]

$$Z^*(\omega) = Z_{HF} + \frac{Z_{LF} - Z_{HF}}{[1+(j\omega\tau)^{\alpha}]^{\beta}} \tag{5.109}$$

can be ascribed as the D-C representation for $\alpha = 1$ with Z_{HF} = high frequency ($\omega \to \infty$) impedance, and Z_{LF} = low frequency ($\omega \to 0$ or dc) impedance. The generic representation of both H-N and D-C responses, shown in Figure 5.10, depicts the real world situation for the Ni-Cd battery impedance data [22, 41], shown in Figure 5.20 as the H-N response. The same data approximate both D-C and H-N representations within the close range of the value of α. The iteration process of the fitting of each relaxation illustrated was previously employed for the D-C response [22]. For $\beta = 0.6$ three separate values of α provide near identical fitting. Thus, two H-N representations cannot be easily distinguished from the D-C representation for the same data. The error in each fitting was brought to as minimum as possible [41].

The values of Z_{LF}, Z_{HF}, α, and β are obtained by trial and error for the measured data of the Ni-Cd battery previously represented as the D-C dielectric behavior [20–22]. Figures 5.21 and 5.22 represent the ratio of the imaginary to the real impedance of the relaxation domain versus angular frequency for H-N response and D-C response, respectively. For the H-N case, when $\alpha = 0.995$ and $\beta = 0.6$, then $\beta\gamma_{max} = \alpha\beta\pi/2 = 0.9378$ and $\tan(\beta\gamma_{max}) = 1.3628$. For the D-C relaxation when $\alpha = 1$ and $\beta = 0.6$, then $\beta\gamma_{max} = \alpha\beta\pi/2 = 0.9425$ and $\tan(\beta\gamma_{max}) = 1.3764$. Realistically the horizontal tangent on the curve in Figure 5.21 and/or 5.22 yields $\tan(\beta\gamma_{max}) = 1.2449$.This finite difference between the theoretical and the experimental values is attributed to the difference between the tangent of the line containing experimental data points and the tangent of the fitted curve, respectively.

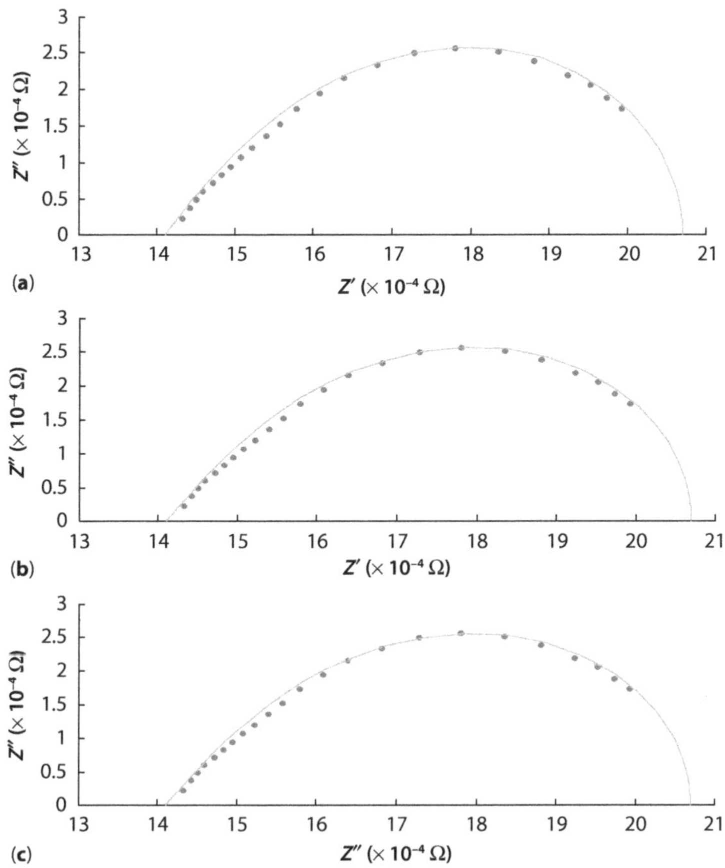

Figure 5.20 Havriliak-Negami representation of the Ni-Cd battery via impedance plot[41] for (a): $\alpha = 1$ and $\beta = 0.6$; (b): $\alpha = 0.999$ and $\beta = 0.6$; and (c): $\alpha = 0.995$ and $\beta = 0.6$.

In order to determine the value of relaxation time τ for the H-N response as well as for the D-C response, equations (5.105) and (5.108) are used. According to equation (5.108), a horizontal line is drawn for $\tan(\beta\gamma_{max}/2)$ which intersects the experimental curve at a point from which a perpendicular line is drawn on the abscissa to extract the relaxation time corresponding to $\tau = 1/\omega$. This gives $\tau = 0.2025$ sec for the H-N case (Figure 5.21) and 0.2009 sec for the D-C case (Figure 5.22). The inset in Figures 5.21 and 5.22 shows enlarged intersecting region of the curve [41].

By setting various values of the product $\omega \times \tau$ in the relaxation function $F^*(\omega)$ in equation (5.75a) it is possible to obtain the impedance function given in equation (5.109). Either of these two functions can be used in plotting the product $\omega \times \tau$ on the abscissa versus imaginary part of the relaxation or impedance function on the ordinate for specific values of α and

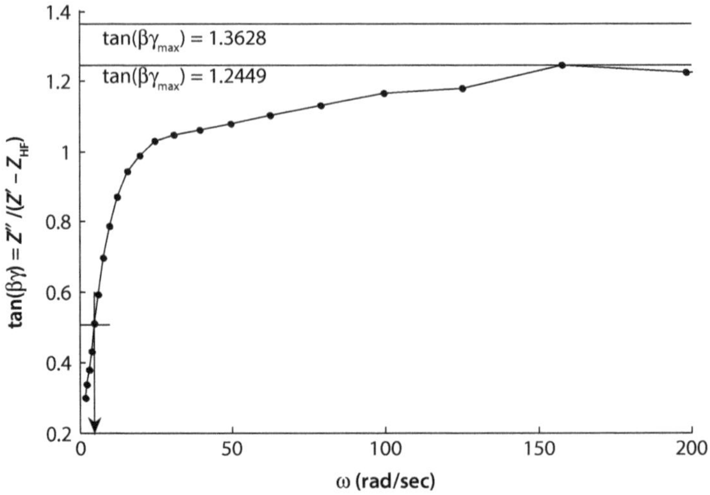

Figure 5.21 H-N response for $\tan(\beta\gamma)$ versus angular frequency ω of the Ni-Cd battery data with $\alpha = 0.995$ and $\beta = 0.6$. Maximum theoretical value of $\tan(\beta\gamma_{max})$ obtained as 1.3628, while experimental value shows 1.2449. Inset shows $\omega = 4.9390$ rad/sec corresponding to $\tau = 0.2025$ sec.

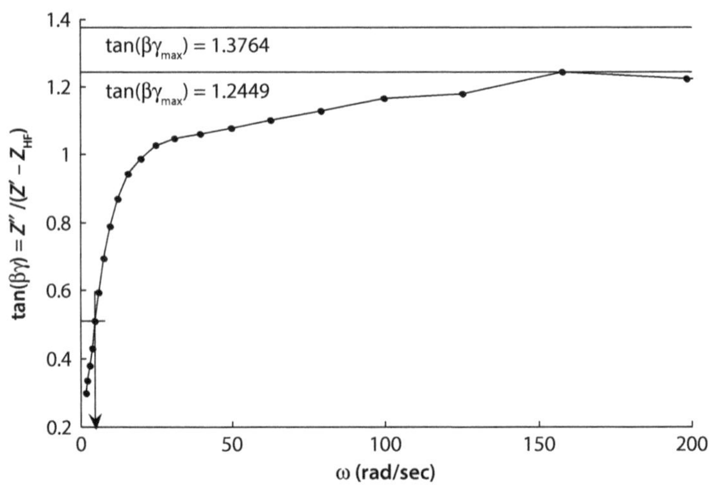

Figure 5.22 D-C response for $\tan(\beta\gamma)$ versus angular frequency ω of the Ni-Cd battery data with $\alpha = 1.0$ and $\beta = 0.6$. Maximum theoretical value of $\tan(\beta\gamma_{max})$ obtained as 1.3764, while experimental value shows 1.2449. Inset shows $\omega = 4.9766$ rad/sec corresponding to $\tau = 0.2009$ sec.

β. Thus, the imaginary part of equation (5.109) can be plotted using equation (5.75a) for Debye, C-C, D-C, and H-N responses. These responses are shown in Figure 5.23 wherein the values of α and β prompted $(\omega\tau)_{peak}$ [41]. It is clear that $(\omega\tau)_{peak} = 1$ only for ideal Debye and ideal non-Debye (i.e.,

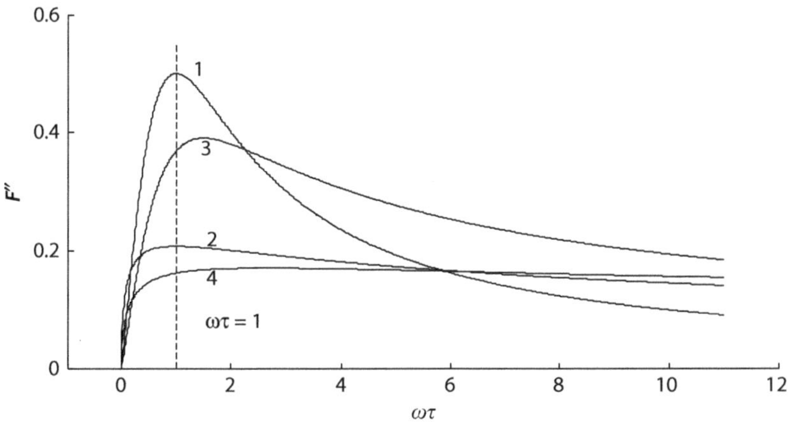

Figure 5.23 The product $\omega \times \tau$ is displayed as a function of the imaginary part of the relaxation function $F^*(\omega)$ for (1) Debye response with $\alpha = 1$ and $\beta = 1$ yielding $(\omega\tau)_{peak} = 1$; (2) Non-Debye (i.e., Cole-Cole) response with $\alpha = 0.5$ and $\beta = 1$ yielding $(\omega\tau)_{peak} = 1$; (3) Davidson-Cole response with $\alpha = 1$ and $\beta = 0.6$ yielding $(\omega\tau)_{peak} = 1.497$; and (4) Havriliak-Negami response with $\alpha = 0.5$ and $\beta = 0.6$ yielding $(\omega\tau)_{peak} = 2.637$. Vertical dotted line indicates $\omega\tau = 1$.

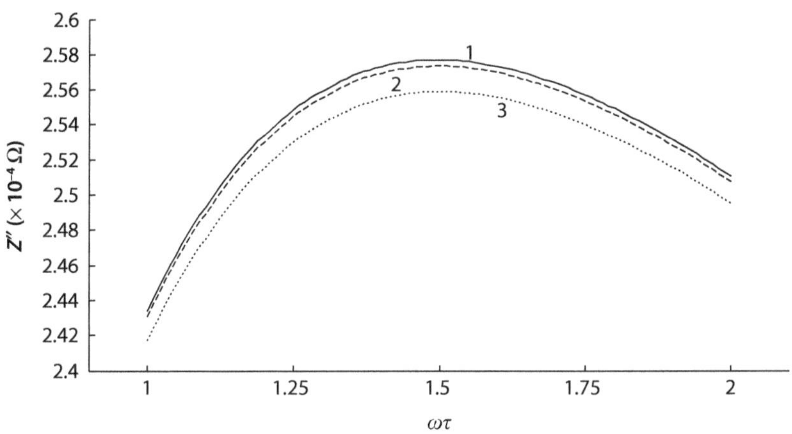

Figure 5.24 The product $\omega \times \tau$ is displayed as a function of the imaginary part of $Z^*(\omega)$ for curve 1: Davidson-Cole response with $\alpha = 1.0$ and $\beta = 0.6$ yielding $(\omega\tau)_{peak} = 1.4970$; curve 2: Havriliak-Negami response with $\alpha = 0.999$ and $\beta = 0.6$ yielding $(\omega\tau)_{peak} = 1.4980$; and curve 3: Havriliak-Negami response with $\alpha = 0.995$ and $\beta = 0.6$ yielding $(\omega\tau)_{peak} = 1.5010$.

Cole-Cole) responses. Further $(\omega\tau)_{peak} \neq 1$ for non-ideal non-Debye D-C and H-N responses.

For the real world situation of Figure 5.20 represented in the F" format in Figure 5.23 that gives $(\omega\tau)_{peak} \neq 1$ (i.e., $(\omega\tau)_{peak} > 1$) is shown in Figure 5.24.

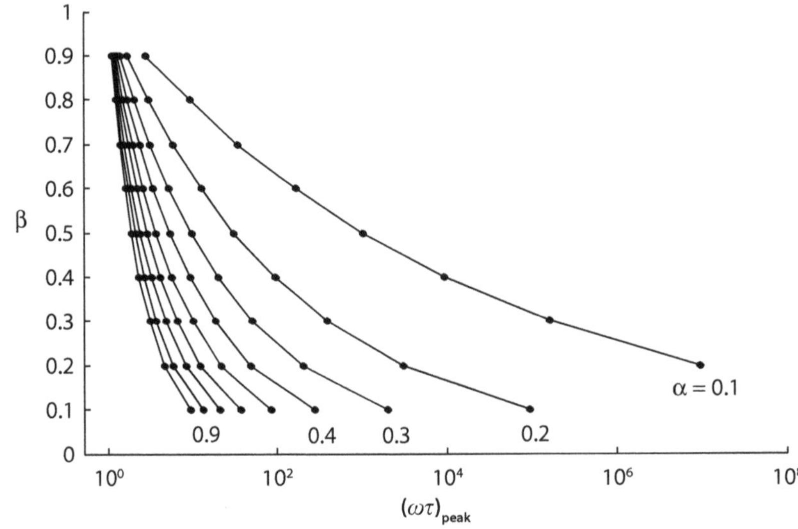

Figure 5.25 The parameter β versus $(\omega\tau)_{peak}$ for various values of α.

The imaginary part F" as a function of the product $\omega \times \tau$ is displayed in Figure 5.23 for Debye response with $\alpha = 1$ and $\beta = 1$ yielding $(\omega\tau)_{peak} = 1$; C-C response with $\alpha = 0.5$ and $\beta = 1$ yielding $(\omega\tau)_{peak} = 1.0$; D-C response with $\alpha = 1$ and $\beta = 0.6$ yielding $(\omega\tau)_{peak} = 1.192$; and H-N response with $\alpha = 0.5$ and $\beta = 0.6$ yielding $(\omega\tau)_{peak} = 1.273$. Figure 5.24 displays the product $\omega \times \tau$ versus $Z''(\omega)$ for both D-C and H-N responses [41]. For D-C response with $\alpha = 1$ and $\beta = 0.6$ yields $(\omega\tau)_{peak} = 1.4970$; for H-N response with $\alpha = 0.999$ and $\beta = 0.6$ yields $(\omega\tau)_{peak} = 1.4980$ and with $\alpha = 0.995$ and $\beta = 0.6$ yields $(\omega\tau)_{peak} = 1.5010$. The values of $(\omega\tau)_{peak}$ are within a close range presumably because of the close proximity of α or β for D-C or H-N response. It is interesting to demonstrate the same Ni-Cd battery data for both D-C and H-N dielectric formalism.

To see the effect of change in the peak value of the product $\omega \times \tau$ on the parameters α and β, different values of α and β are plotted against $(\omega\tau)_{peak}$. Figure 5.25 shows the variation of β with $(\omega\tau)_{peak}$ for various values of α indicating an increase in β with a decrease in α. [41] Figure 5.26 shows the similar variation of α with $(\omega\tau)_{peak}$ for various values of β indicating an increase in α with a decrease in β. [41] In general, both α and β decrease as $(\omega\tau)_{peak}$ increases. Also when the values of α and β approach zero (i.e., highly non-Debye situation), then $(\omega\tau)_{peak}$ becomes an extremely large β. Thus, the extreme non-Debye situation gives $(\omega\tau)_{peak} >> 1$. The values of $(\omega\tau)_{peak}$ for various values of α and β are also determined [41].

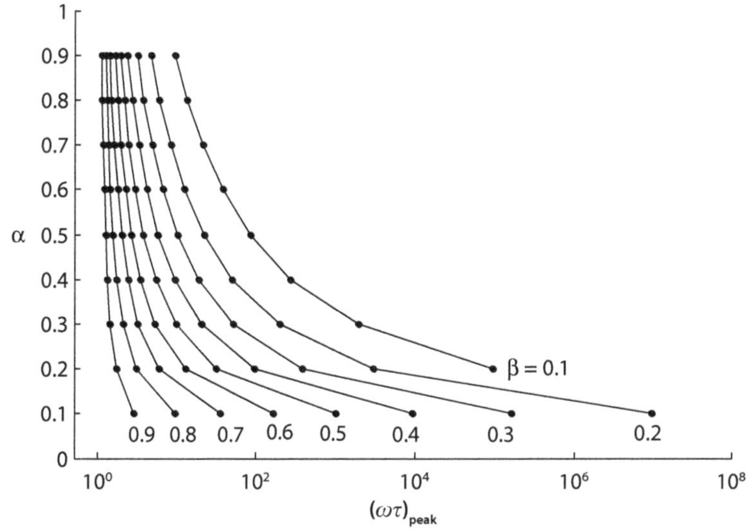

Figure 5.26 The parameter α versus $(\omega\tau)_{\text{peak}}$ for various values of β.

5.12 Checking Generalized Depression Parameter β when α is Real

In order to verify β being a real quantity, it is necessary to test its nature. Thus, considering α as a constant real parameter for β being a complex quantity given by:

$$\beta = \beta_1 + j\beta_2. \tag{5.110}$$

and using it in equation (5.76):

$$F = \frac{1}{(A + jB)^{(\beta_1 + j\beta_2)}} \tag{5.111}$$

$$= X^{-\beta_1} \exp(\beta_2 \gamma) \begin{bmatrix} \{\cos(\beta_2 \ln X)\cos(\beta_1 \gamma) - \sin(\beta_2 \ln X)\sin(\beta_1 \gamma)\} \\ - j\{\sin(\beta_1 \gamma)\cos(\beta_2 \ln X) + \cos(\beta_1 \gamma)\sin(\beta_2 \ln X)\} \end{bmatrix} \tag{5.112a}$$

$$= X^{-\beta_1} \exp(\beta_2 \gamma)\left[\cos(\beta_1 \gamma + \beta_2 \ln X) - j \sin(\beta_1 \gamma + \beta_2 \ln X)\right]. \tag{5.112b}$$

Therefore, equation (5.83) becomes [41]:

$$\frac{F''}{F'} = \tan\left(\beta_1 \gamma + \beta_2 \ln X\right) \tag{5.113}$$

From equations (5.103) and (5.104), it is seen that γ_{max} (when $\omega \to \infty$) is a finite quantity. Similarly taking the logarithm of equation (5.80b) and imposing the limiting case:

$$\lim_{\omega \to \infty} \ln X = \frac{1}{2} \lim_{\omega \to \infty} \ln\left[1 + (\omega\tau)^{2\alpha} + 2(\omega\tau)^{\alpha} \cos\left(\frac{\alpha\pi}{2}\right)\right] \to \infty. \tag{5.114}$$

Taking the limit on both sides of equation (5.113) and using equation (5.114), it is seen that no definite finite limit exists. However, in reality the observed data indicate finite ratio F″/F′ when $\omega \to \infty$ as depicted in Figures 5.10 and 5.19. This gives

$$\beta_2 = 0 \tag{5.115}$$

which in turn yields β as a real quantity. In the foregoing D-C analysis [22] it is noted that the parameter β is the complex quantity for $\alpha = 1$. However, β is a real quantity for the H-N analysis. If the same analytical approach applied to the D-C analysis [22], β will turn to a real quantity. The difference for the parameter β between this analysis and others [45] is due to the fact that when the linear least square fitting equations are used, the independent and dependent variables are real against the complex quantities considered earlier.

5.13 Checking Generalized Depression Parameter α when β is Real

In order to verify α being a real quantity, it is necessary to test its nature. Thus, considering β as a constant real parameter for α being a complex quantity given by

$$\alpha = \alpha_1 + j\alpha_2. \tag{5.116}$$

and using it in equation (5.76):

$$F = \frac{1}{\left[1 + (j\omega\tau)^{(\alpha_1 + j\alpha_2)}\right]^{\beta}} \tag{5.117}$$

$$= \frac{1}{\left[1+(\omega\tau)^{\alpha_1}\exp\left(-\frac{\alpha_2\pi}{2}\right)\cos\phi + j(\omega\tau)^{\alpha_1}\exp\left(-\frac{\alpha_2\pi}{2}\right)\sin\phi\right]^{\beta}} \tag{5.118a}$$

$$= \frac{1}{(C+jD)^{\beta}} \tag{5.118b}$$

where,

$$\phi = \frac{\alpha_1\pi}{2} + \alpha_2\ln\omega\tau, \tag{5.119}$$

$$C + jD = X\exp(j\psi), \tag{5.120}$$

$$\tan\psi = \frac{D}{C} = \frac{(\omega\tau)^{\alpha_1}\exp\left(-\frac{\alpha_2\pi}{2}\right)\sin\phi}{1+(\omega\tau)^{\alpha_1}\exp\left(-\frac{\alpha_2\pi}{2}\right)\cos\phi} \tag{5.121a}$$

$$= \frac{\sin\phi}{(\omega\tau)^{-\alpha_1}\exp\left(\frac{\alpha_2\pi}{2}\right)+\cos\phi} \tag{5.121b}$$

$$= \frac{\sin\left(\frac{\alpha_1\pi}{2}+\alpha_2\ln\omega\tau\right)}{(\omega\tau)^{-\alpha_1}\exp\left(\frac{\alpha_2\pi}{2}\right)+\cos\left(\frac{\alpha_1\pi}{2}+\alpha_2\ln\omega\tau\right)}, \tag{5.121c}$$

and

$$X^2 = C^2 + D^2. \tag{5.122}$$

Then

$$F = \frac{1}{\left(Xe^{j\psi}\right)^{\beta}} = X^{-\beta}e^{-j\beta\psi} = X^{-\beta}\left(\cos\beta\psi - j\sin\beta\psi\right). \tag{5.123}$$

Now equation (5.83) becomes [41]:

$$\frac{F''}{F'} = \tan\beta\psi. \tag{5.124}$$

Taking the limiting case for equation (5.121b) as well as using equation (5.119) and (5.121c):

$$\lim_{\omega\to\infty} \tan\psi = \lim_{\phi\to\infty} \tan\phi \tag{5.125}$$

which does not exist, however, from the observed data and equation (5.103):

$$\lim_{\omega\to\infty} \frac{F''}{F'} = \tan\left(\frac{\alpha\beta\pi}{2}\right) = \text{finite value}. \tag{5.126}$$

Therefore, using equations (5.121c), (5.124), and (5.126) it is seen as:

$$\alpha_2 = 0 \tag{5.127}$$

in order to have a finite limit as $\omega \to \infty$.

5.14 Effect of α and β on the H-N Distribution Function

The H-N distribution function is given by [20, 21, 41–43, 46, 47]:

$$g_{HN}(x) = \frac{1}{\pi}\frac{x^{\beta}\sin(\beta\theta)}{\left[x^{2a} + 2x^{a}\cos(a\pi) + 1\right]^{\beta/2}} \tag{5.128}$$

where

$$\theta = \tan^{-1}\left[\frac{\sin(a\pi)}{x^{a} + \cos(a\pi)}\right] \tag{5.129}$$

and

$$x = \tau / \tau_{HN} \quad \text{for} \quad 0 < \tau < \infty, 0 < x < \infty. \tag{5.130}$$

τ is the continuous relaxation time within a limiting range and τ_{HN} is the fitted relaxation time obtained from the same data. So far, the relaxation time τ_{HN} has been referred to as τ. When $\alpha \to 1$ then $g_{HN} \to g_{DC}$, where g_{DC} is the D-C distribution function. When g_{HN} is written as a function of $\ln(\tau)$, the following properties can be identified as:

$$g_{HN}(\ln \tau) > 0, \text{ for } -\infty < \ln \tau < \infty, \text{ and } \int_{-\infty}^{\infty} g(\ln \tau) d \ln \tau = 1. \quad (5.131)$$

Figure 5.27 displays the effect of β on the H-N distribution function for a fixed value of $\alpha = 0.5$. The inset shows the same response for the fixed value of $\alpha = 0.995$ obtained from the H-N fitting of the Ni-Cd battery data [22, 41]. The curves indicate a shift of the peak of the H-N distribution function as β increases with respect to $x = \ln(\tau/\tau_{HN}) = 0$ line. As β increases, curves become symmetrical about the $x = 0$ line for the H-N distribution function and gradually tend to the C-C distribution function.

The spread of the H-N distribution function (standard deviation) also reduces as it approaches the C-C distribution function. When $\beta = 1$, the H-N distribution function coincides exactly with the C-C distribution function with the peak at $\tau = \tau_{CC}$. Similar behavior is observed at any other value of α.

Figure 5.24 displays the effect of the change in α on the H-N distribution function for a fixed value of $\beta = 0.6$ which was obtained from the H-N fitting of the Ni-Cd battery data. The figure indicates that the increase in α shifts the peak of the H-N distribution function toward the $x = \ln(\tau/\tau_{HN})$

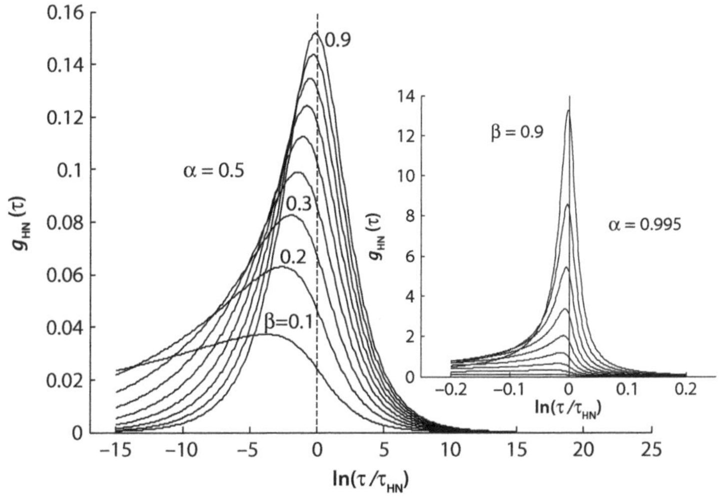

Figure 5.27 Effect of β on the H-N distribution function for $\alpha = 0.5$. For $\tau = \tau_{HN}$.

= 0 line. Each curve becomes more skewed in the negative domain, i.e. larger portion of the curve lies left to the x = $\ln(\tau/\tau_{HN}) = 0$ line. This indicates the H-N distribution function tends to the D-C distribution function. The spread of the H-N distribution function (standard deviation) also reduces as it approaches the D-C distribution function. When $\alpha = 1$, the H-N distribution function coincides exactly with the D-C distribution function which lies to the left of the x = $\ln(\tau/\tau_{HN}) = 0$ line. Similar behavior is also observed at any other value of β.

In Figure 5.27 the breadth of the distribution function becomes narrower as $\beta \to 1$ (i.e., C-C) which indicates a decrease in skewness reflecting near homogeneous distribution of the R-C equivalent circuit values. In Figure 5.28 the breadth of the distribution function becomes narrower as $\alpha \to 1$ (i.e., D-C) indicating asymmetrical distribution of the R-C equivalent circuit values. The value of the resistance is very much limited by the chord of the relaxation loci indicating singular contribution to the relaxation time whereas capacitance is extracted from the y-axis attributing to the skewness. The capacitance is presumed to be nonlinear causing a drift in the $\omega \times \tau = 1$ from the peak of the relaxation arc.

Figure 5.29 shows the nature of the skewed behavior of the H-N distribution function with limiting case of C-C for $\beta = 1$. Here the skewness is quantified as the difference between the expected (mean) value of τ and the parameter τ_{HN}. The expected value of τ from the distribution differs significantly from the fitted parameter τ_{HN} and the difference is dependent on the parameters α and β. The expected value is always less than the fitted

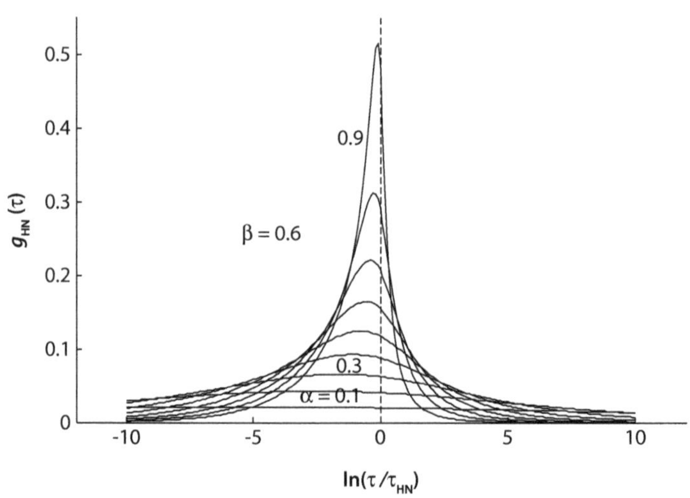

Figure 5.28 Effect of α on the HN distribution function for $\beta = 0.6$.

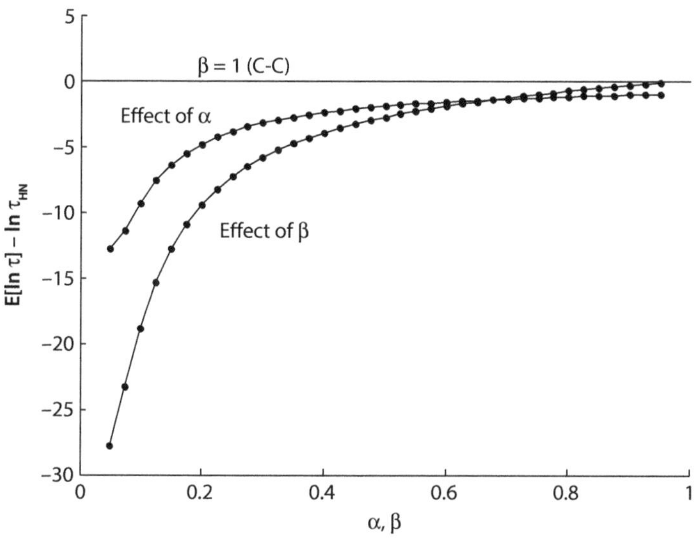

Figure 5.29 Effect of α and β on the difference between expected value of $\ln(\tau)$ and $\ln(\tau_{HN})$ for $\alpha = 0.5$ and $\beta = 0.6$.

value, which indicates the skewed nature of the HN relaxation function. With an increase in β (i.e., H-N → C-C) for a fixed $\alpha = 0.5$, the difference reduces and the expected value of $\ln(\tau)$ coincides with the fitted relaxation time constant at the C-C limit, i.e., Expected[$\ln(\tau)$] → E[$\ln(\tau)$] → $\ln(\tau_{CC})$. Similar behavior is also observed in the case of H-N → D-C situation via the effect of α (for a fixed $\beta = 0.6$). For the limiting C-C case the difference between the expected value of $\ln(\tau)$ and $\ln(\tau_{HN})$ is zero. An increase in α reduces the difference between the expected value and the fitted parameter value of τ_{HN}. However, unlike the C-C limit, the difference does not vanish at the D-C limit condition since the distribution is highly unsymmetrical at the D-C condition [41].

Another way to determine the skewness of the H-N distribution function and the effect of α and β on it is to calculate the skewness statistic of the distribution function. The skewness of the distribution function is defined [46, 47] as:

$$\gamma_1 = \mu_3 / \mu_2^{3/2} \tag{5.132}$$

where μ_2 and μ_3 are the second and third moment about the mean of the distribution, respectively. The skewness of g_{HN} given by γ_1 is plotted in Figure 5.30 as a function of α and β. In reality α is not the skewness but amalgamated with $\beta = 0.6$ in Figure 5.30. The horizontal line shows $\beta = 1$.

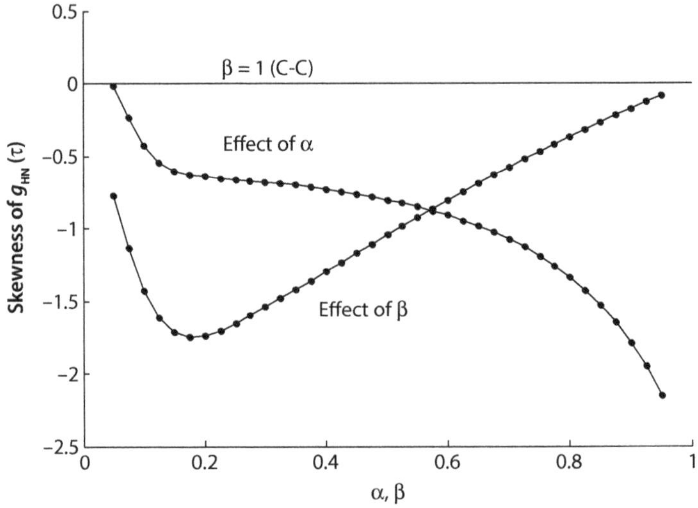

Figure 5.30 Effect of α and β on the skewness of the H-N distribution function for $\alpha = 0.5$ and $\beta = 0.6$.

This figure indicates that the skewness of the H-N distribution function reduces when β increases for a fixed α, and increases when α increases for a fixed β. This is expected since the increase in β indicates H-N → C-C condition, and the distribution becomes more symmetrical; thus, making it more gradual skew positively, i.e., the skewness vanishes when the H-N distribution becomes C-C distribution. On the other hand, the increase in α indicates H-N → D-C condition and the distribution becomes more non-symmetrical, thus making it increasingly negatively skewed distribution.

5.15 Meaning of the Depression Parameters α and β

The straightforward meaning of the depression parameter α implies the presence of the depression angle of the relaxation arc in the absence of β as found in the C-C behavior. This is not the case for the ac electrical data of the Ni-Cd battery displayed in the D-C configuration [35]. The presence of the finite depression angle represents non-ideal conduction processes wherein a considerable degree of non-uniformity is associated with the capacitance term than the resistance term. It also indicates the breadth of the distribution of the relaxation times. The depression parameter α indicates broadness of the frequency dependence of the imaginary part of the impedance ($Z'' \equiv F''$). This is related to the average relaxation time attributing to the reactance. Often the average relaxation time referring to the singular lumped R-C circuit (showing one relaxation time) is termed as the distribution of a set of relaxation times [29, 30]. For all purposes

α lies between 0 and 1 in order to satisfy Debye and C-C responses. The straightforward meaning of the depression parameter β in the D-C behavior implies the breadth and skewness [48] of the relaxation arc for the ac electrical data of the Ni-Cd battery.

The breadth and skewness of the relaxation arc are not necessarily overlapping with the behavior of the depression angle for the relaxation process resulting skewness in the relaxation arc. For the D-C behavior using the Ni-Cd data the relaxation arc is found to be skewed when $\alpha = 1$ and $\beta = 0.6$. The equivalent circuit model of the D–C behavior may be viewed considering both $\omega \times \tau = 1$ corresponding to the maximum of the y-axis parameter and $\omega\tau \neq 1$ corresponding to the maximum of the relaxation function $F^*(\omega)$. Another way to view the equivalent circuit is to refer to Figure 5.19 where $\omega \times \tau = 1$ does not correspond to the maximum of the y-axis parameter but takes the value on the low-frequency side.

Considering the role of both α and β causing the H-N behavior it is understood that these two depression parameters are not isolated from each other. The frequency domain product $\alpha \times \beta$ for the H-N parameters is related to the time domain KWW (Kohlrausch-Williams-Watts) parameter [42, 43]. Any further interpretation of the combined α and β may be extended to both diffusion and activation controlled dielectric behavior of the DUT. The parameters α and β are not optimized as far as H-N representation is concerned. The analytical approaches presented herein reflect that the Ni-Cd battery system is surely too complex to comprehend the underlying conduction processes. Based on the present analysis, it is understood that the internal configuration of the Ni-Cd battery is responsible for the evolution of the parameters α and β. The solution to the problems associated with the internal construction and overall content of the Ni-Cd batteries may not further simplify the parameters α and β.

The previous analyses [22] indicate that there is a chance to view the present H-N representation in the D-C format. This dual complexity requires clarity from the internal construction viewpoint of the Ni-Cd battery. Another study [49] also indicates a similar situation for the Ni-Cd batteries though curve fitting procedure for the identical complex plane analysis is not conducted. Nevertheless, it is clear that the Ni-Cd battery is too complex for the impedance spectroscopy to simplify the conduction processes. It is possible to use the information of the parameters α and β to achieve better performance characteristics. Perhaps in this way impedance spectroscopic approaches can provide improved functions of the Ni-Cd batteries.

The analysis of the ac small-signal electrical data in the H-N dielectric representation in the impedance plane allows understanding of the depression parameters α and β for the Ni-Cd batteries. These two parameters are real quantities and cannot be separated to display independent role for each of them for the H-N formalism. The breadth and skewness are associated with these two parameters primarily at high frequencies when the data are displayed in the impedance plane. The product $(\omega \times \tau)_{peak}$ corresponding to the maximum value of the imaginary part of the complex function $F^*(\omega)$ is not equal to unity for the H-N representation. This product depends on the parameters α and β and the extreme non-Debye situation which is characterized by $(\omega\tau)_{peak} > 1$. The geometrical relationships among various parameters of the H-N relaxation function can be used to determine the relaxation parameter τ_{HN}. Based on the behavior of the two depression parameters, it can be concluded that the optimized values to be determined in the future may yield useful battery performance concerning the reduction of the memory effect.

5.16 Relaxation Function with Respect to the Depression Parameters α and β

The relaxation functions (immittance functions) of Debye, Cole-Cole, Davidson-Cole, and Havriliak-Negami are given by equations 5.9, 5.13, 5.15, and 5.16, respectively. Plots are generated for C-C, D-C, and H-N relaxation functions with respect to the product $\omega Í\tau$ to comprehend the depression parameters α and β. These plots are presented in Figures 5.31 through 5.34 where it is shown how the product $\omega Í\tau$ is located with respect to the peak value of the relaxation function.

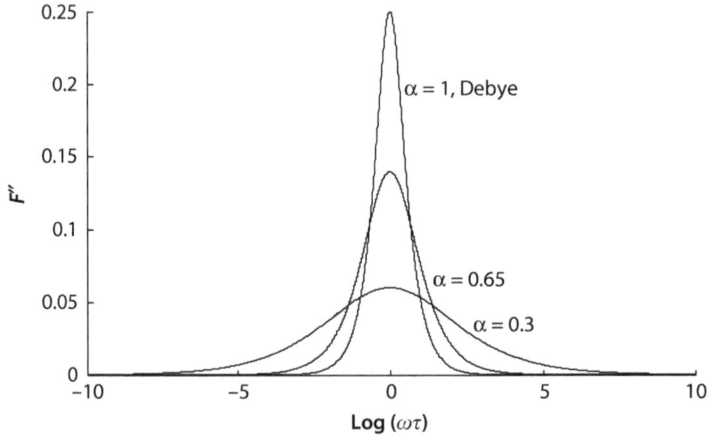

Figure 5.31 Imaginary part of C-C immittance function (F") versus log($\omega\tau$) for different values of the depression parameter α.

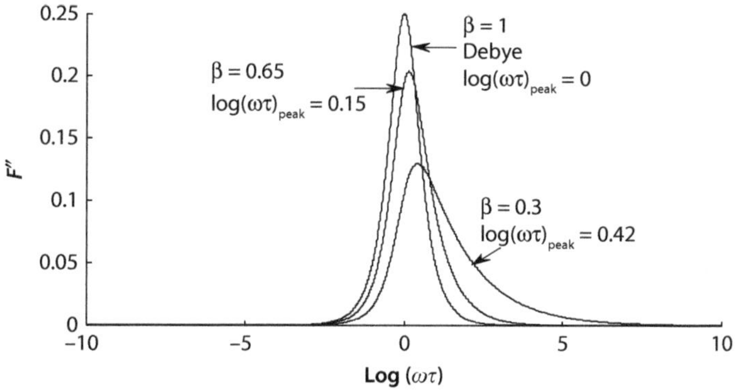

Figure 5.32 Imaginary part of D-C immittance function (F”) versus $\log(\omega\tau)$ for β = 0.3, 0.65 displaying the peak-values of $\log(\omega\tau)$.

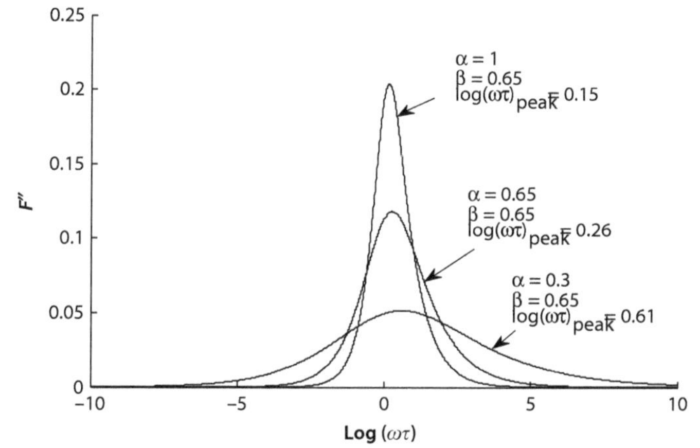

Figure 5.33 Imaginary part of H-N immittance function (F”) versus $\log(\omega\tau)$ for α = 0.30, 0.65 and 1 for fixed value of $\beta = 0.65$ displaying the peak-values of $\log(\omega\tau)$.

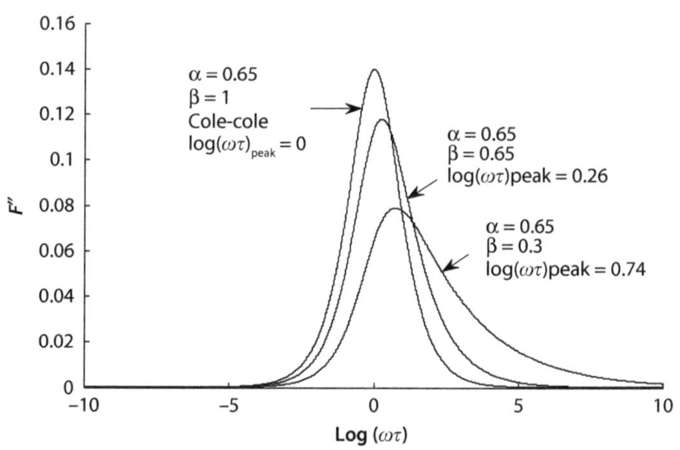

Figure 5.34 Imaginary part of H-N immittance function (F”) versus $\log(\omega\tau)$ for β = 0.30, 0.65 and 1 for fixed value of $\alpha = 0.65$ displaying the peak-values of $\log(\omega\tau)$.

Problems

1. Why the depression angle or depression parameter disappear for solid state devices at increased voltages? What mechanism allows this to happen?
2. Why do single-crystal based devices not normally exhibit a depression angle?
3. In general, liquid systems exhibit a depression angle in the Z^*-plane which is associated with the capacitance or reactance. Can this depression angle disappear at increased voltages? Explain the answer.
4. Write impedance and admittance of each circuit shown below. Take choice of value for each element in each circuit.

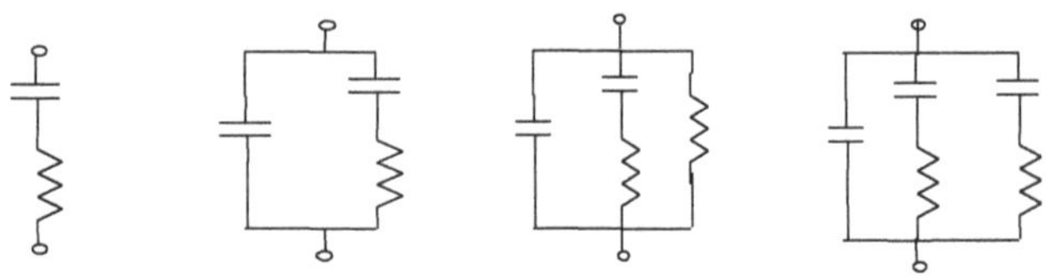

5. Use the following equation for depression angle = 15° for the semicircular relaxation to calculate R. Take choice of reasonable value for C with $\tau = 10^{-6}$ sec. Convert this R-C series circuit to the R-C parallel combination using the depression parameter.

$$C^* = +\frac{C}{1+(j\omega\tau)^{1-\alpha}}.$$

6. Generate the data taking L-C-R series circuit using the values 10 μH, 10 nF, and 10 Ω, respectively. Display the data in both C^*- and Y^*-planes. Describe the observations.

References

1. A. R. von Hippel; *Dielectrics and Waves*, John Wiley & Sons N.Y. (1954).
2. C. P. Smyth; *Dielectric Behavior and Structure*, McGraw-Hill Book Company N.Y., 1955.
3. Cole, K. S.; Cole, R. H.; "Dispersion and Absorption in Dielectrics I. Alternating Current Characteristics," *Journal of Chemical Physics*, 9, 341–351, 1941.
4. M. A. Alim; "Electrical Characterization of Engineering Materials," *Active and Passive Electronic Components*, 19, 139–169, 1996.

5. R. Resnick and D. Halliday; *Physics I & II*, John Wiley & Sons, N.Y., 1966.
6. J. D. Irwin and R. M. Nelms; *Basic Engineering Circuit Analysis*, 10th Edition, John Wiley & Sons, Inc., New York, 2011.
7. S. Bhattacharjee and M. A. Alim; "Frequency Dependent Impedance and Admittance (Immittance) Data-Handling and Interpretation using Complex Plane Formalisms via Nonlinear Regression Analysis for Smart Electronic Materials and Devices: Overview and Case Studies," *Journal of Materials Science: Materials in Electronics*, 26, 4521–4540, 2015.
8. F. A. Grant; "Use of Complex Conductivity in the Representation of Dielectric Phenomena," *Journal of Applied Physics*, 29(1), 76–80, 1958.
9. V. V. Daniel; *Dielectric Relaxation*, Academic Press, New York, 1967.
10. I. Bunget and M. Popescu; *Physics of Solid Dielectrics*, Elsevier, Amsterdam, 1984.
11. I. M. Hodge, M. D. Ingram, and A. R. West; "Impedance and Modulus Spectroscopy of Polycrystalline Solid Electrolytes," *Journal of Electroanalytical Chemistry*, 74, 125–143, 1976.
12. I. M. Hodge, M. D. Ingram, and A. R. West; "A New Method for Analyzing the A. C. Behavior of Polycrystalline Solid Electrolytes," *Journal of Electroanalytical Chemistry*, 58, 429–432, 1975.
13. D. P. Almond and A. R. West; "Impedance and Modulus Spectroscopy of 'Real' Dispersive Conductors," *Solid State Ionics*, 11(1), 57–64, 1983.
14. M. A. Seitz; "Study of Heterogeneous Composite Materials via Lumped Parameter/Complex Plane Analysis," *International Journal of Hybrid Microelectronics* (I.S.H.M.), 3(1), 1–7, 1980.
15. J. J. Ackmann and M. A. Seitz; 'Methods of Complex Impedance Measurements in Biologic Tissue," *CRC Critical Reviews in Biomedical Engineering*, 11(4), 281–311, 1984.
16. R. Coelho; *Physics of Dielectrics for the Engineer*, Elsevier Scientific Publishing Company, Amsterdam, 1979.
17. M. A. Alim, M. A. Seitz and R. W. Hirthe; "Complex Plane Analysis of Trapping Phenomena in Zinc Oxide Varistors," *Journal of Applied Physics*, 63(7), 2337–2345, 1988.
18. D. W. Davidson and R. H. Cole; "Dielectric Relaxation in Glycerine," *Journal of Chemical Physics*, 18(10), 1417–1417, 1950.
19. D. W. Davidson and R. H. Cole; "Dielectric Relaxation in Glycerol, Propylene Glycol, and n-Propanol," *Journal of Chemical Physics*, 19(12), 1484–1490, 1951.
20. S. Havriliak and S. Negami; "A Complex Plane Analysis of Dispersions in some Polymer Systems," *Journal of Polymer Science: Part C*, 14, 99–117, 1966.
21. S. Havriliak and S. Negami; "A Complex Plane Representation of Dielectric and Mechanical Relaxation Processes in some Polymers," *Polymer*, 8, 161–210, 1967.

22. M. A. Alim, S. R. Bissell, A. A. Mobasher; "Analysis of the AC Electrical Data in the Davidson–Cole Dielectric Representation," *Physica B*, 403, 3040–3053, 2008.
23. J. Li, S. Li, P. Cheng, and M, A. Alim; "Advances in ZnO–Bi_2O_3 Based Varistors," *Journal of Materials Science: Materials in Electronics*, 26, 4782–4809, 2015.
24. M. A. Alim, S. Li, F. Liu, and P. Cheng; "Electrical Barriers in the ZnO Varistor Grain Boundaries," *Physica Status Solidi (A)*, 203(2), 410–427, 2006.
25. A-. M. Azad, L. L. W. Shyan, and M. A. Alim; "The AC Electrical Characterization of the Solid-State Reaction Derived $CaSnO_3$," *Journal of Materials Science*, 14, 3375–3396, 1999.
26. A-. M. Azad, L. L. W. Shyan, and M. A. Alim; "Immittance Response of $CaSnO_3$ Prepared by Self-Heat-Sustained Reaction," *Journal of Materials Science*, 14, 1175–1187, 1999.
27. M. A. Alim; "Influence of Intrinsic Trapping on the Performance Characteristics of ZnO-Bi_2O_3 Based Varistors," *Active and Passive Electronic Components*, 17, 99–118, 1994.
28. M. A. Alim; "Influence of Multiple Trapping Phenomena on the Applications Criteria of ZnO-Bi_2O_3 Based Varistors," *Active and Passive Electronic Components*, 17, 57–65, 1994.
29. M. A. Alim; "Admittance-Frequency Response in Zinc Oxide Varistor Ceramics," *Journal of the American Ceramic Society*, 72(1), 28–32, 1989.
30. M. A. Alim and M. A. Seitz; "Singular Nature of Preferential Conducting Paths at High Electric Fields in ZnO Based Varistors," *Journal of the American Ceramic Society*, 71, C246-C249 (1988).
31. M. A. Alim, S. Bhattacharjee, S. Khanam, S. Runa, and N. Muna; "The Behavior of the ZnO-Bi_2O_3-Based Varistor Capacitance," *Journal of Active and Passive Electronic Devices*, 9(2), 157–184, 2014.
32. M. Bartkowiak and G. Mahan, "Nonlinear Currents in Voronoi Networks," *Physical Review B*, 51, 10825 - 10832 (1995).
33. G. E. Pike and C. H. Seager; "The dc Voltage Dependence of Semiconductor Grain-Boundary Resistance," *Journal of Applied Physics*, 50(5), 3414–3422, 1979.
34. G. E. Pike; "Semiconductor Grain-Boundary Admittance: Theory," *Physical Review B*, 30(2), 795–802, 1984.
35. J. P. Holman; *Experimental Methods for Engineers*, 7th edition, McGraw-Hill, N.Y., (2001).
36. E. O. Doebelin; *Measurement Systems: Application and Design*, 5th edition, McGraw-Hill, N. Y., (2004).
37. S. T. Bishay; "Numerical Methods for the Calculation of the Cole-Cole Parameters," *Egyptian Journal of Solids*, 23(2), 179–188, 2000.

38. J. R. Macdonald; "Impedance Spectroscopy: Models, Data Fitting, and Analysis," *Solid State Ionics,* 176, 1961–1969, 2005.
39. A. Sadkowski; "CNLS Fits and Kramers-Kronig Validation of Resonant EIS Data," *Journal of Electroanalytical Chemistry*, 573, 241–253, 2004.
40. J. Macutkevic, J. Banys, and A. Matulis; "Determination of the Distribution of the Relaxation Times from Dielectric Spectra," *Nonlinear Analysis: Modelling and Control*, 5(1), 75–88, 2004.
41. M. A. Alim, S. Bhattacharjee, S. Khanam, N. Muna, and S. Runa; "Ni-Cd Batteries for Automotive Applications: The ac Electrical Data of the Ni-Cd Battery in the Havriliak-Negami Dielectric Representation," in *Autonomous Vehicles: Intelligent Transport Systems and Smart Technologies*, N. Bizon, L. Dascalescu, and N. M. Tabatabaei edited, ISBN 978-1-63321-324-1, Nova Science Publishers, Inc., New York, Chapter 17, 453–483, 2014.
42. F. Alvarez, A. Algeria, and J. Colmenero; "Relationship Between the Time-Domain Kohlrausch-Williams-Watts and Frequency-Domain Havriliak-Negami Relaxation Functions," *Physical Review B*, 44(14), 7306–7312, 1991.
43. F. Alvarez, A. Alegria, and J. Colmenero; "A New Method for Obtaining Distributions of Relaxation Times from Frequency Relaxation Spectra," *Journal of Chemical Physics*, 103(2), 798–806, 1995.
44. M. Sluyters-Rehbach; "Impedances of Electrochemical Systems: Terminology, Nomenclature and Representation - Part I: Cells with Metal Electrodes and Liquid Solutions," *Pure and Applied Chemistry,* 66(9), 1831–1891, 1994.
45. Y. Morioka, S. Narukawa, and T. Itoh; "State-of-the-art of Alkaline Rechargeable Batteries." *Journal of Power Sources,* 100, 107–116, 2001.
46. R. Zorn; "Logarithmic Moments of Relaxation Time Distributions," *Journal of Chemical Physics,* 116(8), 3204–3209, 2002.
47. R. Zorn; "Applicability of Distribution Function for the Havriliak-Negami Spectral Function," *Journal of Polymer Science, Part B: Polymer Physics,* 27, 1043–1044, 1997.
48. S. Havriliak and S. J. Havriliak; "A Rigorous Comparison of Theoretical Autocorrelation Functions with Dielectric Relaxation Parameters for the Alpha Process of Polymers," *Journal of Polymer Science, Part B: Polymer Physics*, 35, 1887–1897, 1997.
49. A. Hammouche, E. Karden, and R. W. De Doncker; "Monitoring state-of-charge of Ni–MH and Ni–Cd batteries using impedance spectroscopy," *Journal of Power Sources*, 127, 105–111, 2004.

6

Modeling and Interpretation of the Data

6.1 Equivalent Circuit Model for the Single Complex Plane (SCP) Representation

Available texts [1–4] to date essentially emphasize single complex plane (SCP) plot using Z^*-plane although other three complex planes are mentioned. This is a limitation of the scope of the immittance spectroscopy for handling the measured ac small-signal electrical data. It may be effective for certain material systems to use Z^*-plane as the SCP route for the analysis [3] but justification or purpose is not adequately addressed. The SCP analysis is not a template format for the liquid electrochemical systems as the underlying mechanisms may run all over the DUT. It is not necessarily true that the Z^*-plane analysis answers all the questions concerning the investigation. No explanation is noticed whether other complex planes are exploited for any possible missing interpretation. Bode plane analysis (BPA) does not answer all the questions as it is sometimes identified to be a limiting tool for the measured immittance data. To some extent BPA confirms some of the observations made via complex plane analysis. The most recent text [4] exercised SCP analysis though dielectric phenomena are demonstrated. As a beginning text [2], a breakthrough was made in the scientific and technological arena though the Z^*-plane was demonstrated and emphasized heavily for the solid state materials. The most recent texts [1, 4] noted the use of other complex planes but the demonstration or necessary explanation is not adequate or sufficient. For some unknown reasons, walking outside the fence of the SCP, implying Z^*-plane, remained restricted. As a technical know-how, SCP became the root solution to all problems.

As a pioneer of the complex plane analysis several questions can come about the classic Cole-Cole plot [5]. Why did complex plane analysis (CPA) begin with the ε^* (or C^*) plot? Why was the use of the Z^*-plane totally absent there? Why was there no mention of other complex plane analysis then? Before the recent so-called popularity of the Z^*-plane representation, Y^*-plane [6] and M^*-plane [7, 8] were used but neither drew much attention. Somehow Z^*-plane analysis has become a sudden massive viable analytical tool of the immittance data that ultimately turned into a rapid template for many investigators or neo-investigators. Is Z^*-plane giving all the necessary information about an unknown DUT – liquid or solid? Certainly Z^*-plane has limitations not only for the solid materials but also for the biological and liquid electrochemical systems. Often the Z^*-plane plot is referred to as the Cole-Cole plot though the plotting components are very different than the components of C^*. The direction of measurement frequency along the semicircular arc or the form of complex quantity equation or expression for Z^* $(= Z' - j Z'')$ or C^* $(= C' - j C'')$ is not adequate to justify the Z^*-plane plot as the Cole-Cole plot. Overall, identifying Z^*-plane analysis as the Cole-Cole plot is a total misnomer as the extracted equivalent circuit has no similarity between the two complex planes. The meaning of the presence of the depression angle for the semicircular relaxation for any of these two complex planes is completely different. Thus, the Z^*-plane and ε^*-plane (or C^*-plane) do not have the same feature or meaning for the extracted equivalent circuit.

It is a general tendency to use the measured electrical data via SCP analysis usually exploiting only the Z^*-plane or occasionally the ε^*-plane in lieu of the C^*-plane. This is presumably done because of the measurement frequency running in the increasing anti-clockwise direction in these two complex planes. Neither the Z^*-plane nor any of the complex planes other than the ε^*-plane is the Cole-Cole plot. However, the C^*-plane can be referred to as the Cole-Cole plot as it is not normalized using the geometrical factors. Resemblance of the anti-clockwise direction for the increasing measurement frequency must not dilute the original feature of the classic Cole-Cole plot [5].

For any SCP plot showing semicircular relaxation yields an equivalent circuit. Thus, four complex planes provide four different types of equivalent circuits. Each of these circuits gives lumped behavior of the unknown DUT. If two resistors have similar magnitude then these two resistors in the series form or in the parallel form will yield as a single resistor in the SCP. This means that the single resistor is the outcome of the behavior of those two resistors. It is likely that those two resistors may be attributed to the similar mechanisms involving conduction due to electron or hole or ion. These two resistors cannot be separated or isolated from each other via complex plane analysis as two distinct resistors for the same branch in the equivalent circuit. In that case these two resistors will emerge as a single resistor which is called the lumped effect. A similar situation can happen for the capacitors as well.

In general, the ac small-signal electrical data do not always give semicircular relaxation in each of the four complex planes in a simultaneous fashion. If a semicircular response is obtained in one complex plane then the same data may appear as unfamiliar curve depicting vertical linear or slant linear or some sort of curve in another plane. Sometimes the same data can show semicircular behavior in more than one complex plane too. This situation gives complexity on the nature of the equivalent circuit model but can yield meaningful interpretation for the observed behavior of the DUT. For an equivalent circuit, a resistor represents a conductive path whereas a capacitor or an inductor represents some sort of polarization process. In the equivalent circuit, relaxation is identified from the combination of these circuit elements.

6.2 Models and Circuits

Multiple equivalent circuits may be extracted for three relaxations in the Z^*-plane known as Voigt, Maxwell, and Ladder circuits. As usual Voigt is straightforward with three R-C parallel segments in series. Maxwell reflects two R-C series in parallel with another R-C parallel circuit. Ladder is a bit complicated as it is a combination of Voigt and Maxwell circuits [1, 2]. One R-C series circuit implies occurrence of a series event such as trapping and de-trapping. Altogether the equivalent circuit must have validity for the elements identified. The choice of a Ladder circuit can be regarded as the Macdonald circuit [1, 2]. This is a tedious type of circuit to consider for the underlying operative mechanisms. The selection of the equivalent circuit must satisfy the DUT for the same set of immittance data. More detailed discussions about these circuits are available elsewhere as noted by Macdonald in several publications [1, 2]. These models are depicted in Figure 6.1.

Several plausible circuit situations for the immittance data are displayed in Figure 6.2 for the Z*-plane. Figure 6.2(a) represents an intercept on the left side of the semicircular relaxation. Figure 6.2(b) represents two consecutive semicircular relaxations yielding the Voigt circuit. As a rule of thumb, the features of the Z*-plane never allows the dc blocking configuration of the equivalent circuit. It means that at zero frequency or at the dc condition implying $f \rightarrow 0$ Hz the equivalent circuit will never allow blocking of the dc current flow. As the measurement angular frequency ω increases along the semicircular arc for Figure 6.2(c) or 6.2(d), the capacitance begins to trace gradual exhaustion yielding peak-frequency corresponding to the time constant τ_1 satisfying:

$$\omega R_1 C_1 = \omega \tau_1 = 1 \quad \text{with} \quad \tau_1 = R_1 C_1. \tag{6.1}$$

Once C_1 is completely exhausted at increased frequency, then starts the next semicircular relaxation process for which again the capacitance C_2

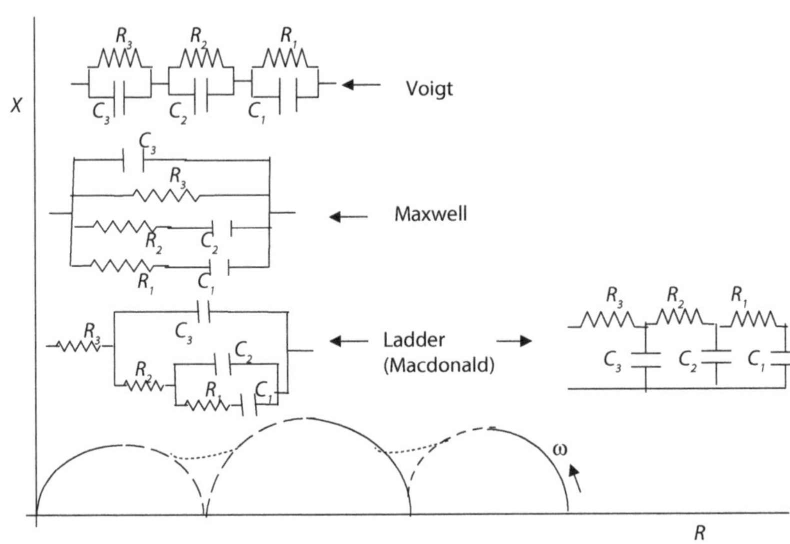

Figure 6.1 Three successive relaxations obtained for a single set of ac small-signal immittance data acquired as a function of frequency displayed in a single complex plane represented by three different equivalent circuits: (1) Voigt model containing three R-C parallel circuits in series; (2) Maxwell model containing two R-C series circuits in parallel with another R-C parallel circuit in parallel; and (3) Ladder or Macdonald model displayed in two arrangements where each relaxation occurred for the R-C series circuit.

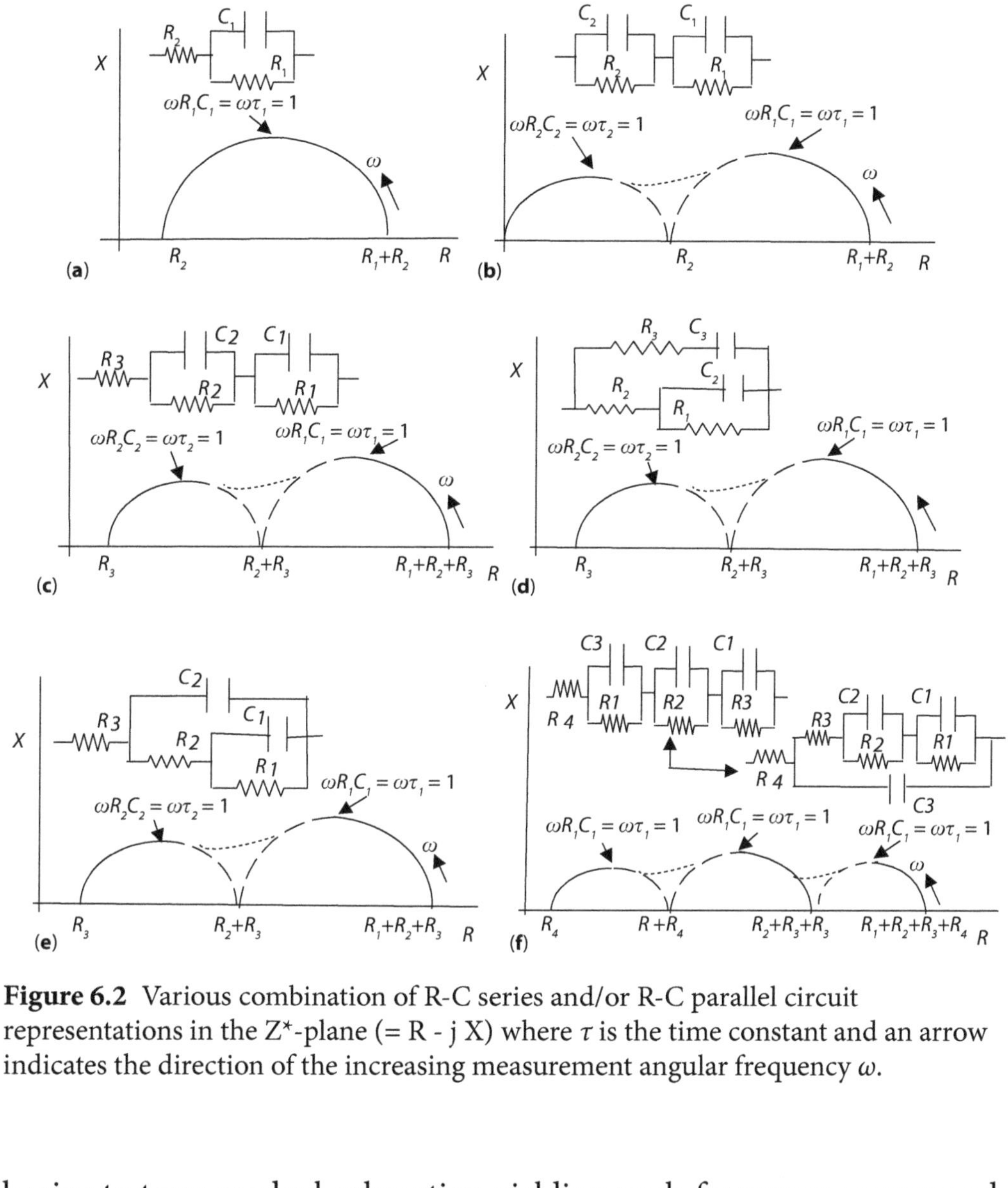

Figure 6.2 Various combination of R-C series and/or R-C parallel circuit representations in the Z*-plane (= R - j X) where τ is the time constant and an arrow indicates the direction of the increasing measurement angular frequency ω.

begins to trace gradual exhaustion yielding peak-frequency corresponding to the time constant τ_2 satisfying:

$$\omega R_2 C_2 = \omega \tau_2 = 1 \qquad \text{with} \qquad \tau_2 = R_2 C_2. \tag{6.2}$$

When C_2 is completely exhausted at an increased frequency termed as the high frequency then the intercept R_3 appears. This feature ascertains that under any circumstance there is no blocking of the dc conduction processes between the electrodes unless some other blocking feature exists within the DUT. For the pure blocking feature invariably a capacitance must appear in series as does in the M*-plane plotting.

The equivalent circuit of Figure 6.2(d) represents partly Ladder model which is slightly different than the complete Ladder model of Figure 6.2(e) though each circuit contains the same number of resistors and the same number of capacitors. Figure 6.2(f) displays two equivalent circuits for three consecutive relaxations in the Z^*-plane. These two circuits essentially represent Voigt and part Ladder model having minor difference in the arrangement with respect to the circuit depicted in Figure 6.1. At this conjecture it may be noted that the variation within the Ladder model represents very complicated nature of the circuit.

Another feature may be noted when two consecutive semicircles are obtained in the Z^*-plane as displayed in Figure 6.2(b) or in 6.2(c) or in 6.2(d) or in 6.2(e). For two semicircles without intercept often rapid decision is taken referring to grain and grain boundary behavior for the DUT if it is polycrystalline. What causes the intercept in Figure 6.2(c) or in 6.2(d) or in 6.2(e) then? Is this intercept related to the electrode? Is this intercept related to the contact leads? How Voigt model can validate grains and grain boundaries having finite high frequency intercept? Can Ladder model be introduced? What mechanisms are likely to tie these together? Answers to these concerns are not difficult to ascertain as they each require addressing one at a time. The lead or electrode can be verified using certain experimental concepts via using conductive electrode matching proper work function of the electrode material and using shorter lead length can eliminate the problem. Some immittance analyzer manufacturers supply these contact terminals. The rest needs proper investigation.

It may be true that the physical regions (grains and grain boundaries) of the DUT are responsible for the two relaxations. But this conclusion or decision has no confirmation if the trapping and de-trapping is predominant. Why the grain and the grain boundary would have the resistance on the same order of magnitude? If the DUT is a polycrystal, particularly with added cations, then there is a high chance of existing large density of defect states across the grain boundaries. Where are the contributions of these defect states in the form of traps in the Z^*-plane feature? Can Z^*-plane accommodate those trapping features? How can these contributions be diagnosed? Sometimes traps can be suppressed if the relevant magnitudes of the series R-C elements are matched enough with respect to the R-C parallel elements attributing to the grains and grain boundaries.

Since trapping and de-trapping are always series events then R-C series feature cannot be ascertained in the Z^*-plane or M^*-plane semicircle or via the presence of the intercept on the left side of the semicircle. In that case the data should be examined or verified in the Y*-plane or C*-plane. When Z*-plane is the only complex plane showing semicircular relaxation then the conclusion may be partially complete by keeping the feature of the defect states away from the discussion. In most SCP presentations, defect states are either outside the discussion or never brought in the scenario to make easy route or passage to draw conclusion. This is what many investigators do by eliminating or distancing further understanding on the rest of the complex plane analysis for the polycrystal DUT. Otherwise full exploitation of the CPA can delineate all aspects of the DUT.

It is worthy to note common features between the single semicircular relaxations obtained in the Z*- and M*-planes with the intercept on the left side of the semicircle. If the DUT allows conduction under the dc condition the feature observed in the M*-plane is not useful. Yet the question remains as to why the immittance data were yielding semicircular relaxation in both the complex planes having intercept in each of them. In this case the circuit elements are easily convertible from the dc blocking feature of the M*-plane to the non-blocking or conduction feature of the Z*-plane or vice versa. This convertible feature is only useful if the relaxation is Debye type. However, the conversion of the circuit elements for the Z*- and M*-planes is only a verification of the two valid circuits. One should find out the reason as to why the dc blocking feature is coming from the data. This means that the DUT must be understood thoroughly with respect to the microstructural configuration. When there is no clue for the dc blocking capacitance in the M*-plane then the only acceptable reasoning could be matching values of the elements that caused the immittance data. A similar situation can happen with other complex planes with respect to the Z*-plane analysis. Thus, only Z*-plane analysis or single complex plane analysis does not provide complete sense about the equivalent circuit. Without examining the data in all the complex planes it is not worthy to draw conclusion about the DUT.

If two semicircles are obtained in the Z*-plane for a single crystal based DUT as shown in Figure 6.2(b) or in 6.2(c) then one can ascertain that there may be no influence of trapping effect. Perhaps the depletion region in conjunction with the bulk effect and/or electrode effect may have contributed for one of the semicircles. Usually low frequency relaxation is attributed to the electrode effect whereas high frequency relaxation is tied with the bulk effect. In order to ascertain bulk, electrode, and depletion region contributions there must have some sort of overlap that ultimately yielded two semicircles. For such a situation all four complex planes must be used to confirm the conclusion about the DUT.

An intercept on the left side of the semicircle on the x-axis in the Z*-plane appears as a series resistor with the R-C parallel combination. The parallel R-C combination usually represents parallel event occurring simultaneously within the DUT as noted for the depletion layer of the p-n junction. This junction represents concurrent event for the capacitance due to the layer formed at the junction between the p- and n-regions, and the conduction taking place across the depletion layer.

A Ladder or Macdonald model can appear in a complicated form. Two semicircles with intercept on the extreme left side on the x-axis is a common feature for three distinct circuits provided in Figure 6.2(c), 6.2(d), and 6.2(e). Both Figure 6.2(d) and 6.2(e) contain Ladder but the steps are slightly different. In Figure 6.2(d) there is a R_3-C_3 series configuration where R_2-C_2 relaxation is noted to occur. If trapping is dominant then certainly R_3-C_3 series combination will get the privilege of trapping contribution. In Figure 6.2(d) the intercept is inclusive in the circuit whereas it is not inclusive in Figure 6.2(e). The interpretation and the choice absolutely depend on the investigator based on the associated mechanisms operative within the DUT. In each case the magnitude of the resistors R_1, R_2, and R_3 are of the same order while the magnitude of C_1 and C_2 are far apart, possibly order differences, from each other as the time constant determines the magnitude. The magnitude of each capacitance should indicate the type of operative mechanism. Two different types of Ladder or partly Macdonald circuits are shown for three relaxations displayed in Figure 6.2(f). Depending on the situation, the investigator has to make the choice of the equivalent circuit matching the underlying operative mechanisms within the DUT.

Several circuit scenarios for the immittance behavior displayed in the Y*-, C*-, and M*-planes are shown in Figures 6.3, 6.4, and 6.5, respectively. Bauerle [6] displayed the circuits of Figure 6.3(a), 6.3(b), 6.3(c), 6.3(d), and 6.3(f) for the investigation of $(ZrO_2)_{0.9}$-$(Y_2O_3)_{0.1}$ solid electrolyte system. Two usual blocking circuits are provided in Figure 6.3(c) and 6.3(d). Two separate circuits of Figure 6.3(e) and 6.3(f) satisfy two consecutive relaxations with an intercept in the Y*-plane. The circuit in Figure 6.3(f) represents a combination of Maxwell and Ladder models for the current dividing nature. As usual simplistic identical circuits in Figures 6.3(a) and 6.4(a) are dc blocking in nature. Existence of the intercept brings a conducting circuit in Figure 6.3(b) but dc blocking in Figure 6.4(b). The influence of the capacitance in parallel with the R-C series circuit brings distortion in the high frequency portion of the single semicircle of Figure 6.3(c) whereas perfect semicircle with intercept obtained at high-frequency on

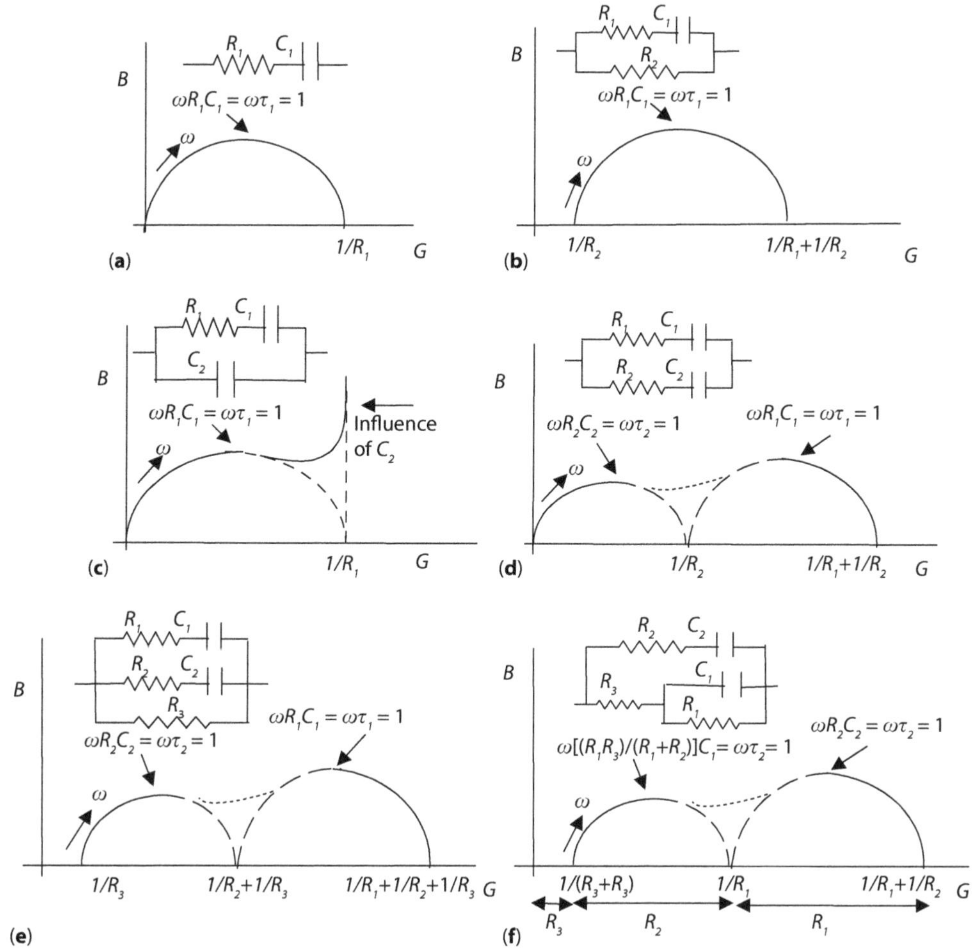

Figure 6.3 Various combination of R-C series and/or R-C parallel circuit representations in the Y*-plane (= G + j B) where τ is the time constant and an arrow indicates the direction of the increasing measurement angular frequency ω.

the left side of the semicircle in the C*-plane of Figure 6.4(b). Grant [9] originally displayed the same circuit both in Y*- and C*-planes, and demonstrated for two material systems: $Ni_{0.4}Zn_{0.6}Fe_2O_4$ and $BaTiO_3$.

Usually semicircular relaxation in the C*-plane with or without intercept yields a dc blocking circuit. The influence of the parallel conductance in the equivalent circuit provides distortion in the semicircular loci at low frequencies in Figure 6.4(d). The intercept in the C*-plane always corresponds to the high frequency. Figure 6.4(f) is an example of a conductive circuit as opposed to the blocking circuit of Figure 6.4(e) for the same two consecutive relaxations having an intercept.

Every relaxation with intercept in the M*-plane yields a dc blocking circuit. Any series resistance in the equivalent circuit obtained in the

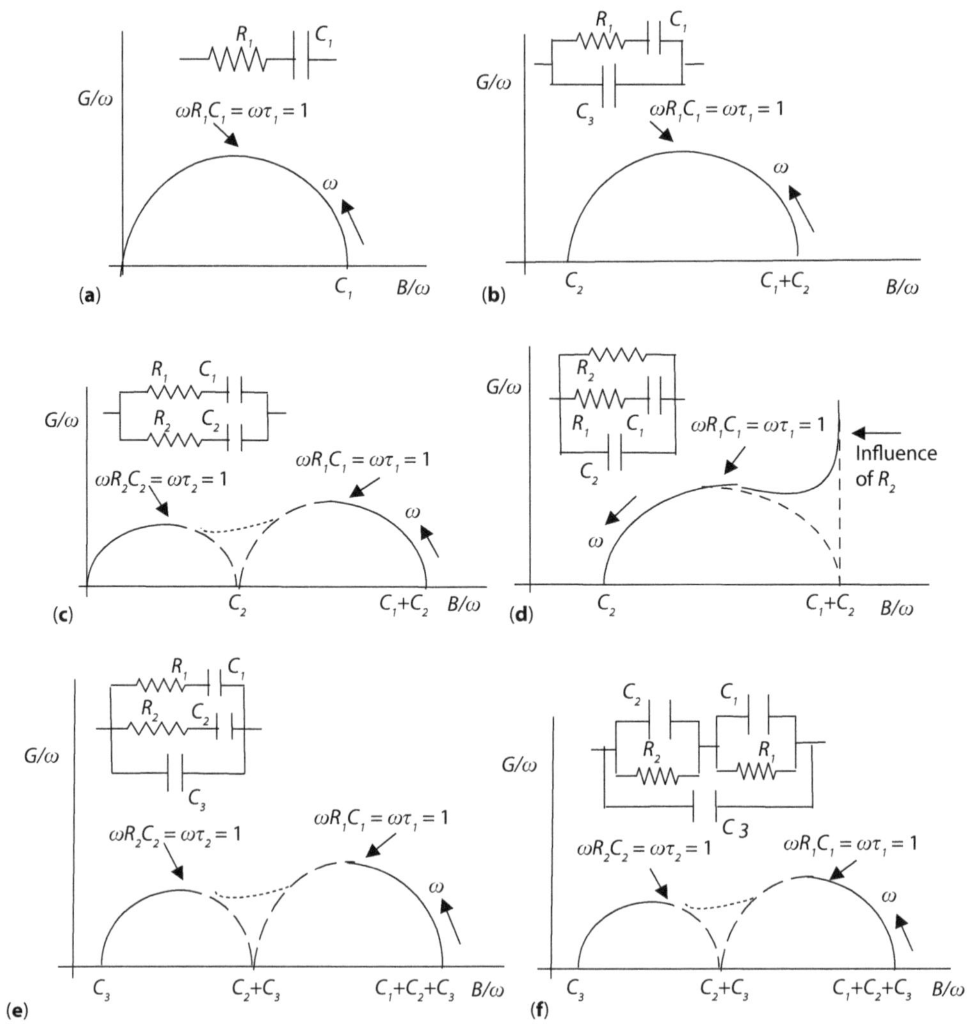

Figure 6.4 Various combination of R-C series and/or R-C parallel circuit representations in the C*-plane (= $B/\omega - j\ G/\omega$) where τ is the time constant and an arrow indicates the direction of the increasing measurement angular frequency ω.

M*-plane gives distortion at low frequencies as shown in Figure 6.5(b), 6.5(d), and 6.5(e). Non-intercept semicircular relaxation is demonstrated in the M*-plane along with the Z*-plane for the solid electrolytes [7, 8].

6.3 Nonconventional Circuits

Sometimes equivalent circuit elements may appear in a non-conventional way. For example, heavy trap concentration can dominate the depletion layer capacitance-resistance parallel combination. In this case R-C series

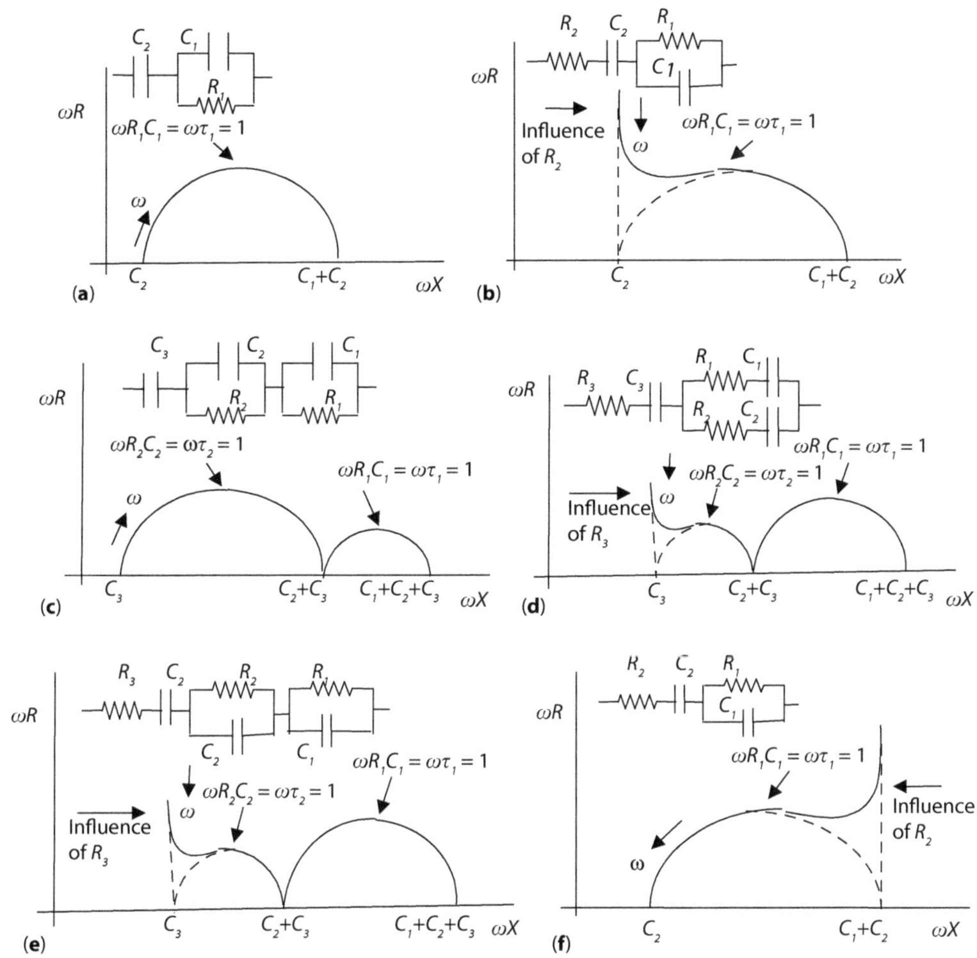

Figure 6.5 Various combination of R-C series and/or R-C parallel circuit representations in the M*-plane (= ωX - j ωR) where τ is the time constant and an arrow indicates the direction of the increasing measurement angular frequency ω.

combination is the dominating feature against the depletion layer feature. Therefore, the Z*-plane semicircular feature with intercept may be represented by series R_1-C_1 in parallel with the intercept resistance R_2 shown in Figure 6.6(a). Sometimes this behavior may be seen in the Y*- or C*-plane. This is a rare case but requires proof for weak influence of the depletion layer. Figure 6.6(b) shows influence of the series capacitor C_2 as distortion at low frequencies but very much convincing as a dc non-blocking type circuit. Sometimes this response coincides with that of the M*-plane. Though electrical data of the DUTs in the Z*-plane is conductive in nature but sometimes conclusive blocking circuit may appear. It is likely that the depletion layer and trapping may appear as parallel features via Z*-plane

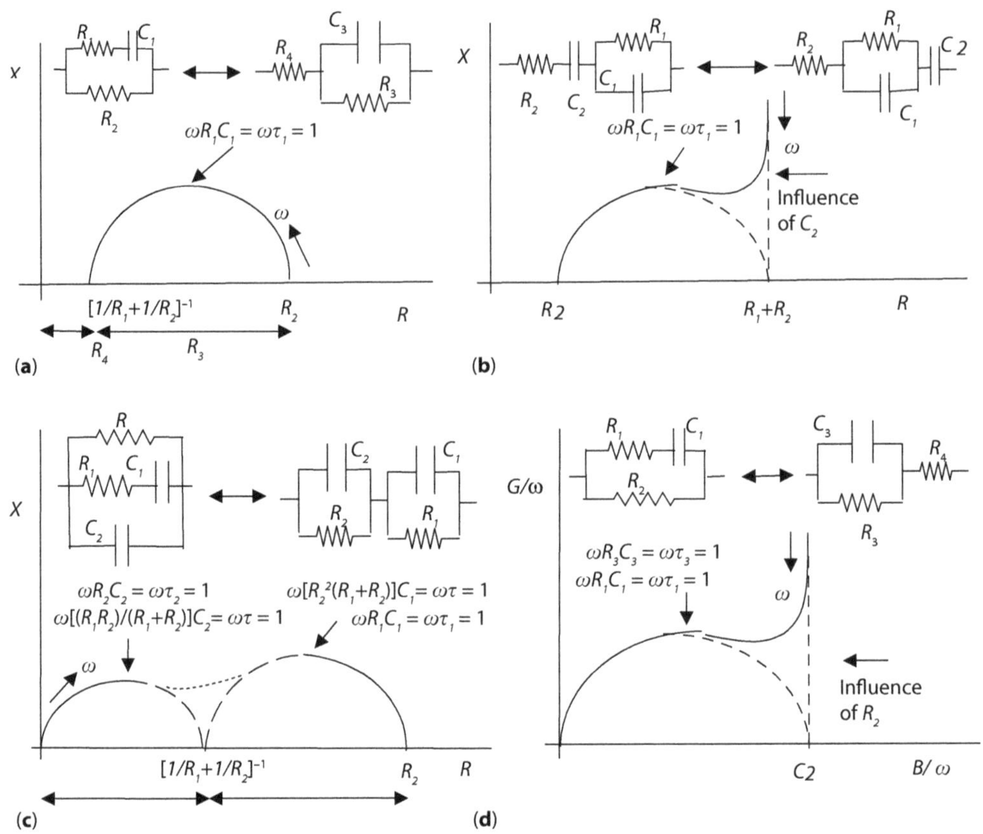

Figure 6.6 Various R-C series and/or R-C parallel combination displayed as nonconventional circuits in the complex planes where τ is the time constant and an arrow indicates the direction of the increasing measurement angular frequency ω.

as shown in Figure 6.6(c). The converse discussion of Figure 6.6(a) can reflect Figure 6.6(d).

Bauerle's circuit [6] appears in Figure 6.7(a). Two possible dc blocking models are presented in Figure 6.7(b) and 6.7(c) as series events but in two complex planes. The dc blocking model circuit cannot be valid for the leakage conduction of the DUT. The same feature is also presented in Figure 6.7(d) in the M*-plane. Often M*-plane relaxation with an intercept on the left side is difficult to interpret for the existing leakage conduction.

6.4 Multiple Equivalent Circuits for Multiple Relaxations in a Single Complex Plane

Derivation of multiple mathematically possible equivalent circuits for two consecutive distinct relaxations in a single complex plane is possible. These

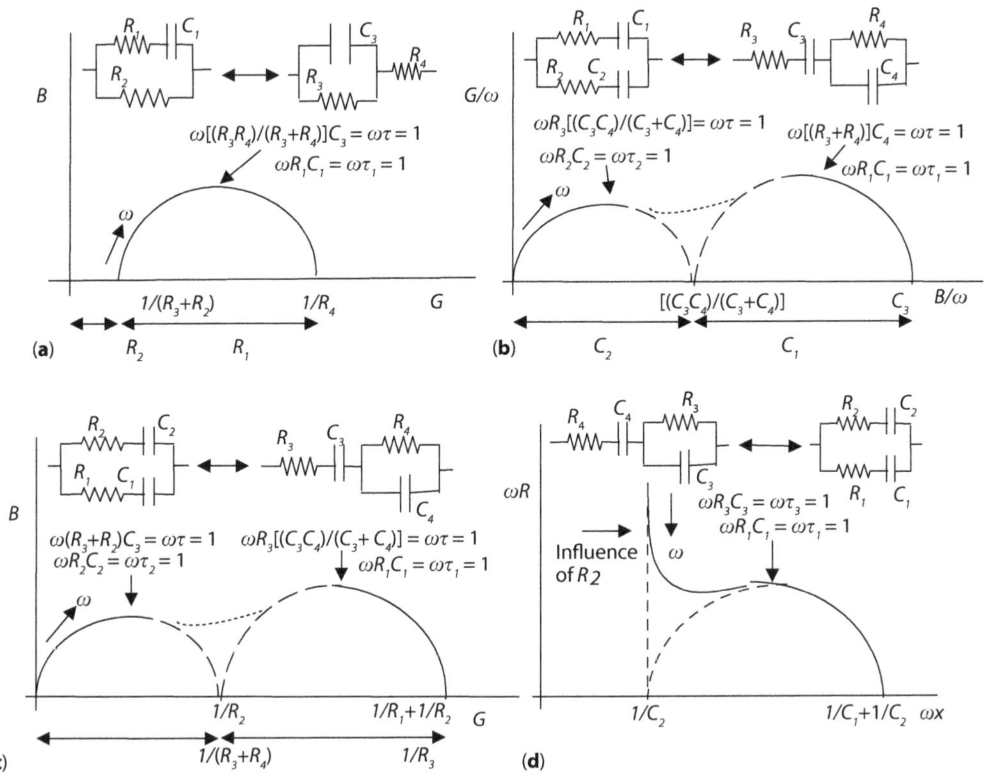

Figure 6.7 Various R-C series and/or R-C parallel circuits in the complex planes for the nonconventional circuits where τ is the time constant, and an arrow indicates the direction of the increasing measurement angular frequency ω.

circuits may include Voigt, Maxwell, and Ladder configurations. Selection of the ultimate equivalent circuit is a challenge as thorough understanding of the DUT can pave the road. Enough care is necessary for adopting an equivalent circuit in this case as other circuits can be mathematically equivalent or valid. Further tests are necessary for confirming the interpretation. Four circuits are displayed for two consecutive Z*-plane relaxations in Figure 6.8. For three consecutive relaxations in the same plane can be more tedious to resolve for an equivalent circuit unless the DUT is simplified.

6.5 Single Equivalent Circuit for Multiple Complex Planes

A single circuit shown in Figure 6.9(a) can emerge from multiple complex planes. The electrical data for this circuit can exhibit two relaxations in the Z*-plane as displayed in Figure 6.9(b). The coinciding point for

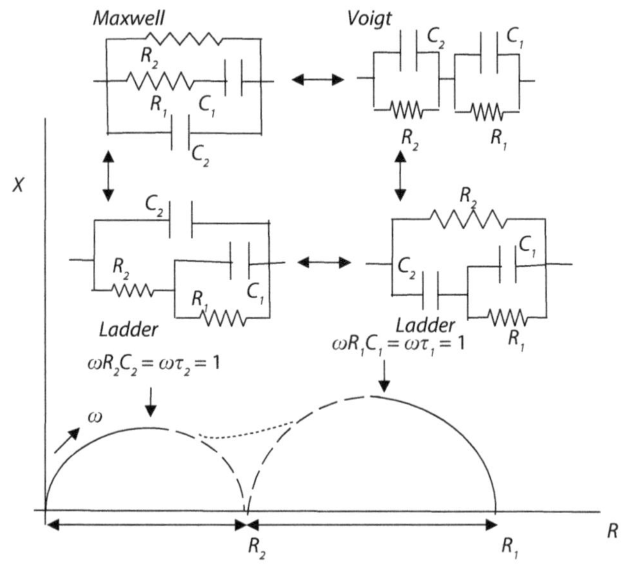

Figure 6.8 Four mathematically possible equivalent circuits for two consecutive distinct relaxations include Voigt, Maxwell, and Ladder configurations where τ is the time constant, and an arrow indicates the direction of the increasing measurement angular frequency ω.

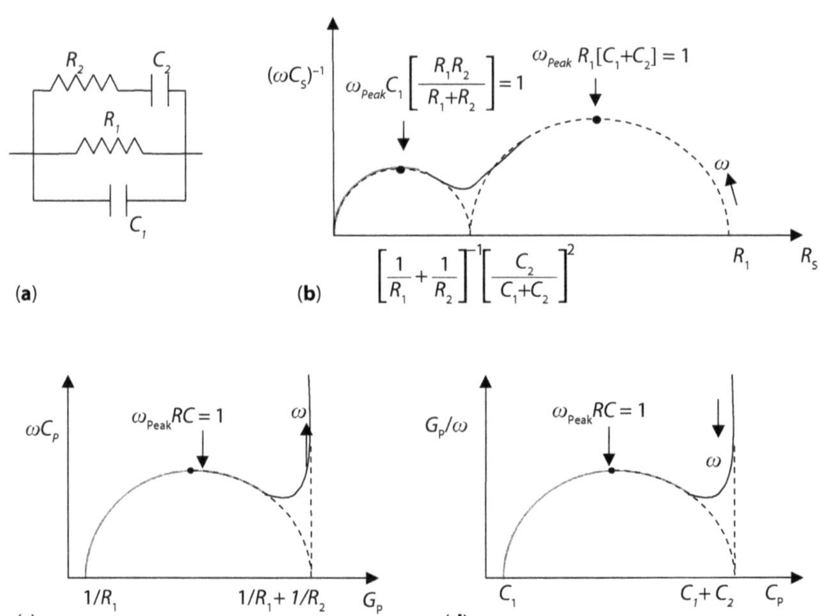

Figure 6.9 Single circuit of (a) satisfies two semicircular relaxations in the Z*-plane of (b) whereas the same circuit satisfies the relaxation of Y*- and C*-planes in (c) and (d), respectively, per Grant [9].

two semicircular relaxations on the real-axis determines the value of the chord as the resistance for the high-frequency semicircular relaxation. This resistance may have more than one face from the circuit shown in Figure 6.9(a). By the same token the intercept on the right side of the low-frequency semicircle determines the value of the chord as resistance for the relaxation. Selection of this circuit is a challenging concept from the type of material being investigated. Enough care is necessary for adopting this circuit as the Z*-plane depicts two distinct semicircles. After satisfying the interpretations, this circuit should be confirmed. The same circuit is demonstrated by Grant [9] for the Y*- and the C*-plane as shown in Figure 6.9(c) and 6.9(d), respectively, and the same is supported by the Z*-plane relaxations. Based on logic and sense this is the feature that the Z*-plane can delineate for the possible trap-containing system.

6.6 Equivalent Circuit for Resonance

Two separate resonance circuits are shown in Figure 6.10(a) and (b) satisfying Y*-plane relaxation. A perfect circle appears for the R-L-C series circuit. Two different chord values are noted for the lumped resistance as structured in Figure 6.10(b). One chord is for the semicircular relaxation in the positive domain before resonance, and the other chord is reflecting the post-resonance semicircle.

6.7 Single Equivalent Circuit from Z*- and M*-Planes

Simple circuits of Figure 6.11 satisfy both Z*- and M*-planes [7, 8] for various material systems. Therefore, concurrent representation of the complex planes for a single set of electrical data has strong ground for establishing

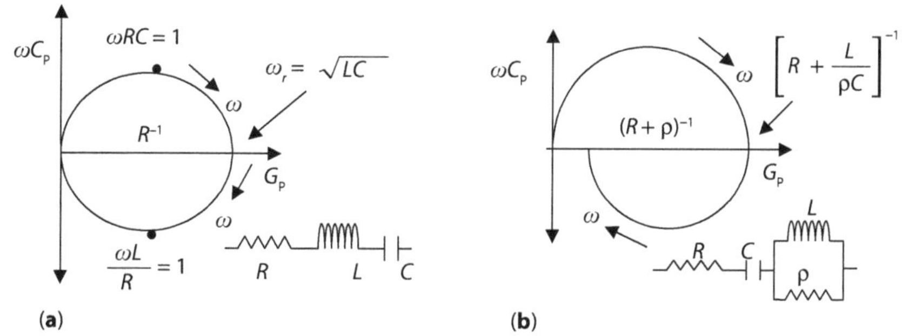

Figure 6.10 Two different resonance circuits can be obtained from the Y*-plane.

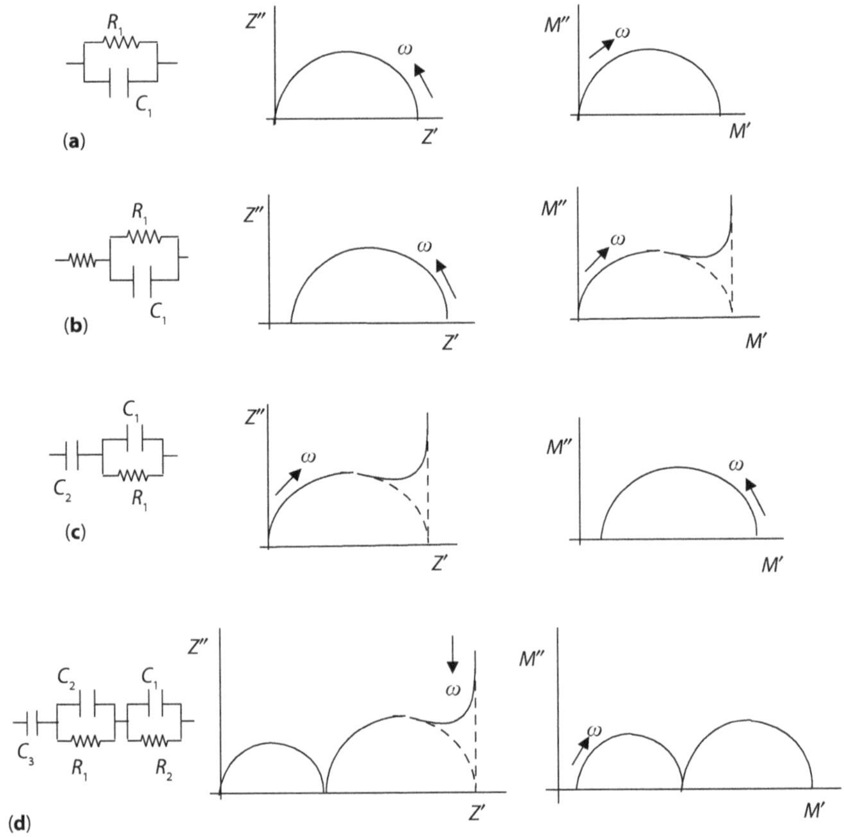

Figure 6.11 Z*- and M*-plane plots of four basic equivalent circuits that are sometimes found as dual representation of the same measured electrical data.

underlying operative mechanisms that yield a nice single equivalent circuit. It is a convention that the M*-plane depicts dc blocking feature, but when the same data are presented in the Z*-plane then the non-blocking feature is strongly validated. For leakage conduction, DUTs satisfying M*-plane feature is achieved via distortion in the semicircular loci.

6.8 Temperature and Bias Dependence of the Equivalent Circuit Modeling

Elevated temperature and/or dc biasing of the DUT can influence level of conduction. As a result, carriers can increase and causes reduction in the effect of the depletion layer. Thus, R-C combination can be altered significantly and become dominating contributors with respect to room temperature behavior. Thermal activation energies for R and C can easily dominate the thermal behavior of trapping and de-trapping within the DUT. Usually a polycrystalline device exhibits drastic changes in the equivalent circuit model at elevated temperatures and dc biasing compared to room temperature and zero bias of the DUT [10–12]. Thorough investigation is necessary to determine precise equivalent circuit at various conditions.

6.9 Equivalent Circuit: Zinc Oxide (ZnO) Based Varistors

Simultaneously operative multiple phenomena in ZnO-based varistors can be used in establishing an equivalent circuit model [10–13]. The semicircular arc in the C*-plane is an indication of a series R-C circuit [9, 14]. Thus, each distinct trapping relaxation may be interpreted as such, and modeled as a parallel event with respect to the barrier layer capacitance (C_{BL}). Thus, three distinct trapping branches are designated as series R_2-C_2, R_3-C_3, and R_4-C_4 in parallel. The low frequency admittance Y_{LF} is unknown via unresolved arrow-cut series resistance and capacitance of the parallel branches. Basically Y_{LF} is given by:

$$Y_{LF} = G_{LF} + j\,\omega C_{LF} \tag{6.3}$$

for the branch representing low frequency. All these parallel trapping branches are effectively in parallel with the dc resistance R_{dc} of the device. The bulk capacitance (C_{ZnO}) and resistance are effectively in parallel which is in series with the grain boundary elements and the inductance (L_r) predominantly related to the external leads. This scenario of the equivalent circuit model [10–13] is developed from the immittance data displayed [2] in the C*- and Y*-plane. This circuit is summarized in Figure 6.12 for electric field $0 \leq E \leq 80\%\ E_{1mA/cm^2}$ in the temperature range 26 °C ≤ T ≤ 90 °C where electric field E_{1mA/cm^2} corresponds to the current density 1 mA/cm^2. It is very difficult to visualize this kind of equivalent circuit displaying entire features of the ZnO-based varistor as a device. In reality the measurement conducted in the frequency range 10^{-2} Hz ≤ f ≤ 10^9 Hz that revealed trapping contributions and series resonance behavior via C*-plane plotting which in turn provided the lumped response in the Z*-plane plotting.

The resonance behavior was confirmed via Y*-plane plotting in the frequency range 10^6 Hz ≤ f ≤ 10^9 Hz that contributed to the series L_r-C_r-R_r resonance circuit [10, 13, 15–17]. The parameter C_6 identified as C_{ZnO} from the C*-plane after the resonating behavior was identified. Only the value of R_r obtained from the Y*-plane matched with the small resistance obtained

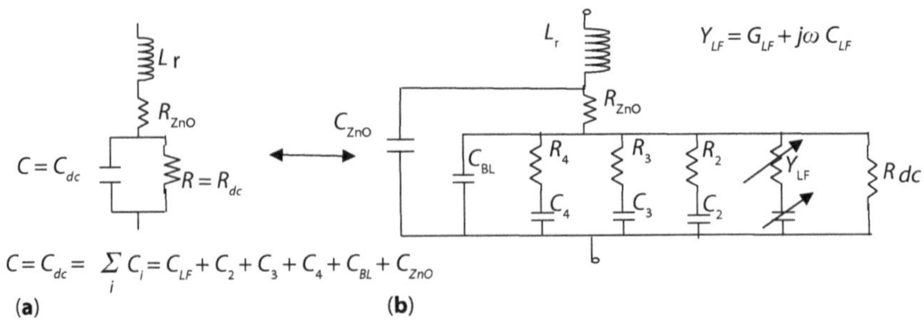

Figure 6.12 Equivalent circuit model of the ZnO-based varistor in the frequency range 10^{-2} Hz $\leq f \leq 10^9$ Hz for electric field $0 \leq E \leq 80\%E_{1mA/cm^2}$ in the temperature range 26 °C $\leq T \leq$ 90 °C where electric field E_{1mA/cm^2} corresponds to the current density 1 mA/cm^2.

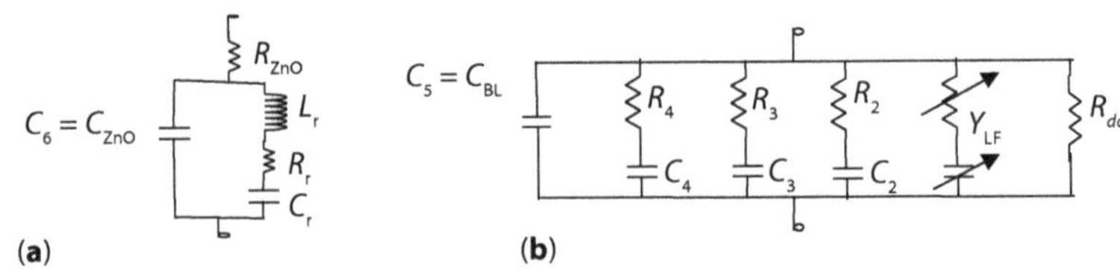

Figure 6.13 Resonance circuit in (a) in the frequency range 10^6 Hz $\leq f \leq 10^9$ Hz, and trapping circuit in (b) in the frequency range 10^{-2} Hz $\leq f \leq$ 13 MHz for ZnO-based varistor for electric field $0 \leq E \leq 80\%E_{1mA/cm^2}$ in the temperature range 26 °C $\leq T \leq$ 90 °C where electric field E_{1mA/cm^2} corresponds to the current density 1 mA/cm^2.

from the left-side intercept in the Z*-plane [10–12]. However, this resistance value is likely to coincide with the combined resistance of the lead and contact electrode for the varistor beside the lumped grain resistance. Thus, R_r and R_{ZnO} appeared alike those obtained from the left-side intercept of the Z*-plane and chord of the Y*-plane circle. The capacitance C_6 or C_{ZnO} is placed in parallel with the series L_r-C_r-R_r circuit [10, 13, 15–17] with the R_{ZnO}. The resonating circuit is shown in Figure 6.13(a) and the trapping related circuit is shown in Figure 6.13(b). Upon combining these two circuits and using legitimacy or validity of the equivalent circuit model Figure 6.12(b) is presented whereas Figure 6.12(a) results from the Z*-plane [10–12] without the inductor. The inductor is placed with circuit as the resonance is believed to be associated externally with the lead.

The Z*-plane relaxation conducted at zero dc bias at room temperature reveals R-C parallel combination where [10, 20]:

$$C = C_{dc} = \sum_i C_i = C_{LF} + C_2 + C_3 + C_4 + C_{BL} + C_{ZnO} \tag{6.4}$$

and

$$R = R_{dc}, \tag{6.5}$$

with R_{ZnO} as the intercept of the left side of the semicircle. At the dc condition all the capacitances should add from the parallel branches regardless of the resistance in each branch. Resistance in each branch of Figure 6.12(b) cannot exceed the value of the dc resistance which is obtained as the highest value from the chord of the Z*-plane semicircular relaxation.

At elevated temperature and dc bias the immittance data were plotted in the Z*-plane that provided lumped R_{im}-C_{im} parallel combination along with a series resistor designated as R_{ZnO} where electric field E_{1mA/cm^2} corresponds to the current density 1 mA/cm^2. The subscript "im" stands for the impedance plane parameter. The equivalent circuit model depicted in Figure 6.12 is visible between room temperature and less than 90 °C, and dc biasing less than 80%E_{1mA/cm^2} [11, 12]. Above 90 °C the room temperature model disappears as the Z*-plane display becomes dominant yielding simplified R_{IM}-C_{IM} parallel combination with series R_{ZnO} element. Likewise biasing above 80% E_{1mA/cm^2} to less than E_{1mA/cm^2} the same circuit is obtained via the same plane [11, 12]. These circuits satisfying experimental situation are shown in Figure 6.14. Essentially C_{IM} reflects summation of six contributing capacitances (lumped together) and R_{IM} reflects R_{dc} shown in Figure 6.12. No trapping resistance appears in this scenario because each branch shown in Figure 6.12(b) is blocked under the dc condition.

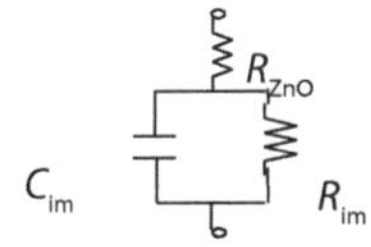

(a) Equivalent circuit of ZnO varistor at zero *dc* bias at T ≥ 90 °C.

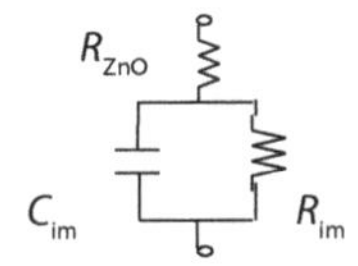

(b) Equivalent circuit of ZnO varistor at *E* ≤ 80% E_{1mA/cm^2} and at T ≥ 90°C.

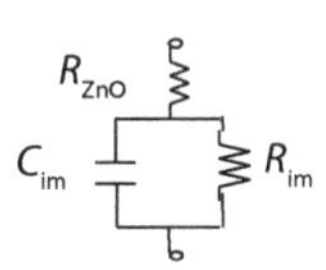

(c) Equivalent circuit of ZnO varistor at *E* ≤ 80% E_{1mA/cm^2} at room temperature.

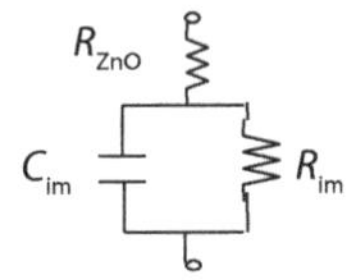

(d) Equivalent circuit of ZnO varistor in the range 80% E_{1mA/cm^2} ≤ *E* at room temperature.

Figure 6.14 Equivalent circuit model of ZnO-based varistor at various experimental conditions.

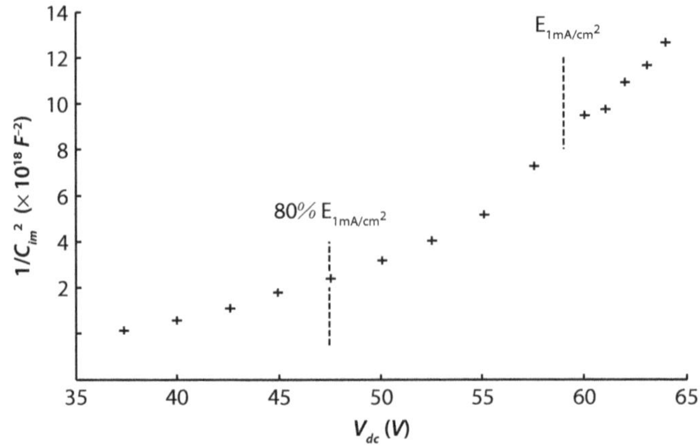

Figure 6.15 The Mott-Schottky plot using equivalent circuit capacitance C_{im} obtained in the Z*-plane semicircular relaxation for the ZnO-based varistor.

The existence of the back-to-back Schottky barrier [10–13] in ZnO varistor is a fundamental concept because of the nonlinear current-voltage response. The circuit capacitance C_{im} when plotted in the Mott-Schottky plane as C_{im}^{-2} versus V_{dc} instead of a straight line a curvature is obtained as shown in Figure 6.15. This curvature does not represent a straight line in the nonlinear range of the current-voltage response. Therefore, circuit C_{im} is not the Schottky barrier related capacitance.

Under certain restricted experimental conditions Faradaic impedance like response was monitored in the data for the ZnO varistor. The low frequency immittance did not reveal clear trapping response [7] but exhibited distorted curve in the C*-plane at low frequencies. This portion was modeled as Y_{LF} in 1988 which was alternately termed as low frequency admittance [7] in the equivalent circuit.

Since the unresolved admittance Y_{LF} remained unknown for the equivalent circuit model, curios effort led to the use of the elimination process of the rest of the data leaving only for the equivalent circuit branches combined as R_{dc} and Y_{LF} as parallel form. The branches constituting R_{dc} and Y_{LF} were extracted from the measured immittance data by subtracting the rest of the branches of the equivalent circuit model, and plotted in the Z*-plane. Surprisingly, a convincing semicircular relaxation exhibiting finite depression angle was achieved. Rest of the complex planes did not show relaxation. Bode plots using |Z|, R_{LF}, and X_{LF} (= $1/(\omega C_{LF})$) versus Log(f) revealed certain strange pieces of information. Based on the information, the impedance of this semicircle can be expressed as:

$$
\begin{aligned}
Z_F = Z_{LF} &= \frac{1}{Y_{LF}} = R_{LF} + X_{LF} \\
&= A\,\omega^{-\beta_1} + B\,\omega^{-\beta_2} = A\,\omega^{-.0.71} + B\,\omega^{-0.59}
\end{aligned}
\tag{6.6}
$$

where R_{LF} and X_{LF} denote apparent Faradaic resistance and reactance, respectively. For this impedance, A and B are unequal, and also the exponents β_1 and β_2 are unequal while for Warburg, these are equal making 0.5. The non-Faradaic nature of the impedance expressed in equation (6.6) is the result of the inequality in the exponents of β_1 and β_2. This is not an ideal Faradaic polarization. A theoretical treatment leading to unequal exponents is not available yet. Since equation (6.6) yields a semicircular fit in the Z*-plane, all derived parameters in this plane may be represented in the form of Cole-Cole type expression:

$$
Z_F = Z_{LF} = \frac{1}{Y_{LF}} = R_{High-LF} + \left(\frac{R_{LF}}{1 + (j\,\omega\,\tau_{LF})^{(1-\phi)}} \right), \tag{6.7}
$$

where

$$
\tau_{LF} = R_{LF}\,C_{LF} \tag{6.8}
$$

and the depression angle

$$
\theta_{LF} = \frac{\phi\pi}{2} \tag{6.9}
$$

where ϕ is the depression parameter. There is some dc bias dependence of ϕ or θ_{LF} having two linear ranges for which the meaning is unclear. The dc biasing does not indicate decrease in depression angle but the semicircle disappears because of impossible measurement for the exclusive C*-plane plot. These low frequency parameters are strongly influenced by dc biasing and temperature for which the exponents remained unequal nearly in the same domain.

The room temperature dc bias (or current) dependence of R_{LF} and C_{LF} indicated nearly exponential decay and nearly exponential rise, making the time constant τ_{LF} nearly constant with bias, but a definite cause for these responses remains inclusive because of the unequal exponents. Equation (6.9) represents highly complicated situation though a semicircle has been achieved in the Z*-plane that yielded equation (6.7). What should be the status of A and B? This is truly a complicated aspect of the ZnO-based varistor. There is no clue yet for the parameters A, B, β_1, and β_2 as there is no mention anywhere about them.

6.10 Equivalent Circuit: Lithium Niobate ($LiNbO_3$) Single Crystal

The Lithium Niobate (LN) crystal is neither a perfect insulator nor a perfect conductor. Therefore, in reality there is no dc blocking capacitor and significant leakage resistor (or conductor) between the two opposite electrodes. That's why the ac small-signal electrical data in the measurement frequency range $10\ Hz \leq f \leq 10^6\ Hz$ provided semicircular behavior in both C*- and M*-planes for the LN crystal in a narrow temperature range 59 °C≤ T ≤ 73 °C [18]. These two complex planes attribute to the blocking equivalent circuit in the presence of an intercept on the left side of the semicircular relaxation. Since the LN crystal cannot have the blocking feature, it is essential to derive a reasonable equivalent circuit that works under the dc condition. Accordingly an equivalent circuit model is developed for the LN crystal.

There was no visible distortion in the semicircular loci for the frequency range used. It means there may be a limitation in the window of measurement. The distortion at low frequencies in the C*-plane provides the influence of the dc conduction as found in ZnO-based varistors [10–12]. This kind of behavior is noted by Coelho [19]. This low frequency identification in the C*-plane was not obvious in the M*-plane. Therefore, a reasonable assumption is made to establish the equivalent circuit model for the LN crystal [18].

A semicircular relaxation with the left-side intercept either in C*- or in M*-plane provides a blocking equivalent circuit shown in Figure 6.16(a) and (b), respectively. Apparently the LN crystal indicates any one of these circuits is possibly valid because of the observed feature in the C*- or in M*-plane. Since this crystal cannot be a perfect insulator and due to the limitation in the low frequency measurement it is assumed that both circuits possess leakage resistor for the dc conduction. It is apprehended that the low frequency behavior for the dc resistor in the C*-plane causes a shunt resistor for Figure 6.16(a) whereas in the M*-plane causes a series resistor. With this assumption the equivalent circuit is developed which is presented in Figure 6.17.

As a first step the modified configuration of Figure 6.16(a) is shown in Figure 6.17(a) with shunt R_2, and then Figure 6.16(b) is shown in Figure 6.17(b) with series R_4. Figure 6.16(a) is reproduced in Figure 6.17(c)

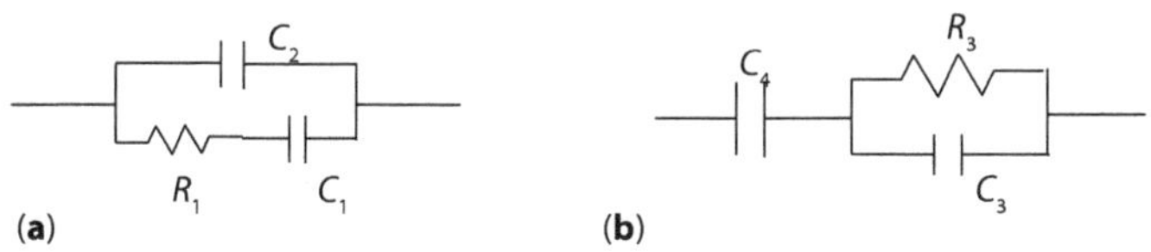

Figure 6.16 Equivalent circuit model corresponds to (a) C*-plane semicircular relaxation and (b) M*-plane semicircular relaxation.

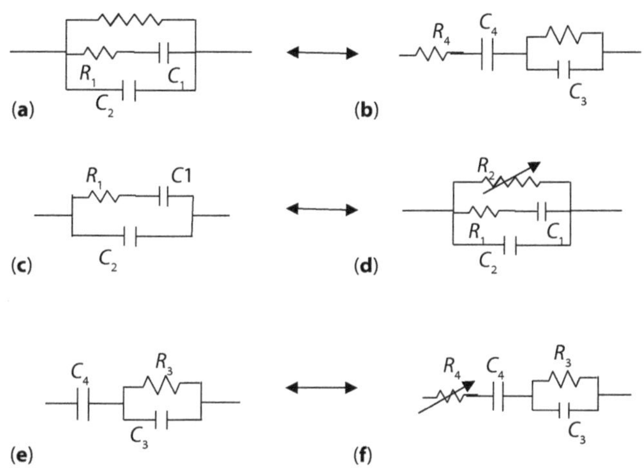

Figure 6.17 Equivalent circuit model in (a) and in (b) obtained from the C*- and M*-planes, respectively, for the $LiNbO_3$ single crystal in the temperature range 59 °C ≤ T ≤ 73 °C. These two equivalent circuits represent idealized forms for single semicircular relaxation where the presence of R_2 and R_4 are partly recognized via skewed response attributing to the limited window of data acquisition. Equivalent circuits in (c) and in (d) are the same when the skewed portion of the data representing R_2 (arrow-cut) is eliminated to achieve blocking behavior. Equivalent circuits in (e) and in (f) are the same when the skewed portion of the data representing R_4 (arrow-cut) is eliminated in achieving blocking behavior. Equivalent circuit elements R_2 or R_4 represent dc resistance in respective complex plane.

with a highlighted arrow-cut in Figure 6.17(d) for R_2. Similarly Figure 6.16(b) is reproduced in Figure 6.17(e) with a highlighted arrow-cut in Figure 6.17(f) for R_4. The dotted box indicates that the parameter R_2 or R_4 is not visible in the measurement frequency range. Surely the circuit in Figure 6.16(b) or in Figure 6.17(f) is not valid as it portrays the dc blocking feature whereas the circuit in Figure 6.17(a) or in Figure 6.17(d) is valid.

The series equivalent circuit elements R_1–C_1 in Figure 6.17(a, c, and d) are the successive events of trapping and de-trapping. This event causes resistance for de-trapping and capacitance for trapping indicating in-phase and out-of-phase conduction. For the trapping of the carriers the capacitance effect is realized while the de-trapping effect provides conduction originating from free carriers. These equivalent circuit elements are extracted from the C*-plane that portray point defects related traps as well as other associated defect states or defect disorder within the LN crystal. For a number of material systems, whether single crystal or not, the R–C series combination obtained either in the C*-plane or in the Y*-plane represents a trapping and de-trapping response of the carriers. This type of response is demonstrated with several material systems [10, 20–27].

The value of C_2 represents lumped bulk and space charge capacitance in series implying the equivalent combinations of C_3 and C_4 listed in Figure 6.18. Thus, C_2 matches with $C_{M3,4}$ and given by:

$$C_2 \equiv C_{M3,4} = \frac{C_3\, C_4}{C_3 + C_4}. \tag{6.10}$$

The underlying meaning of each element invariably needs serious attention to unravel reciprocating relationships. Overall, the equivalent circuit for each of the two complex planes is straightforward and the corresponding elements are delineated considering the highly resistive dc conducting situation and listed in Figure 6.18. The dc condition requires the assumption of non-existence or near absence of the lumped bulk resistance R_3. If the single crystal is totally devoid of defects then the resistance is expected to be a minimum. Thus, justification of the equivalence of C_2 and $C_{M3,4}$ appears reasonable verification.

The resistances R_2 and R_4 are identical and referred to the dc resistance of the LN crystal. In the C*-plane R_2 can be realized from the distorted

Temperature (°C)	C_1 from C*-plane (pF)	C_2 from C*-plane (pF)	C_3 from M*-plane (pF)	C_4 from M*-plane (pF)	Equivalent capacitance of series C_3 and C_4 $C_{M3,4} = \frac{C_3 C_4}{C_3 + C_4} \equiv C_2$ (pF)
59	331.8	39.4	46.9	363.9	40.8
63	42.1	27.3	44.9	69.4	27.3
69	12.9	17.3	40.4	30.2	17.3
73	7.9	13.8	37.4	21.7	13.7

Figure 6.18 Comparison of equivalent circuit parameters obtained simultaneously from the C*- and M*-planes.

semicircular loci similar to that demonstrated by Grant [9] at low frequencies as was observed in the case of the ZnO varistor [10], whereas the same effect is presumed to be as R_4 in the M*-plane on the left side of the semicircular relaxation. This implies that the two resistors R_2 and R_4 are equivalent and shown within the arrow-cut in Figure 6.17(d) and (f). Overall the blocking behavior of the LN crystal presumably led to the duality between the two complex plane relaxations. Though Grant demonstrated duality of the Y*- and C*-plane [9] (that is σ* and ε*, respectively) relaxations for the same equivalent circuit, it is unique to achieve for the LN crystal two near blocking equivalent circuits for the relaxations in the C*- and M*-planes. The dc resistance R_2 or R_4 is very large and visualized from the immittance data reflected by the small range distortion at the low-frequency end. The capacitance C_2 represents lumped behavior of the bulk and space charge capacitance whereas C_4 represents space charge capacitance. The exact magnitude of the dc resistance is not quantified from the C*- and M*-planes. This is because of the constituting quantity of the abscissa parameter in each plane. However, the abscissa parameter yields worthy information about the trapping capacitance (C_1) and absolute bulk capacitance (C_3). The ordinate parameter in conjunction with the relaxation times (τ_C and τ_M) yield trapping resistance (R_1) and lumped bulk resistance (R_3) when the crystal is either perfect or devoid of defects.

It is not reasonable to visualize the barrier layer capacitance in the LN single crystal as there is no physical boundary with distinct phases. The left-side intercept in the M*-plane semicircular relaxation denoted as C_4 refers to the space charge capacitance which is possible in the LN crystal due to possible dislocations or planar defects in the array of atomic layers. This plausible concept is realized as a very high value at 59 °C and then sharply decreases at 63 °C; thereafter, as temperature increases this value decreases. In the light of the data-handling/analyzing criteria systematic equivalent circuit analog is developed using associated operative phenomena within the LN crystal. The dual complex-plane analytical approach via single semicircular relaxation in the C*- and M*-planes appears to be useful when the meaning of C_2 (lumped bulk and space charge capacitance) is translated to series combination of C_3 (absolute bulk capacitance) and C_4 (space charge capacitance) is understood. Likewise the meaning of R_2 or R_4 stands as the total dc resistance.

The arrow-cut resistances R_2 and R_4 in Figure 6.17(d and f) represent the dc resistance. Overall, two separate equivalent circuit models appear to provide identical information although both indicate near blocking feature of the LN crystal. This interacting feature is unique to this crystal in the narrow range of ambient temperature. The rapid decrease of space charge capacitance (C_4) with increase in temperature indicates discharge of charges in the crystal. The same analogy applies to the trapping capacitance (C_1). Two equivalent circuit models are quite different in appearance; however, establishing the interacting relationship among the elements remains tedious.

Since the LN crystal is piezoelectric and ferroelectric, it cannot be ruled out the above stated plausible explanations for the equivalent circuit elements, derived from the dielectric representation of modulus, due to the presence of the defective dipolar charge distribution in the narrow temperature range (59 °C $\leq$ T $\leq$ 73 °C). The evolution of these charges and defect clusters' interactions are also not eliminated. Overall, it is interesting to delineate the equivalent circuit elements obtained via semicircular relaxation in the C*- and M*-planes for the same set of the ac small-signal electrical data.

6.11 Equivalent Circuit: Polycrystalline Yttria (Y_2O_3)

The ac small-signal electrical data were examined in three complex planes (Z*, M*, and Y*) where semicircular relaxation was observed for both ZrO_2 added (doped) Y_2O_3 as well as undoped Y_2O_3 samples [21, 22]. The Z*-plane representation exhibited two relaxations that provided two R-C parallel circuits in series as shown in Figure 6.19(a). This is possible because R_1 and R_2 are of the same order of magnitude. These two relaxations are attributed to the lumped response of the grains and grain boundaries containing traps, including possible electrode/sample effects if there is any.

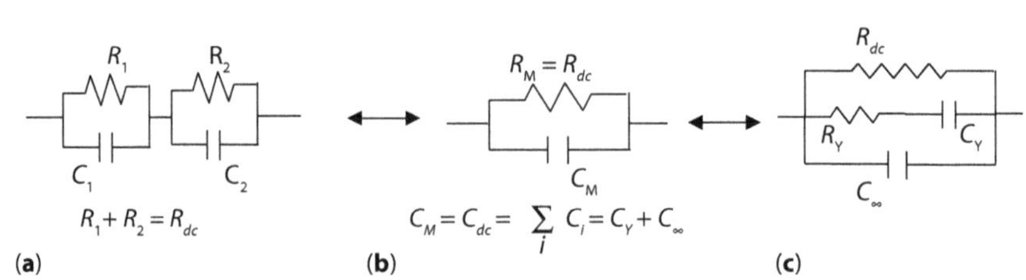

Figure 6.19 (a) Equivalent circuit model corresponding to the two semicircular relaxations in the Z*-plane where R_1-C_1 and R_2-C_2 parallel combinations in series constitute the low-frequency relaxation associating grain-boundaries and high-frequency relaxation associating grains, respectively. (b) Equivalent circuit obtained in the M*-plane where one semicircle was obtained. The series combination of R_Y- C_Y constitutes the low-frequency relaxation. R_{dc} and C_∞ are the dc resistance and high-frequency limit capacitance, respectively.

The same data provided one relaxation in the M*-plane where one lumped R-C parallel circuit is observed as depicted in Figure 6.19(b). This single-like relaxation is close representative of the high-frequency relaxation in the Z*-plane attributing to the dominant contribution of the grains. Essentially lumped R_M is the summation of R_1 and R_2 of Figure 6.19(a) which is the dc resistance R_{dc} shown in Figure 6.19(c) obtained from the left-side intercept of the Y*-plane semicircle. The capacitance C_M is the direct transformation from the circuit of Figure 6.19(a) which is the summation of C_∞ and C_M of Figure 6.19(c). The R_Y-C_Y series combination is the information related to trapping whereas C_∞ denoted geometric capacitance related to lumped behavior of bulk grain. C_∞ indicates high frequency limit which coincides with C_1 of the Z*-plane [21, 22].

Two relaxations of the Z*-plane representation gave the dc resistance obtained from the M*-plane relaxation as:

$$R_1 + R_2 = R_{dc} = R_M, \tag{6.11}$$

and the corresponding capacitance as:

$$C_M = C_{dc} = \sum_i C_i = C_Y + C_\infty, \tag{6.12}$$

where C*-plane yielded summation of barrier layer capacitance, C_∞, and trapping capacitance, C_Y per Figure 6.19(c) [21, 22].

6.12 Equivalent Circuit: Polycrystalline Calcium Zirconate ($CaZrO_3$)

The ac small-signal electrical data were examined in three complex planes (Z*, M*, and Y*) where semicircular relaxation was observed in $CaZrO_3$ [23]. Two semicircles were observed in the Z*-plane for which the equivalent circuit is shown in Figure 6.20 (a). When the same data were displayed in the M*-plane, a single-relaxation-like semicircle was obtained. From the Y*-plane plot, relevant electrical parameters were determined and they are found to match those obtained from the Z*-and M*-planes. The M*-plane circuit is shown in Figure 6.20(b) whereas the Y*-plane circuit is shown in Figure 6.20(c).

Two relaxations of the Z*-plane representation gave the dc resistance obtained from the M*-plane relaxation as:

$$R_1 + R_2 = R_{dc} = R_M, \tag{6.13}$$

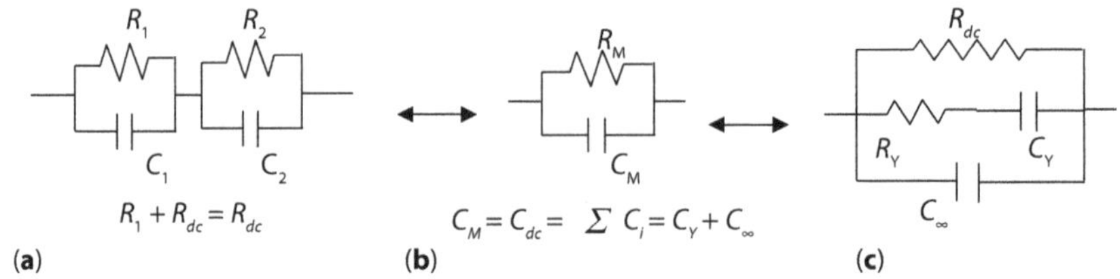

Figure 6.20 (a) Equivalent circuit model corresponding to the two semicircular relaxations in the Z*-plane where R_1-C_1 and R_2-C_2 parallel combinations in series constitute the low-frequency relaxation associating grain-boundaries and high-frequency relaxation associating grains, respectively. (b) Equivalent circuit obtained in the M*-plane where one semicircle was obtained. The series combination of R_Y- C_Y constitutes the low-frequency relaxation. R_{dc} and C_∞ are the dc resistance and high-frequency limit capacitance, respectively.

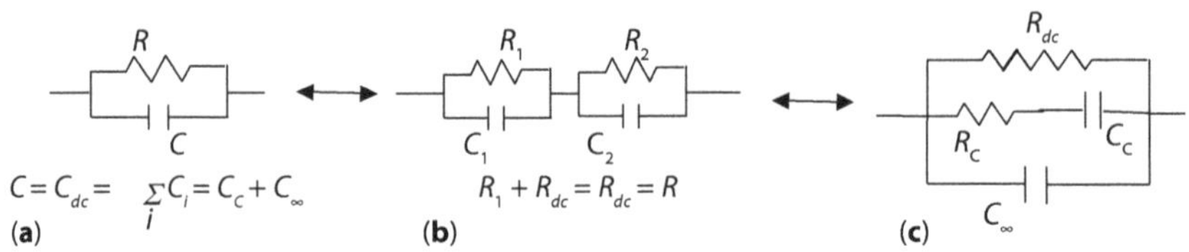

Figure 6.21 (a) Equivalent circuit model corresponding to the semicircular relaxations in the Z*-plane, (b) equivalent circuit representation of the M*-plane, and (c) the series R-C combination comes from the C*-plane.

and the corresponding capacitance as:

$$C_M = C_{dc} = \sum_i C_i = C_Y + C_\infty, \tag{6.14}$$

where C*-plane yielded summation of barrier layer capacitance, C_∞, and trapping capacitance, C_Y per Figure 6.20(c) [23].

6.13 Equivalent Circuit: Polycrystalline Calcium Stannate ($CaSnO_3$)

The ac small-signal electrical data were examined in three complex planes (Z*, C*, and M*) where semicircular relaxation was observed in $CaSnO_3$ [24, 25]. Lumped behavior observed via single relaxation in the Z*-plane provided R-C parallel combination shown in Figure 6.21(a). The M*-plane revealed two semicircles yielding the circuit of Figure 6.21(b). The same data in the C*-plane provided Figure 6.21(c). The choice goes with the meaning of each circuit representation. Obviously trapping component shows up as R-C series combination in Figure 6.21(c). The rest of the meaning of the circuit elements is identical to the ZnO-based varistor.

Two relaxations of the M*-plane representation gave the dc resistance obtained from the Z*-plane relaxations as:

$$R_1 + R_2 = R_{dc} = R \tag{6.15}$$

and the corresponding capacitance as:

$$C = C_{dc} = \sum_i C_i = C_C + C_\infty, \tag{6.16}$$

where C*-plane yielded summation of barrier layer capacitance, C_∞, and trapping capacitance, C_C per Figure 6.21(c) [24, 25].

6.14 Equivalent Circuit: Polycrystalline Titanium Dioxide (TiO_2)

Electrical characterization was performed on the sintered body comprising of anatase TiO_2 added with Y_2O_3 or ZrO_2 was evaluated as potentially gas sensing material [26, 27]. The ac small-signal electrical data exhibited some meaningful response in the Z*-plane as well as in the C*-plane that provided an equivalent circuit model shown in Figure 6.22. Two semicircles are observed for the effect of ZrO_2 addition [26] whereas one semicircle is observed for the effect of Y_2O_3 addition [27] to the anatase in the Z*-plane. There was no C*-plane relaxation for the ZrO_2 added samples but Y_2O_3 addition exhibited relaxation.

To some extent the equivalent circuit has resemblance with the matured ZnO varistor when the τ_4-relaxation [10] is masked and the resonance

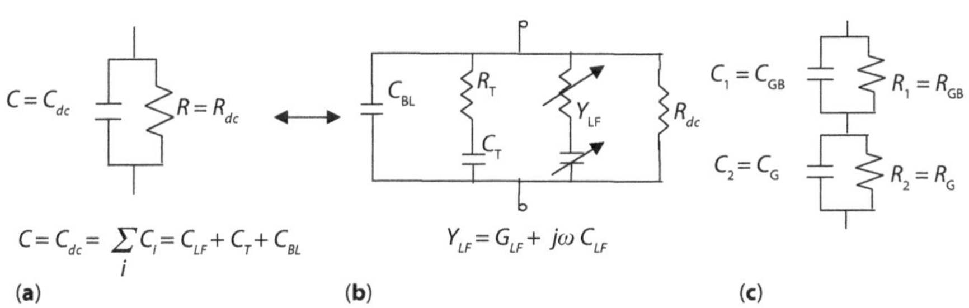

Figure 6.22 Equivalent circuit model corresponding to the (a) one relaxation in the Z*-plane for the effect of the Y_2O_3 addition, (b) relaxation in the C*-plane for the effect of the Y_2O_3 addition, and (c) two relaxations in the Z*-plane for the effect of the ZrO_2 addition.

behavior is difficult to distinguish [13]. The equivalent circuit obtained in the C*-plane converges to two R-C parallel combination in series as displayed in Figure 6.22(a). The extracted components are characterized to satisfy spectroscopic analysis which is identical to ZnO varistors. The C*-plane feature provides:

$$C = C_{dc} = \sum_{i} C_i = C_{LF} + C_T + C_{BL}, \tag{6.17}$$

because of parallel branches in Figure 6.22(b). C_{GB} and C_G are grain boundary capacitance and grain capacitance, respectively, whereas R_{GB} and R_G are grain boundary resistance and grain resistance, respectively, as shown in Figure 6.22(c).

One relaxation of the Z*-plane representation gave the dc resistance for the Y_2O_3 addition as:

$$R_{dc} = R, \tag{6.18}$$

and the corresponding capacitance as:

$$C_{dc} = C. \tag{6.19}$$

Two relaxations, which were evident in the Z*-plane for the ZrO_2 addition, gave the dc resistance as:

$$R_1 + R_2 = R_{dc} = R, \tag{6.20}$$

and no relaxation was observed in the C*-plane [26, 27].

6.15 Equivalent Circuit: Multi-Layered Thermoelectric Device (Alternate SiO_2/SiO_2+Ge Thin-Film)

The non-Debye equivalent circuit model for the 200-layered alternate SiO_2 and SiO_2+Ge thin-film based ion-bombarded and unbombarded thermoelectric devices is developed from the conventional Bode plot in conjunction with the Z*- and Y*-plane plots [28]. Figure 6.23(a, b, and c) displays the R-C parallel combination for the Z*-plane relaxation. At lower bombardment doses two R-C parallel circuits in series emerged whereas at a higher bombardment dose single R-C parallel circuit was obtained. This implies that two R-C parallel circuits in series were lumped into one R-C parallel circuit. Small depression angle was observed in each semicircle in the Z*-plane. The dc resistance R_{dc} of

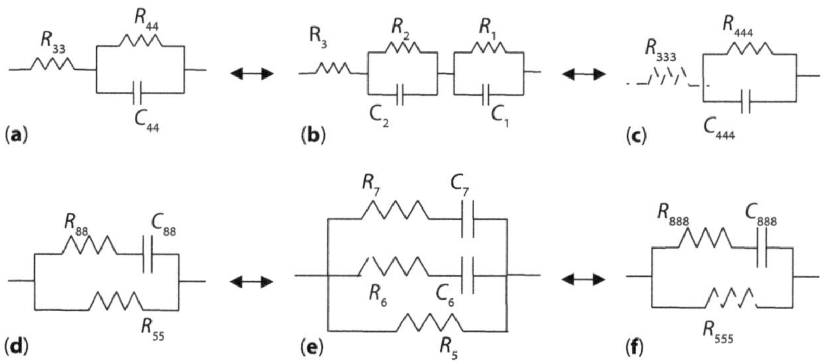

Figure 6.23 Equivalent circuit model of the (a) unbombarded and bombarded (1×10^{14} ions/cm^2) 200-multilayered thermoelectric devices as obtained in the Z*-plane, (b) bombarded (5×10^{12} ions/cm^2, 1×10^{13} ions/cm^2, and 5×10^{13} ions/cm^2) 200-multilayered thermoelectric devices as obtained in the Z*-plane, (c) bombarded (1×10^{14} ions/cm^2) 200-multilayered thermoelectric devices in the Z*-plane having dotted resistance R_{333} obtained as the intercept on the left side of the semicircle [the dotted resistance often may be negligible as it is too small to obtain left-intercept for the semicircle upon using CNLS fitting procedure], (d) unbombarded and bombarded (1×10^{14} ions/cm^2) 200-multilayered thermoelectric devices as obtained in the Y*-plane, (e) bombarded (5×10^{12} ions/cm^2, 1×10^{13} ions/cm^2, and 5×10^{13} ions/cm^2) 200-multilayered thermoelectric devices as obtained in the Y*-plane, (f) bombarded (1×10^{14} ions/cm^2) 200-multilayered thermoelectric devices in the Y*-plane having dotted resistance R_{333} obtained as the intercept on the left side of the semicircle [the inductance is external to the multilayered device and omitted from the equivalent circuit model].

each type of the device is extracted from the Z*-plane analysis which is given by [28]:

$$R_{dc} = R_{33} + R_{44} = R_3 + R_2 + R_1 = R_{333} + R_{444}. \tag{6.21}$$

The resistors R_{33}, R_3, and R_{333} are obtained from the left-side intercept of the semicircle in the Z*-plane. The dotted resistance R_{333} appeared as tiny compared to the chord in the Z*-plane fitting. The resistance R_{dc} obtained from Z*-plane and Y*-plane is very close to each other. There is a little mismatch in determining capacitance for each of these complex planes. This happened due to the data points in the overlapped region of the semicircular loci as well as the conversion using the Debye equation [28].

In this case of an ideal situation, the parallel combination of capacitances C_6 and C_7 [i.e., $C_{equivalent} = C_6 + C_7$] obtained from the Y*-plane must be equal to the series combination of capacitances C_1 and C_2 [i.e., $1/C_{equivalent} = (1/C_6) + (1/C_7)$] obtained from the Z*-plane. The overlapped region caused heavy distortion both in the geometric and CNLS fitting of the semicircular curve [28].

The non-Debye equivalent circuit model is presented in Figure 6.23(d, e, and f) for the same devices obtained from the Y*-plane analysis. At lower bombardment doses two R-C series circuits in parallel emerged whereas at a higher bombardment dose a single R-C series circuit was obtained. This implies that two R-C series circuits in parallel were lumped into one R-C series circuit. A small depression angle was observed in each semicircle in the Y*-plane. The dc resistance R_{dc} of each device is obtained from the Y*-plane analysis which is given by:

$$R_{dc} = R_{55} = R_{5} = R_{555}. \tag{6.22}$$

The resonant frequency is obviously the same for both Z*- and Y*-plane because of the measured data. The associated inductance calculated from both these planes is close but there is a finite difference. This is due to the approximate nature of the semicircular loci in the negative domain and the overlapped relaxation process. The inductance is attributed to the electrode lead for the 200-multilayered device.

The equivalent circuit model confirms the existence of the dc resistance. The relaxation is attributed to the series or parallel R-C circuit based on the Y*-plane or on the Z*-plane semicircle, respectively. Such a situation verifies operative conductive path between the two electrodes. The dc resistance or conductance is highly visible in the form of distorted or skewed semicircle for the unbombarded device as well as for the bombarded device. This kind of influence of the dc resistance was demonstrated by Grant [9] and Coelho [19]. The entire fitting process is in close conformity via Debye conversion among these two complex planes despite the presence of the small depression parameter. An ideal Debye response makes a perfect match between the two complex plane relaxation processes [28].

6.16 Equivalent Circuit: Polycrystalline Tungsten Oxide (WO_3)

Electrical characterization was performed on the sintered body comprising of WO_3 pellets for the surface electronic properties [29]. The ac small-signal electrical data exhibited meaningful response in the Z*-plane that provided an equivalent circuit model for the surface electronic response shown in Figure 6.24. The relaxation is attributed to the R_2-C_2 parallel combination for the lumped grain boundaries on the surface. There was an ultra-small intercept on the left-side of the semicircle that yielded the

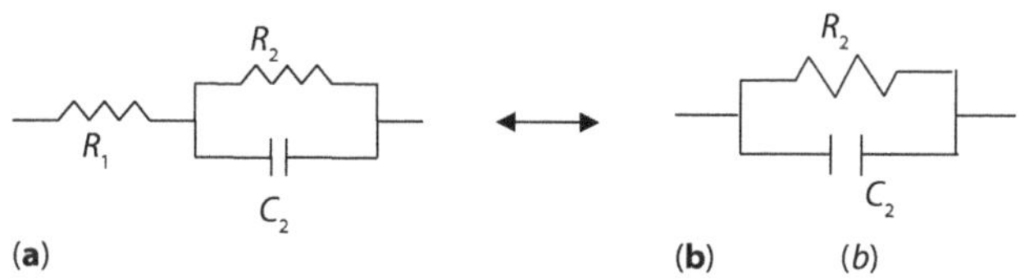

Figure 6.24 Equivalent circuit model of the polycrystalline tungsten oxide pellet surface showing total resistance $R_{total} = R_1 + R_2 \approx R_2$ since R_1 is detected as too small via the left-intercept i.e., intercept on the left side of the semicircular relaxation in the Z*-plane.

lumped surface grain resistance R_1. Basically the circuit in Figure 6.24(a) and (b) are the same because of the ultra-small value of R_1. The resistance R_2 is very close to the dc resistance. Therefore, R_1 can be neglected for all practical purposes.

6.17 Equivalent Circuit: Biological Material – *E. Coli* Bacteria

Escherichia Coli (*E. Coli*) bacteria-suspended brain heart infusion (BHI) (hereafter *E. Coli*) is investigated out of curiosity to find the non-blocking feature of this conductive sample [30]. This experiment served localized verification of the previous findings [31–39] done elsewhere. These findings asserted R-C series combination as the equivalent circuit model for either similar or the same type of biological systems. Such an equivalent circuit is absolutely erroneous for the conductive biological samples. How is it possible to represent a conductive biological sample by the dc blocking R-C series circuit? To answer this question, local interest was taken to verify the series of so-called assertive validity [30–38].

The non-Debye equivalent circuit model developed from the Bode plot in conjunction with the Y*-plane as well as C*-plane relaxations is presented in Figure 6.25. The resonance behavior was ascertained via Y*- and C*-plane representations and is similar to that of ZnO varistor [10, 13] found in 1988. In these two complex planes non-Debye behavior was evident due to the presence of the small depression parameter. This circuit is in strong agreement among these two complex planes in conjunction with the Bode plane representation of the same ac small-signal electrical data. In fact, the resulting circuit shown in Figure 6.25 is established by Grant [9] and verified for the *E. Coli* bacteria samples.

In the Y*-plane semicircular relaxation, the CNLS curve fitting yielded a very tiny intercept on the left side of the semicircle that substantiates the skewed response in the C*-plane. Both complex planes confirmed

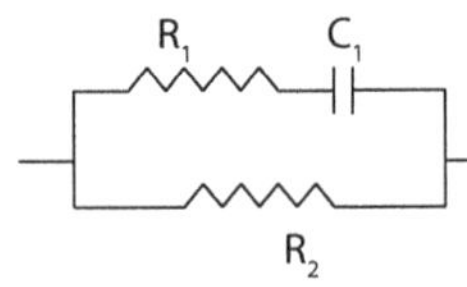

Figure 6.25 Non-Debye equivalent circuit model observed for the sterile BHI or the *E. Coli* bacteria suspended BHI system. The shunt resistance R_2 for the blocking R_1-C_1 series circuit obtained from the Bode plot in conjunction with the complex plane representation.

the presence of the dc conductance in parallel with the R_1-C_1 series combination which is extracted from the semicircular relaxation. The resistance R_1 is verified with the horizontal straight line in the high frequency domain of the Bode plot which is cross verified from the chord of the Y*-plane semicircle. The dc resistance denoted as R_2 is extracted from the slant straight line via extrapolation and good eye-estimated in the low frequency domain as there was no existence of the horizontal line. Thus, R_2 is the shunt resistor confirming as the dc resistance, which is at least 10 or more times larger than R_1. This R_1 constitutes the blocking branch in series with C_1 in the equivalent circuit model depicted in Figure 6.25.

Invariably this equivalent circuit verifies the operative conductive path between the two electrodes for the *E. Coli* bacteria. The shunt resistance (or conductance) is highly visible in the form of a distorted or skewed semicircle at low frequencies in the C*-plane. This kind of influence of the shunt resistance for the blocking R_1-C_1 circuit was demonstrated by Grant [9] and Coelho [19]. It is a valid observation that the blocking nature does not exist in the *E. Coli* bacteria in the suspended system.

Problems

1. Why did LN single crystals exhibit dc blocking feature in both C*- and Y*-planes although it had some leakage behavior? What is the dominating factor that led to this behavior?
2. Consider the following two circuits. Under what circumstance can each circuit be obtained from the Z*-plane as well as from the C*-plane? Under what circumstance can each circuit be obtained from the M*-plane as well as from the Y*-plane? Choose own symbol for each element.

3. Consider the following two circuits. Under what conditions can each of these circuits satisfy two or more complex planes concurrently?

References

1. E. Barsoukov and J. R. MacDonald; *Impedance Spectroscopy: Theory, Experiment, and Applications,* Edited, John Wiley & Sons, Inc., Hoboken, New Jersey, 2005.
2. J. R. MacDonald; *Impedance Spectroscopy: Emphasizing Solid materials and Systems*, Edited, John Wiley & Sons, Inc., Hoboken, New Jersey, 1987.
3. M. E. Orazem and B. Tribollet; *Electrochemical Impedance Spectroscopy*, John Wiley & Sons, Inc., Hoboken, New Jersey, 2008.
4. V. F. Lvovich; *Impedance Spectroscopy: An Application to Electrochemical and Dielectric Phenomena,* John Wiley & Sons, Inc., Hoboken, New Jersey, 2012.
5. R. H. Cole and K. S. Cole; "Dispersion and Absorption in Dielectrics: I. Alternating Current Characteristics," *Journal of Chemical Physics*, 9, 341–351, 1941.
6. J. E. Bauerle; "Study of Solid Electrolyte Polarization by a Complex Admittance Method," *Journal of Physics and Chemistry of Solids,* 30, 2657–2670, 1969.
7. I. M. Hodge, M. D. Ingram, and A. R. West; "A New Method for Analyzing the A. C. Behavior of Polycrystalline Solid Electrolytes," *Journal of Electroanalytical Chemistry,* 58, 429–432, 1975.
8. I. M. Hodge, M. D. Ingram, and A. R. West; "Impedance and Modulus Spectroscopy of Polycrystalline Solid Electrolytes," *Journal of Electroanalytical Chemistry*, 74, 125–143, 1976.
9. F. A. Grant; "Use of Complex Conductivity in the Representation of Dielectric Phenomena," *Journal of Applied Physics*, 29(1), 76–80, 1958.
10. M. A. Alim, M. A. Seitz and R. W. Hirthe; "Complex Plane Analysis of Trapping Phenomena in Zinc Oxide Varistors," *Journal of Applied Physics*, 63(7), 2337–2345, 1988.
11. M. A. Alim, M. A. Seitz and R. W. Hirthe; "High-Temperature/Field Alternating Current Behavior of ZnO-Based Varistors," *Journal of the American Ceramic Society,* 71, C52–C55 (1988).

12. M. A. Alim and M. A. Seitz; "Singular Nature of Preferential Conducting Paths at High Electric Fields in ZnO-Based Varistors," *Journal of the American Ceramic Society,* 71, C246-C249 (1988).
13. M. A. Alim; "High-Frequency Terminal Resonance in ZnO-Bi_2O_3-Based Varistors," *Journal of Applied Physics,* 74(9), 5850–5853, 1993.
14. D. H. Eaton and C. T. Sah; "Series Equivalent Circuit Representation of SiO_2—Si Interface and Oxide Trap States," *Solid State Electronics,* 16, 841–846, 1973.
15. A. R. von Hippel; "*Dielectrics and Waves,*" John Wiley & Sons, N.Y. (1954).
16. A. R. von Hippel; "*Dielectric Materials and Applications,*" John Wiley & Sons, N.Y. (1954).
17. A. R. von Hippel; "*Molecular Science and Molecular Engineering,*" John Wiley & Sons, N.Y. (1959).
18. M. A. Alim, A. K. Batra, S. Bhattacharjee, and M. D. Aggarwal; "Complex Capacitance in the Representation of Modulus of the Lithium Niobate Crystals," *Physica B,* 406, 1088–1095, 2011.
19. R. Coelho; *Physics of Dielectrics for the Engineer,* Elsevier Scientific Publishing Company, Amsterdam, 1979.
20. M. A. Alim; "Electrical Characterization of Engineering Materials," *Active and Passive Electronic Components,* 19, 139–169, 1996.
21. C. C. Wang, V. D. Patton, S. A. Akbar, and M. A. Alim; "Effect of Zirconia Doping on the Electrical Behavior of Yttria," *Journal of Materials Research,* 11, 422–429, 1996.
22. V. D. Patton, C. C. Wang, S. A. Akbar, and M. A. Alim; "The ac Electrical Behavior of Polycrystalline Yttria," *Journal of Applied Physics,* 78, 1757–1762, 1995.
23. C. C. Wang, W. H. Chen, S. A. Akbar, and M. A. Alim; "High-Temperature a.c. Electrical Behaviour of Polycrystalline Calcium Zirconate," *Journal of Materials Science,* 32, 2305–2312, 1997.
24. A-. M. Azad, L. L. W. Shyan, and M. A. Alim; "Immittance Response of $CaSnO_3$ Prepared by Self-Heat-Sustained Reaction," *Journal of Materials Science,* 14, 1175–1187, 1999.
25. A-. M. Azad, L. L. W. Shyan, and M. A. Alim; "The AC Electrical Characterization of the Solid-State Reaction Derived $CaSnO_3$," *Journal of Materials Science,* 14, 3375–3396, 1999.
26. A. M. Azad, S. A. Akbar, L. B. Younkman, and M. A. Alim; "High Temperature Immittance Response in Anatase-Based Sensor Materials," *Journal of the American Ceramic Society,* 77(12), 3145–3152, 1994.
27. A. M. Azad, L. B. Younkman, S. A. Akbar, and M. A. Alim; "Characterization of TiO_2-Based Sensor Materials by Immittance Spectroscopy," *Journal of the American Ceramic Society,* 77(2), 481–486, 1994.

28. M. A. Alim, S. Budak, and S. Bhattacharjee; "The AC (Alternating Current) Electrical Behavior of Multi-layered Thermoelectric Devices," *Journal of Electronic Materials,* 45(11), 5588–5599, 2016.
29. A. K.Batra, J. R. Currie, M. A. Alim, and M. D. Aggarwal; "Impedance Response of Polycrystalline Tungsten Oxide," *Journal of Physics and Chemistry of Solids,* 70, 1142–1145, 2009.
30. Samples of *E. Coli* bacteria-suspended brain heart infusion (BHI) were prepared at the Department of Food and Animal Science at Alabama A & M University, and investigated at the Electrical Characterization Laboratory in the Department of Electrical Engineering and Computer Science.
31. L. Yang; "Electrical Impedance Spectroscopy for Detection of Bacterial Cells in Suspensions Using Interdigitated Microelectrodes," *Talanta*, 74, 1621–1629, 2008.
32. L. Yang, C. Ruan, and Y. Li; "Detection of Viable Salmonella Typhimurium by Impedance Measurement of Electrode Capacitance and Medium Resistance," *Biosensors and Bioelectronics*, 19, 495–502, 2003.
33. L. Yang and R. Bashir; "Electrical/Electrochemical Impedance for Rapid Detection of Foodborne Pathogenic Bacteria," *Biotechnology Advances,* 26,135–150, 2008.
34. L. Yang, Y. Li, C. L. Griffis, and M. G. Johnson; "Interdigitated Microelectrode (IME) Impedance Sensor for the Detection of Viable Salmonella Typhimurium," *Biosensors and Bioelectronics*, 19, 1139–1147, 2004.
35. L. Yang, Y. Li, and G. F. Erf; "Interdigitated Array Microelectrode-based Electrochemical Impedance Immunosensor for Detection of *Escherichia Coli* O157:H7," *Analytical Chemistry*, 76, 1107–1113, 2004.
36. R. Gomez-Sjöberg and R. Bashir; "Impedance Microbiology-on-a-Chip: Microfluidic Bioprocessor for Rapid Detection of Bacterial Metabolism, *Journal of Microelectromechanical Systems*, 14(4), 829–838, 2005.
37. R. Gomez, R. Bashir, and A. K. Bhunia; "Microscale Electronic Detection of Bacterial Metabolism," Sensors and Actuators B, 86, 198–208, 2002.
38. R. Gomez, R. Bashir, A. Sarikaya, M. R. Ladisch, J. Sturgis, J. P. Robinson, T. Geng, A. K. Bhunia, H. L. Apple, and S. Werely; "Microfluidic Biochip for Impedance Spectroscopy of Biological Species," *Biomedical Microdevices,* 3(3), 201–209, 2001.
39. Y.-S. Liu, P. P. Banada, S. Bhattacharya, A. K. Bhunia, and R. Bashir; "Electrical Characterization of DNS Molecules in Solution Using Impedance Measurements," *Applied Physics Letters*, 92, 143902 (2008).

7

Data-Handling and Analyzing Criteria

7.1 Acquisition of the Immittance Data

The usefulness of impedance spectroscopy lies in the ability of distinguishing individual contributions of the equivalent circuit elements of a DUT. For this purpose the acquired ac small-signal electrical data must be valid before these are analyzed and interpreted. If the measured data are not valid then an analysis or interpretation is next to useless. For this purpose many precautionary procedures are documented in the texts [1–5]. These procedures are apart from the calibration of the instrument. Users must be knowledgeable in instrumentation and the data acquisition system.

In order to validate the acquired immittance data as a function of frequency via reproducibility a rapid evaluation can be conducted. Today almost all the machines available are computer-controlled. Each of these is very efficient and versatile displaying plenty of capabilities. Measured data are reliable, accurate, and trustworthy with respect to the standard metal-film resistor or parallel-plate capacitor. Before conducting an investigation of an unknown DUT it is better to check the nature of the data for several immittance values corresponding to the ac small-signal voltage. For a fixed number of arbitrarily selected spot-frequencies the measured parameters must be the same for these signal-voltages in order to realize that the appropriate ac small-signal voltage as applicable, which is essentially non-destructive.

These arbitrary spot-frequencies can lie from low to high range. Checking in this way the nature of the data can be validated for the acquired ac small-signal electrical data. If the ac small-signal voltage is too low, only then there is a chance where the data may not be the same for other short range or short interval small-signal voltages. Only when the immittance data are the same for the ac small-signal voltages corresponding to the spot-frequencies then the data acquisition can be considered as valid. The electrochemical liquids require ultra-low signal-voltage to begin with, whereas for the solid system the same voltage may not work. For the single crystal based single-junction system the signal-voltage may be too low, while the polycrystalline solid may require a little higher signal-voltage.

Software can be developed to acquire the data for almost all the modern machines. Sometimes these machines include data acquisition software capabilities. All modern machines are computer-controlled and, therefore, the data acquisition cannot be troublesome. Investigators have to select the machine based on several pieces of technical information such as tan(δ) handling, conducting and dielectric samples handling, frequency range covering, etc. that are conducive to the measurement for the DUT. Since analysis depends on the investigators, preference of the customized software development is no longer tedious. A sample guide is provided in this text using geometric and CNLS fittings.

7.2 Lumped Parameter/Complex Plane Analysis (LP/CPA)

Acquisition of ac small-signal electrical data as a function of applied frequency and then displaying it in the Z^*-plane to obtain semicircular relaxation followed by the extraction of an equivalent circuit is not *impedance spectroscopy*. This process can simply be termed as the lumped parameter/complex plane analysis (LP/CPA). In order to accomplish impedance spectroscopy several aspects of analysis must be done to confirm the characteristics of the DUT. A sort of near complete spectroscopic approach is demonstrated for the ZnO-based varistor [6, 7] where detailed analyses are found for the extracted parameters such as τ_2 and τ_3 relaxations along with other parameters while the equivalent circuit is developed. Thus, the term *impedance spectroscopy* bears a broad meaning for the investigation of the DUT. Concerning demonstration of the data, concurrent evaluation was conducted via three complex planes so that a sound picture was available.

For a polycrystalline varistor only one pronounced semicircle is obtained in the Z*-plane along with a tiny intercept on the left side of the semicircle [7]. This observation does not conform with usual conclusion that two semicircles must be there, having one for the lumped grain boundaries in the low frequency domain and the other for the lumped grains in the high frequency domain. Why for the ZnO varistor only one semicircle is pronounced having a small left-intercept? This left-intercept constituted the lumped resistive behavior of the grains. The fact is that the capacitive behavior of the lumped grains is so small compared to the lumped trapping capacitance that the low-frequency semicircle did not unmask the high frequency semicircle. The left-intercept referring to the lumped resistance appeared as a tiny intercept on the x-axis which is several orders smaller, 10^8 Ω versus 10 Ω. Polycrystalline DUTs contain plenty of defect states, discontinuities of structures, variety orientations of grains, possible other phases, etc. Trapping contributions are identified in the C*-plane via multiple relaxations beside the resonance behavior. Upon revealing the resonance behavior barrier layer contribution along with the lumped ZnO capacitance are identified. Exploring ZnO varistor with this much investigation is not making complete scenario of the *impedance spectroscopy.*

When each trap is investigated or characterized thoroughly using one external variable such as temperature, dc biasing, variation of recipe, and processing steps, etc. at a time then a complete scenario is obtained from the LP/CPA approach. Of course Bode plots can be used for cross-verification purposes. For ZnO varistor, none of the Bode plane plots was useful concerning the development of the equivalent circuit. Once all aspects are completed only then *impedance spectroscopy* is performed for the ZnO varistor. If these trap findings are further supported by a DLTS (deep-level transient spectroscopy) experiment [8, 9] then confirmation can be obtained for increased confidence. In reality *impedance spectroscopy* is a meaningful term in a broad sense for the unknown DUT as *spectroscopy* was once related to the atom of an element to identify with characteristic frequency. Characterization of an element via *spectroscopy* is similar to that of the *impedance spectroscopy* for the unknown DUT. Therefore, acquisition of the frequency-dependent small-signal electrical data or representing the same data in the impedance plane is not the *impedance spectroscopy.*

Macdonald's perspective of *impedance spectroscopy* has a separate viewing window [10, 11], which is definitely much different and broader than many usual investigators'. Several problems [12] on using *impedance* *spectroscopy* were noted but rectification never observed among the users. In many cases diluted definitions along with strange practicing styles are observed. This is perhaps due to insufficient short-cut effort to comprehend the insight of *impedance spectroscopy.* The amount of pioneering work by Cole-Cole [13] in 1941 and Bauerle [14] in 1969 provided the concept of the equivalent circuit which is massive to validate the findings, while Grant [15] confirmed the use of two concurrent complex planes in 1958 ascertaining a single equivalent circuit. This is the beginning of the usage of concurrent complex planes for analytical purposes to derive one valid equivalent circuit. Despite these milestones, the template representation of the immittance data in the Z*-plane has taken over overwhelmingly, particularly after 1970, in the name of *electrochemical impedance spectroscopy.* Very soon the constant phase element (CPE) was imported and circulated or introduced widely instead of using the angle $\alpha\pi/2$ or the parameter α that Cole-Cole [13] identified. Nevertheless, Cole-Cole [13] demonstrated $\alpha\pi/2$ as the depression angle via multiple complex plane plots in 1941, and discussed α as the parameter referring to depression parameter. Precisely Cole-Cole [13] used $(1 - \alpha)\pi/2$ as the independent of frequency term where $\alpha\pi/2$ is certainly included though it is the depression angle. None of the parameters $(1 - \alpha)\pi/2$, $\alpha\pi/2$, and α were noted as the constant phase element (CPE) nor provided indication of an "element" that bears constant phase with frequency. Rather the proper sentence was: "This impedance has the striking property that, while the resistance and reactance components are functions of the frequency, the phase angle is a constant independent of the frequency and equal to $\alpha\pi/2$." So the depression angle noted as the phase angle is not the constant phase element or CPE. Cole-Cole [13] referred the so-called would not be "Debye resistance" as the "complex impedance" implied from the expression [13]

$$\varepsilon^* = \varepsilon_\infty + \frac{(\varepsilon_o - \varepsilon_\infty)}{1 + (j\omega\tau)^{1-a}} \tag{7.1}$$

as

$$Z = \frac{\tau_{peak}\left(j\omega_{peak}\,\tau_{peak}\right)^{-a}}{\varepsilon_o - \varepsilon_\infty} \tag{7.2}$$

where $j = \sqrt{-1}$. Thus, equation (7.2) is completely a complex quantity, and does not have the feature of a Debye resistance. Since impedance is inherently a complex quantity, Cole-Cole [13] considered the term impedance in lieu of the Debye or the non-Debye resistance. This aspect of the nomenclature impedance is referred to as the "complex impedance" instead of erecting the Debye or the non-Debye resistance.

Figure 7.1(a) illustrates a depressed semicircle having left-intercept in the ε*-plane as how the electrical parameters are extracted, where θ (= $\alpha\pi/2$) is the depression angle, α is the depression parameter, and U_i and V_i are the chords of the semicircular loci. The semicircle fitting is achieved via $\text{Log}(U_i/V_i)$ versus $\text{Log}(\omega)$ plot. For ideal semicircle (θ = 0°) Debye circuit is shown in Figure 7.1(b). The Cole-Cole [13] circuit termed as the non-Debye circuit in Figure 7.1(c) that satisfies equation (7.2). The Debye resistor ρ or R of Figure 7.1(b) is replaced by the frequency-independent impedance known as the non-Debye element shown in Figure 7.1(d). Cole-Cole [13] referred ρ or R to as the resistance. It may be noted that impedance cannot relate to one element but can be considered for more than one element, presumably in series. Thus, the Cole-Cole [13] prescribed frequency-independent impedance of equation (7.2) cannot be defined for a single element. The term "complex impedance" is improper to use

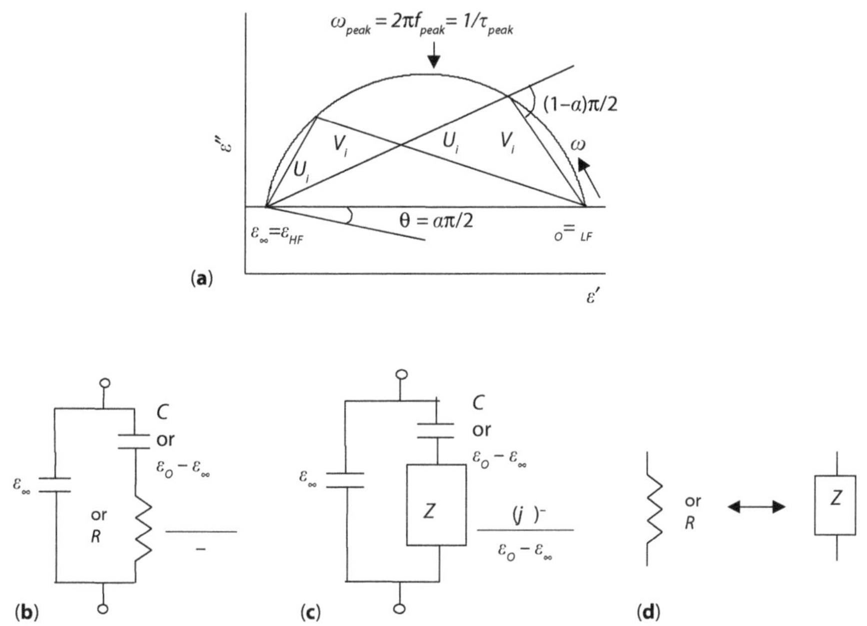

Figure 7.1 (a) Cole-Cole representation in the ε*-plane displaying semicircular relaxation where θ is the depression angle, α is the depression parameter, U_i and V_I are the chords of the semicircular loci; (b) ideal Debye circuit presuming Debye relaxation for (a); (c) Cole-Cole proposed circuit termed as non-Debye circuit for the relaxation in (a); (d) Debye resistor of (b) replaced by the Cole-Cole (non-Debye) frequency-independent impedance.

while referring to simple impedance of a circuit except for the frequency-independent impedance when the Cole-Cole element Z of Figure 7.1(c) is obtained. By the same token, constant phase element or CPE, though it refers to equation (7.2), is likely to be associated with one element, as it is called, which is a simple misnomer or misleading term. In many cases as seen frequently, this so-called CPE in the equivalent circuit is related with the depression parameter per y-axis which is the reactance in the Z^*-plane plot. This CPE has no resemblance with the Cole-Cole proposition [13] for the ε^*-plane plot whether termed as constant phase element. Sometimes it is claimed that the acronym CPE [16] surfaced around 1984 as if it is initiated by Fricke [17] in 1932.

Fricke demonstrated the features of the polarization capacity C_p (capacitance) which varies inversely as the *m*th power of the frequency (means $C_p \propto f^{-m}$), where *m* ranges between 0 and 1 but nowhere noted as the CPE. During 1988 experimentally it was demonstrated [18] that the so-called CPE is no longer a constant element but gradually turned to a transparent Debye element as documented in the equivalent circuit modeling [18, 19]. This means that the Z^*-plane parameter indicated CPE at room temperature and at zero dc bias which in turn provided a clear passive circuit element capacitance, C [18, 19]. Surely Cole-Cole [13] identification of the associated parameter is not related to the capacitance but with the conductance because the y-axis essentially was G_p/ω in 1941. Upon introduction of the Z^*-plane as the template representation the associated parameter is reactance implying capacitance. Fricke's demonstration was presumably based on the impedance revealing R-C series circuit as R_p and C_p were noted in the calculations [16].

The term *electrochemical* or the associated species is not the cause for the presence of the depression angle in the impedance spectroscopy. No matter the troublemaking tendency that arises in a situation with the depression angle, it is more tedious to comprehend when the chunk of the analysis is left behind for the sake of impedance spectroscopy. Invariably for the liquid system the meaning of the depression angle may not be the same as is found in the solid system. Thermal or dc biasing studies for the liquid system are rarely conducted and sometimes impossible for time-dependent ultra-rapid or slow degradation; whereas for the solid system it can be conducted to understand better. The fact is that the presence of the depression angle creates a perennial panic for the liquid system users. If the process of degradation is understood for the liquid system then the meaning of the depression angle becomes somewhat comprehensible.

Also there is no way to understand as to how this depression angle can disappear or disintegrate or exhibit signs of breaking ice or give a clue to visualize the origin of this behavior. In general the liquid system users do not have exposure to the solid system, where similar behavior has a non-destructive breakdown process under non-equilibrium conditions like dc biasing or elevated temperature or simultaneously biasing and temperature. At least these processes give some understanding or clue to penetrate light inside the depression angle. That's why the easy or short-cut way to summarize this behavior is to jump on the rapid inconclusive opinion. Of course a non-equilibrium experimental condition may not maintain the intact status or non-destructive testing situation for the liquid DUT.

Invariably solid system is much different though structurally often may be very complicated. For most situations regardless of the microstructures, number of phases present, grain size or grain size distribution, processing recipe or processing steps, etc. high dc biasing causes the depression angle to decrease and ultimately vanish [6, 7, 18–20] without perturbing the major characteristics of the DUT. When the dc biasing is taken away from the ZnO varistor avoiding subsequent heat development or thermal runaway [21], the device gets back to its usual functional characteristics. Sometimes high dc biasing or elevated temperature switches the window of evaluation from one complex plane to another complex plane at which the depression angle begins to decrease. Before thermal activation can be determined for the depression angle this behavior goes out of the scene as the behavior transforms to generate another window from the presence of the depression angle in a complex plane to another complex plane. This means that the complex plane switches to another complex plane which makes the analysis more critical and difficult [6, 7, 18–20]. The breakdown of the presence of the depression angle is possible at high dc biasing while the elevated temperature shifts the window of observation before any conclusion can be drawn, if systematic study is conducted. The disappearance of the depression angle is termed as the concept of *current localization* [22] for the ZnO varistor. It is doubtful whether this concept existed for the device before 1988. The term current localization was coined in 1995 by replacing the original term *path channeling* [18] for the conduction processes introduced in 1988. Although by 1999, several network simulations were available but nowhere current localization was mentioned per claim [23, 24] and subsequent elaboration [25, 26]. A similar concept for the liquid system is not familiar as degradation dominates.

Initially current localization became a hypothetical concept for the ZnO-based varistor via Voronoi network model [22] because of the near steady nonlinear coefficient α defined by:

$$\alpha = \frac{d\log(I)}{d\log(V)} = \frac{\log(I_2 / I_1)}{\log(V_2 / V_1)}, \tag{7.3}$$

though occasional multi-peaked [27] nature of α is observed. The current-voltage behavior of the ZnO varistor is shown in Figure 7.2 where ac and dc curves are also displayed. Consideration of the Voronoi network [22] in the simulation of varistor conduction processes revealed the concept of current localization for near temperature-independent tunneling and near constant nonlinear coefficient α. However, this concept needed experimental evidence preferably via Debye response justifying uniformity in conduction.

Resistivity in the two ohmic regions (Regions I and IV) depicts two extreme ends of the J-E curve. The nonlinear region is divided into two segments (Regions II and III) because 1 mA/cm^2 is the demarcation that caused the depression angle approaching zero. The entire J-E curve extends about 12 orders of magnitude in current and about five to six orders of magnitude in voltage. The ac and dc responses merge between 0.5 and 1 mA/cm^2. Typically the nonlinear coefficient α exceeds 60 for quality devices satisfying superb energy handling capability.

Though industrial corner used R-C parallel equivalent circuit for the ZnO-based varistor representing the design of the surge protection device since 1970s, but there was no confirmation for the validity of this circuit as a confirmed or proven model. Gradual logic and wisdom justified the rationale of using the R-C parallel circuit for the ZnO varistor where R became the nonlinear resistor and C became the back-to-back double Schottky barrier related capacitance. In reality both concepts are somewhat incorrect [28]. This is because the capacitance of the foregoing equivalent circuit does not provide a Mott-Schottky plot but somewhat satisfies Mott-barrier type behavior. Overall, this presumed equivalent circuit satisfied all aspects of designing the surge protector. Experimentally this circuit was found to be non-Debye type at room temperature and zero bias confirmed via Z*-plane representation [7, 18, 19]. The depression angle remained steady when the dc biasing gradually increased to 80%E_{1mA/cm^2}, where 1 mA/cm^2 is the current density corresponding to the field E_{1mA/cm^2}. Further increase in dc biasing caused depression angle to decrease slowly. This is a clear indication that the device is entering into a uniform conduction process and perhaps this biasing is the onset of

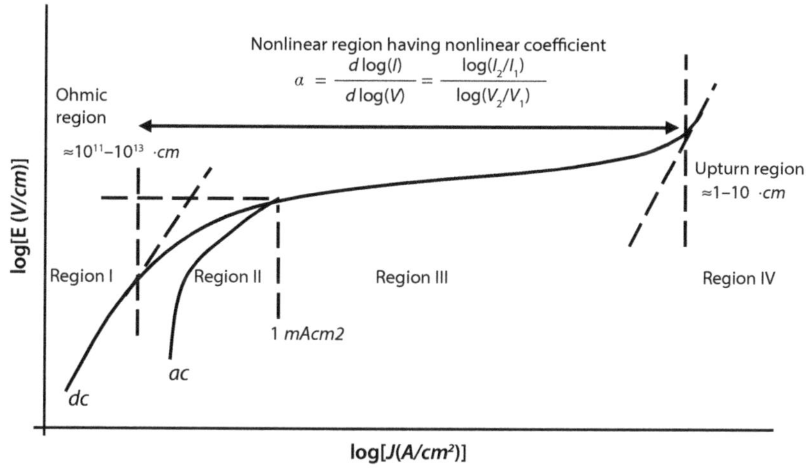

Figure 7.2 The dc and ac current density (J) versus electric field (E) showing four distinct regions for ZnO-based varistor.

the current localization [18, 22]. Certainly this behavior suggests that the depression angle gradually leads toward the uniform conduction associating the reactance component, and which is invariably no longer a CPE as the Debye capacitive element emerged. Eventually at E_{1mA/cm^2} the depression angle totally vanished as Debye semicircle emerged [18, 19]. This observation tells that the conduction becomes uniform between the electrodes where electrical paths cannot be distinguished. The capacitive path that caused the depression angle in the first place reveals preference of the operative paths exhibiting via uniformity of the identical parallel paths. For this capacitive path the corresponding dc resistance is nonlinear considered as the conduction process taking place due to tunneling beside thermal carriers crossing over the barriers.

The presence of the depression angle in the trapping relaxation via C^*-plane and the total lumped grain boundary depression angle via Z^*-plane are invariant with dc biasing up to $80\%E_{1mA/cm^2}$ [6, 7, 18–20]. No measurement is possible for the C^*-plane for $E > E_{1mA/cm^2}$ except for a very short range of about another 5% above E_{1mA/cm^2}. Therefore, homogeneous conduction for the trapping is not a feasible concept based on the experimental situation. Thus, the possible situation for the solid materials may not be exercised for the liquid systems unless a thorough structural knowledge at the atomistic level is exercised, whether the liquid system can be biased in conjunction with the elevated temperature without destruction. Only then extraction of some insight is possible for the so-called CPE. Therefore, LP/CPA is the powerful tool/technique to delineate complicated devices and systems, even for obtaining the origin of the depression angle.

The resonance phenomenon in ZnO varistor was confirmed via C^*- and Y^*-plane circles [7, 29]. In both planes non-Debye feature was evident. At that time the M^*-plane was examined revealing the same number of semicircular shapes displaying near identical mirror-like image of the C^*-plane feature. It was then strongly felt that it reveals the dc blocking circuit which is not logical for the ZnO varistor as it always exhibits some level of leakage current at any given voltage. Due to the practical features of the device, the M^*-plane representation was restricted then, and never felt useful or meaningful.

Farther ahead in late 1990s, the M^*-plane feature was revealed [30] for the valid device where it is presumed that there was no left-side intercept of the semicircle. From the single semicircular relaxation displayed in the Z^*-plane for the same data it appears that the device was somewhat weaker or poorer in 1990s than those investigated in 1980s in the U.S.A. It is noticeable from the magnitude of the resistance obtained via the chord of the semicircle which is also the indicative of watts loss for the device. This means the larger this resistance is, the lower the watts loss is. Nevertheless, this work contained concurrent multi-plane analysis [30] having some merit. It is true that the highest quality of ZnO-based varistor possessing low watts loss was always manufactured in the U.S.A. since the 1970s, though Matsuoka of Japan remains pioneer [31–33] in this technology.

Concerning the data, the demonstration [30] was ridiculous because there was no mention of the nature of the fitting procedures for analysis or interpretation. When two consecutive semicircles appeared in the M^*-plane the reason of the relaxation times remained unidentified though theoretical circuit analysis was done presuming that there was no dc blocking element. Moreover, the single semicircle in the Z^*-plane or two semicircles in the M^*-plane were non-Debye despite relevant equations and conversions used regular Debye style [30]. This type of demonstration is misleading and impractical as it loses the physical basis of the total representation. There was no concurrent report of the C^*- or Y^*-plane feature in this work. However, for the sake of reporting the presence of M^*-plane semicircle this work [30] may be noted.

7.3 Spectroscopic Analysis (SA)

The admittance spectroscopic analysis (SA) of the immittance data is familiar among the investigators for decades as those of the complex plane analysis [34]. It consists of systematic analytical steps for the measured immittance components or phasor form representation under the non-equilibrium experimental conditions such as applied voltages, elevated temperatures, compositional changes, processing or fabrication variables, ambient environment, atmospheric pressure, etc. Certainly this analytical approach is very useful in the cases where the measurement window is limited due to the lack of complete facility of instrumentation. For example, if the relaxation process is not visible, partially or fully, in the complex plane representation at room temperature, then the spectroscopic approach as a function of temperature can be useful, and yield results identical to those obtained from the lumped parameter/complex plane analysis (LP/CPA) or Bode plane analysis (BPA) [34]. The SA technique can be very effective especially when the DUT does not yield a meaningful solution in the LP/CPA and/or BPA, as it involves the evaluation of the as-measured in-phase and out-of-phase parameters for both series and parallel configurations under non-equilibrium conditions.

Figure 7.3 illustrates the correspondence between SA and LP/CPA where demonstration of identical information is given [34]. The SA technique has

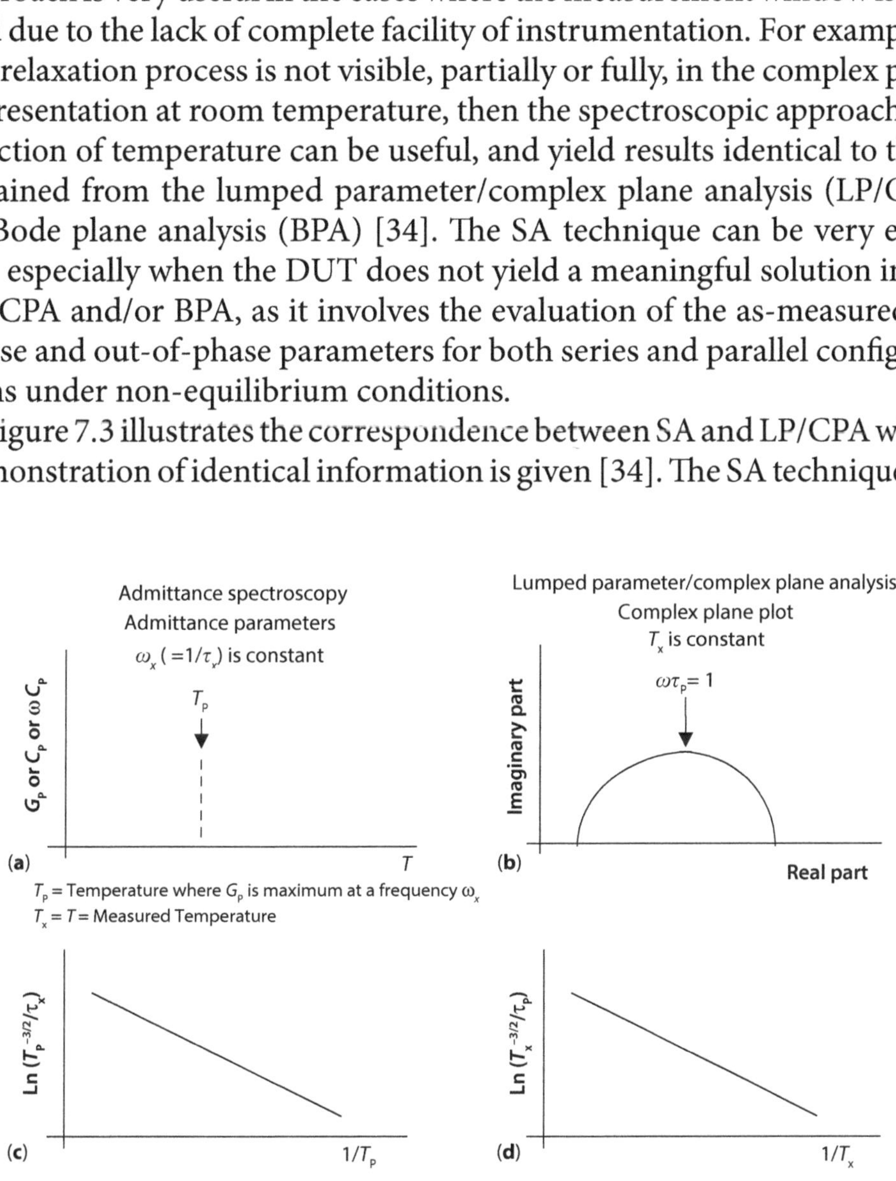

Figure 7.3 Demonstration of identical results obtained via (a) admittance spectroscopic analysis (SA) and (b) lumped parameter/complex plane analysis (LP/CPA) techniques where reciprocal temperature for the peak $1/T_p$ is plotted against spectroscopic term $[Ln(T_p^{-3/2}/\tau_X)]$ in (c) and reciprocal of various temperatures $1/T_X$ is plotted against spectroscopic term $[Ln(T_x^{-3/2}/\tau_p)]$ in (d).

been used in the early 1970s by several investigators [35–38]. Since then it is in successful use for a variety of devices [6, 7, 39–64]. This analytical approach is useful in the cases where the measurement window is limited due to the proper scaling range in the instrumentation. For example, if the relaxation process is not visible at all in the complex planes at room temperature, the spectroscopic approach as a function of temperature can be useful and yields identical results as if obtained via the LP/CPA or BPA technique [34].

A thorough spectroscopic analysis conducted on ZnO-based varistor [6, 7] may be referred to as a near perfect example of employing this technique. In this work one of the trapping relaxations denoted as τ_3 is characterized in detail. Temperature dependence of τ_3 yields thermal activation energy where the constituting components R_3 and C_3 are also thermally active [6], as shown:

$$\tau_3 = R_3 C_3 = \tau_{3o} \exp(E_3 / kT) \tag{7.4}$$

or

$$\begin{aligned} \tau_3 &= [R_{3o} \exp(E_{R3} / kT)][C_{3o} \exp(E_{C3} / kT)] \\ &= R_{3o}\, C_{3o} \exp[(E_{R3} + E_{C3}) / kT] \end{aligned} \tag{7.5}$$

where

$$\tau_{3o} = R_{3o} C_{3o} \tag{7.6}$$

and

$$E_3 = E_{\tau 3} = E_{R3} + E_{C3}. \tag{7.7}$$

It was shown that about 90% of the activation energy of τ_3 is associated with R_3 and about 10% is associated with C_3. The entire thermal activation must be associated with the in-phase component R_3 while the trap occupation or charge storage should not be temperature-dependent; hence C_3 should not be thermally activated [65–68]. Considering this response in conjunction with the presence of the depression angle, it implies that the nature of the τ_3-relaxation is more complicated, and of course suggests a distribution of activation energies and/or relaxation times [69]. The origin of complexity is still not known though Debye equations are still in use lacking understanding [44–46, 70] of the nature of the trapping response of ZnO-based varistor. Identification of a Debye or non-Debye type conduction process is always unobvious in the SA approach. The trapping behavior yielded capture cross-section, capture rate, etc. using spectroscopic equations [7, 68]. Both time constant (τ_3) and depression angle (θ_3) are independent of dc biasing [1] up to 80%E_{1mA/cm^2}. Above this biasing level, measurement for the τ_3-relaxation is not possible as the distortion in the C*-plane semicircle begins.

7.4 Bode Plane Analysis (BPA)

The classical BPA involves plotting of the absolute magnitude of the immittance function as a function of frequency [71]. It is basically a circuit analysis tool that provides understanding on the behavior of the phase angle, as-measured electrical parameters, in-phase and out-of-phase quantities of the immittance function, etc. From the shape of the curve obtained in each Bode plane, an equivalent circuit analog may be developed. In simplified cases, this approach can reveal identical results to that obtained from the LP/CPA. The time-dependent behavior and/or reaction-related dynamic processes are, however, generally not straightforward in the LP/CPA representation. The response is usually distorted and/or overlapped with other mechanisms, and often cannot be resolved due to severely lumped behavior of simultaneous competing phenomena [7, 39–64]. In this case, extraction of an equivalent circuit in a simplified form is not meaningful as it may not match with the information revealed via LP/CPA technique [34]. This situation can be improved using SA or BPA.

The simplest form of the Bode plot for the R-C parallel circuit is shown in Figure 7.4 where absolute magnitude of impedance, ($|Z|$), and phase angle are plotted as a function of frequency beside the Z*-plane plot. The slope of the slant line in Figure 7.4(a) is –1. From the nature of the curve of the Bode plot the equivalent circuit can be determined with proper values. The intersecting horizontal line on the y-axis gives the value of the resistance R shown in Figure 7.4(a). The phase angle plot shows reverse symmetry between the top and bottom segments around 45° as shown in Figure 7.4(b). The intersection of the slant and horizontal lines provides the relaxation-frequency or peak-frequency, ω_{peak}. It gives relaxation time τ shown in Figure 7.4 (a) or (b) or (c). The capacitance C is extracted from τ when the value of R is used. The behavior of the same circuit is displayed in Figure 7.5(a) for the Bode plot using absolute magnitude of admittance, ($|Y|$), versus frequency. The slope of the slant line shown in Figure 7.5(a) is +1. The vertical line is the feature of the capacitance C for the Y*-plane representation shown in Figure 7.5(b). The plots in Figure 7.5 yield the same information as extracted from Figure 7.4. In case of the limited window

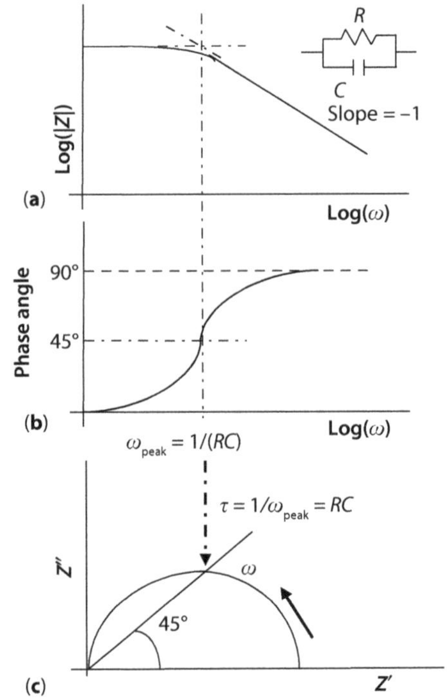

Figure 7.4 (a) Bode plot of absolute impedance versus frequency, (b) Bode plot of phase angle versus frequency, and (c) Z*-plane plot for the R-C parallel circuit.

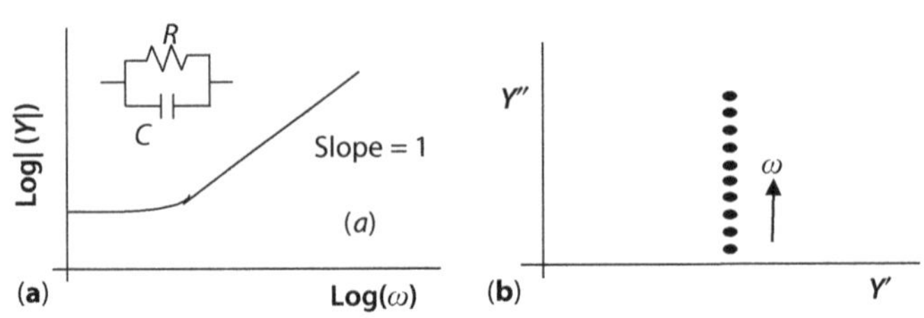

Figure 7.5 (a) Bode plot of absolute admittance versus frequency, and (b) Y*-plane plot for the R-C parallel circuit.

in the low frequency domain, parallel horizontal line of Figure 7.4(a) or Figure 7.5(a) may not be available. In that case exact value of R cannot be assessed. At the same time a small portion of the semicircle may not have the data points although a good amount can be extrapolated efficiently using proper complex nonlinear least squares (CNLS) fitting procedures to obtain the equivalent circuit parameters almost accurately. If at least 50% or more data points cover the semicircular loci in the measured frequency range on either side of ω_{peak} or τ via CNLS fitting procedures in conjunction with the confidence level approaching unity or above 95%,

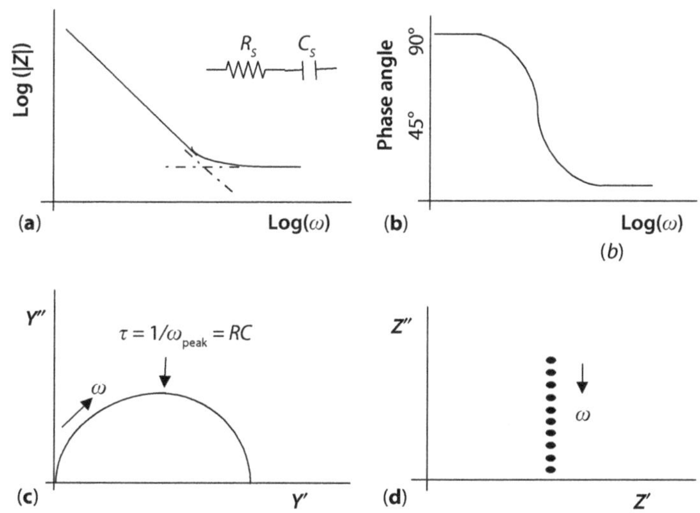

Figure 7.6 (a) Bode plot of absolute impedance versus frequency, (b) Bode plot of phase angle versus frequency, and (c) Z*-plane plot for the R-C parallel circuit.

then acceptance of the extracted parameters are well supported. This is a practical consideration concerning experimental standpoint as this kind of fitting was successfully used first time in establishing high-temperature/field character of the ZnO-based varistor [7, 17, 18] where extracted parameters were reasonable.

The simplest form of the Bode plot for the R-C series circuit is shown in Figure 7.6 where absolute magnitude of impedance, (|Z|), and phase angle are plotted as a function of frequency. The features are different than the R-C parallel circuit. Semicircular relaxation is obtained in the Y*-plane whereas vertical line obtained in the Z*-plane. The Bode plot explicitly shows information for every step of frequency. Thus, impedance can be extracted at two extreme ends of frequencies for $\omega \rightarrow 0$ and $\omega \rightarrow \infty$. Also resistive equivalent circuit parameters can be realized from the Bode plot. Figure 7.7 shows the response of the R-C series and parallel combination [72, 73] having the same value for R and C. Like complex plane plot, the Bode plot representation must not be time-dependent. It can be exploited using absolute magnitude of complex capacitance (|C|) or admittance (|Y|) or modulus (|M|) versus Log(f) or Log(ω) in the same way.

7.5 Misrepresentation of the Measured Data

Figure 7.8 shows the data plotting in the complex plane to obtain equivalent circuit related parameters where the immittance parameter $I^* = I' \pm I''$.

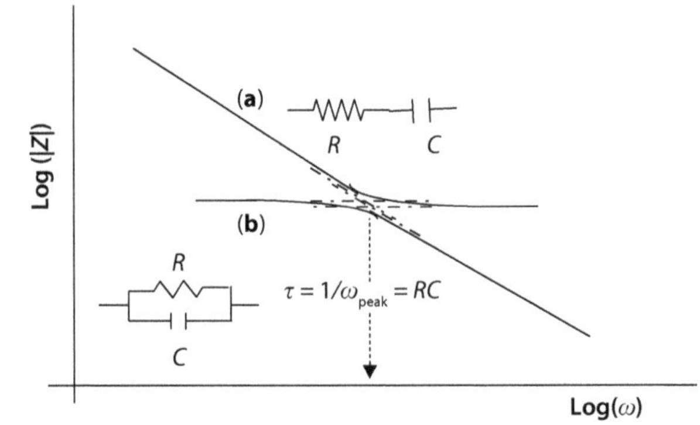

Figure 7.7 Bode plot of R-C series and parallel circuits where horizontal line gives the value of R, and slope of the slant line dictates the value of C from the time constant τ.

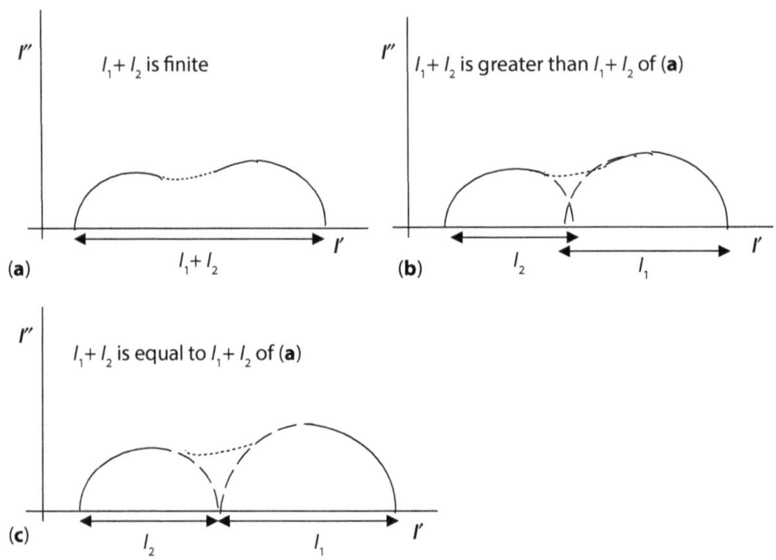

Figure 7.8 Fitting of two consecutive semicircular relaxations in a complex plane where I* is the immittance parameter: (a) as it appears upon plotting, (b) attempted fitting having significant or finite overlapped region, and (c) correct way of fitting having no overlapped region.

The appearance of two consecutive semicircles having overlapped region with two distinct relaxation peaks is displayed in Figure 7.8(a). Without prior understanding many investigators hurriedly use a very convincing semicircle fitting, presumably utilizing a protractor-like method, for each relaxation that never coincides or cuts at a single point for the two relaxations on the real-axis. Thus, the fitting process for two semicircles looks like per Figure 7.8(b). Chords I_1 and I_2 are extracted from the real-axis using the style of fitting for each semicircle. The quantity $I_1 + I_2$ obtained in Figure 7.8(b) is not equal to but greater than $I_1 + I_2$ per Figure 7.8(a). Obviously this process has increased the actual value of $I_1 + I_2$ because of the small overlapped region on the real-axis counted twice. If the fitting for each semicircle is correctly done then $I_1 + I_2$ obtained in Figure 7.8(c) will match with $I_1 + I_2$ of Figure 7.8(a). The error is caused from the small overlapped region located just above the edge of the two overlapping arrows accounted for both I_1 and I_2 in Figure 7.8(b). The correct process of fitting two semicircles is shown in Figure 7.8(c) where two semicircles meet at a single point on the real-axis.

It is worthwhile mentioning about this kind of erroneous fitting procedure as it is expanding with many journals by squaring away reviewers' knowledge. As an example, erroneous information in determining conductivity of tin oxide [74] may be noted. The data were displayed in the Z^*-plane where two overlapped semicircles were observed similar to that demonstrated in Figure 7.8(a). There is no indication of using the CNLS fitting procedures. However, the shape of each semicircular fitting curve is displayed in an overlapped manner [74]. similar to Figure 7.8(b).

There are three types of errors in this fitting work. The first type is the approach like Figure 7.8(b). The second type is the extraction of the values from the chord of the so-called fitted each semicircle in the Z^*-plane. The third type is using each chord for converting the resistance to conductivity using the physical electrode area and the thickness of the DUT.

A comment appeared [75] accordingly at that time in the same journal. This comment noted about the underlying problem of using the extracted values and consequences of practicing the same thereafter. Instead of comprehending the correct approach, in reply [76] the contributors walked through the periphery leaving the central problem, keeping the argument alive for the problem and kept explaining what has been done in that work. The first two types of errors are straightforward to understand. The third type of error involves the physical electrode area and thickness of the DUT. Physical electrode area and thickness of the DUT are not associated with the lumped grain boundary. Therefore, the concept of using physical electrode area and thickness for the lumped grain boundary resistance or conductance is an incorrect approach. This is because physical electrode area and thickness are not the area and thickness of the lumped junctions. For each junction the grain-to-grain contact area is useful along with the depletion width for the same junction. Unless a single-junction study [77, 78] is conducted it is incorrect to determine grain boundary resistivity or conductivity from the erroneous chord of the Z*-plane semicircle [74]. A single-junction study [77, 78] was available before this erroneous [74] extraction methodology was demonstrated. By that time fitting of the two consecutive semicircles [79] was demonstrated in 1982. Also, the discussion on CNLS was available [2] in 1987 besides existed several papers. Despite having access to the necessary pieces of information for the correct way to handle the data or plot, it is hard to understand as to why visible erroneous information generated a paper that skipped the correction process. The same type of extraction of the parameters from the consecutive overlapped semicircles still exists in many journals. To achieve correct fitting, two viable options can be considered. The first option is to employ simultaneous fitting of the two consecutive semicircles as demonstrated [75]. The second option is to fit one semicircle at a time but taking both semicircles by doing an iteration method, which is a common sense or wisdom to overcome the underlying problem. In fact, ZnO varistor semicircles in the C*-plane for two consecutive trapping behaviors employed fitting using the second option [6, 7, 18–20].

7.6 Misinterpretation of the Bode Plot: Equivalent Circuit

For biological liquid samples [80–88] often the equivalent circuit is modeled as the dc blocking circuit using only the conventional Bode plot. Usually these samples are conductive and, therefore, cannot have the dc blocking feature. On each occurrence such a sample is modeled as the R-C series combination supported only via the conventional Bode plot. Use of mass scale identical Bode plot [80–88] ignoring the nature of the dc blocking circuit for the biological liquid samples is not adequate for the conductive biological sample. No complex plane plot or other supporting method has been explored in each investigation [80–88]. Instead of evaluating properly not only for the sake of the material system but also for the sake of verifying the dc blocking circuit, if at all it exists, there had been no effort other than cross-referring the same blocking circuit. This is nothing but enhancing erroneous process multiplication via skipping or ignoring the conceptual error.

To verify the dc blocking circuit, out of curiosity, *Escherichia coli* (*E. Coli*) bacteria-suspended brain heart infusion (BHI) (hereafter *E. Coli*) is investigated using ac small-signal electrical measurements [89]. The BHI broth based material system used as a medium for the *E. Coli* bacteria growth has been investigated [82] via impedance or admittance (immittance) measurements to determine underlying operative mechanisms. The immittance measurement provides a powerful tool to understand the detection process of *E. Coli* as well as the total behavior of the same in the suspended medium. In order to detect bacteria, relative or absolute changes in the constituting immittance parameters are used [78, 81]. The electrical response of the bacterial behavior is ascertained by specific patterns of the data display and subsequent proposition of the equivalent circuit model [83–88, 91–95]. These data are analyzed via complex plane formalisms besides the same conventional Bode plot [89]. From these plots it is revealed that the R-C series combination is shunted to support the non-blocking feature of the dc operative path in *E. Coli*. Thus, the existence of the shunt resistor is verified for the *E. Coli* [89].

The total data are analyzed via Y*-plane as shown in Figure 7.9(a) where a complete circle was observed displaying negative capacitance for the resonance behavior. When the positive portion of the data is displayed in the Y*-plane in Figure 7.9(b) and C*-plane in Figure 7.9(c) then a single semicircle was observed in each of these planes. There is sound evidence

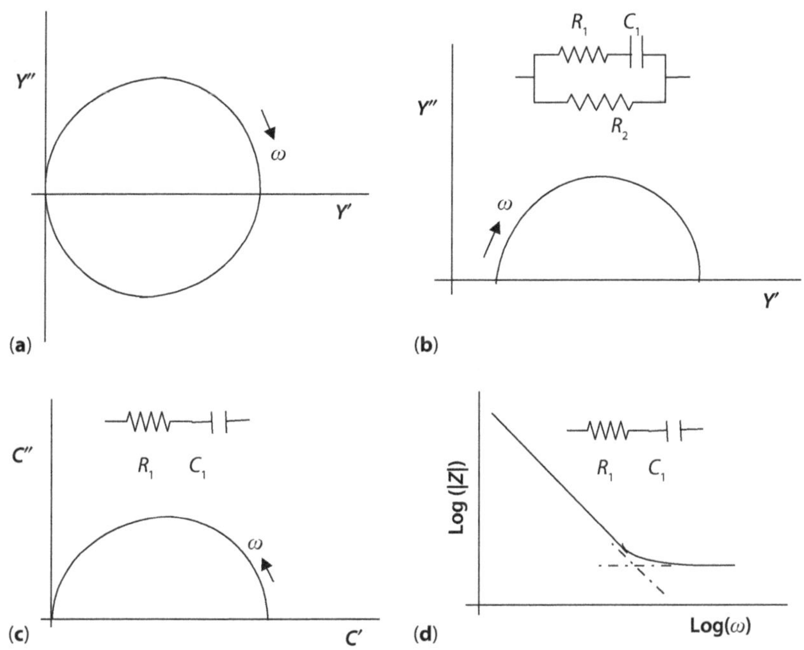

Figure 7.9 (a) Behavior of the total measured data of the *E. Coli* bacteria suspended in the BHI medium in the Y*-plane displaying negative capacitance via resonance, (b) using only the positive portion of the data in the C*-plane, and (c) Bode plot via Log(|Z|) versus Log(ω) showing dc blocking R-C series circuit.

of the influence of the dc conductance in the low frequency region via significant distortion in the semicircular loci of the C*-plane relaxation like that of the ZnO varistor [6, 7, 20]. The behavior observed in both Y*- and C*-planes was noted earlier by Grant [90]. The Y*-plane fitting directly gives the finite value of the shunt resistor which is apprehended as the dc resistance, whereas the C*-plane semicircle strongly suggests the influence of the dc conductance which is in parallel with the R-C series relaxation. This R-C series circuit indicates the relaxation that satisfies both the semicircles in Y*- and C*-planes.

The growth of bacteria causes the increase of admittance between the two electrodes reflecting the decrease in resistance. Both fitted semicircles show the presence of a finite non-zero small left-side intercept means a finite non-zero dc resistance. Both figures indicate the presence of non-Debye relaxation process as depicted by the depressed semicircles with finite depression parameter $\alpha \approx 0.18$ for the *E. Coli*.

No investigator of the biological samples provided a display of the data in the complex plane but always forwarded Bode plot using Log(|Z}) versus Log(ω) plot where only R-C series circuit is revealed. There is no chance to see enough data in the low frequency domain where the dc resistance can emerge. When the same data of the Y*- and C*-planes are plotted in the Bode plane then an identical curve is obtained, as was demonstrated by some investigators [80–88]. Therefore, only R-C series combination is revealed from the Bode plot. There is no way to recognize the shunt resistance in the Bode plane since the frequency window is limited there. This limitation is never realized by the same investigators but kept presenting the R-C series blocking circuit as the only finding. This erroneous presentation has become a problem for many investigators because of the stagnant concept concerning the usage of the Bode plane.

Problems

1. What is admittance spectroscopy? How does it differ from impedance spectroscopy? Can a Bode plot be extended to some form to obtain similar information as possible in admittance spectroscopy?
2. Consider references 35 through 38. What is the definition of admittance spectroscopy there? Why is there no admittance plot in any of these references? Explain.

3. What is lumped parameter/complex plane analysis (LP/CPA)? How does it differ from the admittance spectroscopy?
4. How can two overlapped semicircles be used in extracting the chord for each semicircle? What steps or measures should be taken to achieve correct values of the chord?
5. Consider Lang's 1974 paper where deep-level transient spectroscopy (DLTS) is discussed. How does the DLTS results compare with that of impedance spectroscopy and of admittance spectroscopy?
6. Consider Grant's 1958 paper where four basic circuits are developed. Each circuit is satisfied via relaxations in the Y*- and C*-planes. Likewise develop four basic circuits that satisfy relaxations in the Z*- and M*-planes.

References

1. E. Barsoukov and J. R. MacDonald; *Impedance Spectroscopy: Theory, Experiment, and Applications,* Edited, John Wiley & Sons, Inc., Hoboken, New Jersey, 2005.
2. J. R. MacDonald; *Impedance Spectroscopy: Emphasizing Solid materials and Systems,* Edited, John Wiley & Sons, Inc., Hoboken, New Jersey, 1987.
3. M. E. Orazem and B. Tribollet; *Electrochemical Impedance Spectroscopy,* John Wiley & Sons, Inc., Hoboken, New Jersey, 2008.
4. V. F. Lvovich; *Impedance Spectroscopy: An Application to Electrochemical and Dielectric Phenomena,* John Wiley & Sons, Inc., Hoboken, New Jersey, 2012.
5. G. S. Popkirov and R. N. Schindler; "Validation of Experimental Data in Electrochemical Impedance Spectroscopy," *Electrochimica Acta*, 38(7), 861–867, 1993.
6. M. A. Alim; "Admittance-Frequency Response in Zinc Oxide Varistor Ceramics," *Journal of the American Ceramic Society*, 72, 28–32, 1989.
7. M. A. Alim, M. A. Seitz, and R. W. Hirthe; "Complex Plane Analysis of Trapping Phenomena in Zinc Oxide Based Varistor Grain Boundaries," *Journal of Applied Physics*, 63, 2337–2345, 1988.
8. D. K. Schroder; "*Semiconductor Material and Device Characterization,*" 3rd Edition, John Wiley and Sons, New York, 2006.
9. D. V. Lang; "Deep-Level Transient Spectroscopy: A New Method to Characterize Traps in Semiconductors," *Journal of Applied Physics*, 45(7), 3023–3032, 1974.
10. J. R. Macdonald; "Impedance Spectroscopy," *Annals of Biomedical Engineering*, 20, 289–305, 1992.

11. J. R. Macdonald; "Impedance Spectroscopy," *Encyclopedia of Physical Science and Technology*, pp 279–291, 1992.
12. J. R. Macdonald; "Impedance Spectroscopy: Old Problems and New Developments," *Electrochimica Acta*, 15(10), 1483–1492, 1990.
13. K. S. Cole and R. H. Cole; "Dispersion and Absorption in Dielectrics I. Alternating Current Characteristics," *Journal of Chemical Physics*, 9, 341–351, 1941.
14. J. E. Bauerle; "Study of Solid Electrolyte Polarization by a Complex Admittance Method," *Journal of Physics and Chemistry of Solids*, 30, 2657–2670, 1969.
15. F. A. Grant; "Use of Complex Conductivity in the Representation of Dielectric Phenomena," *Journal of Applied Physics*, 29(1), 76–80, 1958.
16. G. J. Brug, A. L. G. Van Den Eeden, M. Sluyters-Rehbach, and J. H. Sluyters; "The Analysis of Electrode Impedances Complicated by the Presence of a Constant Phase Element," *Journal of Electroanalytical Chemistry*, 176, 275–295, 1984.
17. H. Fricke; "The Theory of Electrolytic Polarization," *Philosophical Magazine*, Series 7, Volume 14, Issue 90, 310–318, 1932.
18. M. A. Alim and M. A. Seitz; "Singular Nature of Preferential Conducting Paths at High Electric Fields in ZnO-Based Varistors," *Journal of the American Ceramic Society*, 71, C246-C249 (1988).
19. M. A. Alim, M. A. Seitz and R. W. Hirthe; "High-Temperature/Field Alternating Current Behavior of ZnO-Based Varistors," *Journal of the American Ceramic Society*, 71, C52-C55 (1988).
20. L. C. Sletson, M. E. Potter, and M. A. Alim; "Influence of Sintering Temperature on Intrinsic Trapping in Zinc Oxide-Based Varistors," *Journal of the American Ceramic Society*, 71(11), 909–913, 1988.
21. M. A. Seitz, R. W. Hirthe, and M. E. Potter; "Thermal Runaway in Metal Oxide Varistors," Materials Research Society Symposia Proceedings: Grain Boundaries in Semiconductors, 5, 405–410, 1982, H. J. Leamy/G. E. Pike/C. H. Seager edited, Elsevier Science Publishing Co., N.Y.,
22. M. Bartkowiak and G. Mahan, "Nonlinear Currents in Voronoi Networks," *Physical Review B*, 51, 10825–10832, 1995.
23. D. R. Clarke; "Varistor Ceramics," *Journal of the American Ceramic Society*, 82(3), 485–502, 1999.
24. Q. Wen and D. R. Clarke; "Modeling the Electrical Characteristics of Polycrystalline Varistors Using Individual Grain Boundary Properties," Ceramic Transactions: Grain Boundaries and Interfacial Phenomena in Electronic Ceramics, 41, 217–230, 1994, L. M. Levinson and S.-I. Hirano edited, *American Ceramic Society*, Westerville, Ohio.
25. A. Vojta, Q. Wen, and D. R. Clarke, "Influence of Microstructural Disorder on the Current Transport Behavior of Varistor Ceramics," *Computational Materials Science*, 6(1), 51–62, 1996.

26. A. Vojta and D. R. Clarke, "Microstructural Origin of Current Localization and 'Puncture' Failure in Varistor Ceramics," *Journal of Applied Physica*, 81(2), 985–93, 1997.
27. M. Bartkowiak, G. D. Mahan, F. A. Modine, M. A. Alim; "Multiple-Peaked Structure in the Nonlinearity of ZnO Varistors," *Japanese Journal of Applied Physics, Part II*: 35(4)A, L414-L417 (1996).
28. M. A. Alim, S. Bhattacharjee, S. Khanam, S. Runa, and N. Muna; "The Behavior of the ZnO-Bi_2O_3-Based Varistor Capacitance," Journal of Active and Passive Electronic Devices, 9(2), 157–184, 2014.
29. M. A. Alim; "High-Frequency Terminal Resonance in ZnO-Bi_2O_3-Based Varistors," *Journal of Applied Physics*, 74(9), 5850–5853, 1993.
30. M. Andres-Verges and A. R. West; "Impedance and Modulus Spectroscopy of ZnO Varistors," *Journal of Electroceramics*, 1-2, 125–132, 1997.
31. M. Matsuoka; "Nonohmic Properties of Zinc Oxide Ceramics," *Japanese Journal of Applied Physics*, 10, 736–746, 1971.
32. M. Matsuoka, T. Masuyama, and Y. Iida; "Voltage Nonlinearity of Zinc Oxide Ceramics Doped with Alkali Earth Metal Oxide," *Japanese Journal of Applied Physics*, 8, 1275–1276, 1969.
33. M. Matsuoka, T. Masuyama, and Y. Iida; "Nonlinear Electrical Properties of Zinc Oxide Ceramics," Proceedings of the 1st Conference of the Solid State Division, Tokyo (1969), Supplement to the *Journal of the Japanese Society of Applied Physics*, 39, 94–101, 1970.
34. M. A. Alim; "Electrical Characterization of Engineering Materials," *Active and Passive Electronic Components*, 19, 139–169, 1996.
35. D. L. Losee; "Admittance Spectroscopy of Deep Impurity Levels: ZnTe Schottky Barriers," Applied Physics Letters, 21 54–56, 1972.
36. D. L. Losee; "Admittance Spectroscopy of Impurity Levels in Schottky Barriers," *Journal of Applied Physics*, 46, 2204–2214, 1975.
37. G. Vincent, D. Bois, and E Pinard; "Conductance and Capacitance Studies in GaP Schottky Barriers," *Journal of Applied Physics*, 46, 5173–5178, 1975.
38. G. Vincent, A. Chantre, and D. Bois; "Electric Field Effect on the Thermal Emission of Traps in Semiconductor Junctions," *Journal of Applied Physics*, 50, 5484–5487, 1979.
39. C. C. Wang, V. D. Patton, S. A. Akbar, and M. A. Alim; "Effect of Zirconia Doping on the Electrical Behavior of Yttria," *Journal of Materials Research*, 11, 422–429, 1996.
40. V. D. Patton, C. C. Wang, S. A. Akbar, and M. A. Alim; "The ac Electrical Behavior of Polycrystalline Yttria," *Journal of Applied Physics*, 78, 1757–1762, 1995.
41. C. C. Wang, W. H. Chen, S. A. Akbar, and M. A. Alim; "High-Temperature a.c. Electrical Behaviour of Polycrystalline Calcium Zirconate," *Journal of Materials Science*, 32, 2305–2312, 1997.

42. A. M. Azad, S. A. Akbar, L. B. Younkman, and M. A. Alim; "High-Temperature Immittance Response in Anatase-Based Sensor Materials," *Journal of the American Ceramic Society*, 77, 3145–3152, 1994.
43. A. M. Azad, L. B. Younkman, S. A. Akbar, and M. A. Alim; "Characterization of TiO_2-Based Sensor Materials by Immittance Spectroscopy," *Journal of the American Ceramic Society,* 77, 481–486, 1994.
44. Y. Shim and J. E Cordaro, "Admittance Spectroscopy of Polycrystalline ZnO-Bi_2O_3 and ZnO-BaO Systems," *Journal of the American Ceramic Society*, 71, 184–188, 1988.
45. J. C. Simpson and J. E Cordaro, "Characterization of Deep Levels in Zinc Oxide," *Journal of Applied Physics,* 63, 1781–1783, 1988.
46. J. E Cordaro, Y. Shim, and J. E. May; "Bulk Electron Traps in Zinc Oxide Varistors," *Journal of Applied Physics*, 61, 4186 (1986).
47. A. K. Jonscher and T. J. McCarthy; "Admittance Spectroscopy of Silicon Zener Diodes," *Semiconductor Science Technology*, 1, 150–160, 1986.
48. A. K. Jonscher; "Admittance Spectroscopy of Systems Showing Low-Frequency Dispersion," *Electrochimica Acta*, 35(10), 1595–1600, 1990.
49. P. Krispin; "Impurity Detection by Admittance-Frequency Dispersion," *Physica Status Solidi* (a), 40, K81-K84 (1977).
50. K. Kobayashi, M. Takata, and Y. Fujimura, and S. Okamoto; "Investigation of Trapping States in a Nb-Doped Rutile by Admittance Spectroscopy," *Journal of Applied Physics*, 60, 4191–4198, 1986.
51. M. D. Giulo, G. Micocci, and A. Tepore; "Admittance Spectroscopy of Traps in Au-InSe Schottky Cells," *Solid State Electronics*, 27, 1015–1019, 1984.
52. L. C. Isett; "Characterization of CdS/CdTe Thin-Film Xolar Cells by Admittance Spectroscopy and Deep-Level Transient Spectroscopy," *Journal of Applied Physics*, 56(12), 3508–3517, 1984.
53. D. L. Polla and G. E. Jones; "Admittance Spectroscopy of Deep Levels in $Hg_{1-X}Cd_XTe$," *Journal of Applied Physics*, 51(12), 6233–6237, 1980.
54. C. X. Qiu and I. Shih; "Investigation of Electrodeposited $CulnSe_2$ Films by Admittance Spectroscopy," *Journal of Applied Physics*, 64, 758–760, 1988.
55. L. V. Hmurcik and R. A. Serway, "Frequency Dispersion in the Admittance of the Polycrystalline Cu_2S/CdS Solar Cell," *Journal of Applied Physics,* 61, 756–761, 1987.
56. J. Versluysa, P. Clauws, P. Nollet, S. Degrave, and M. Burgelman; "DLTS and Admittance Measurements on CdS/dTe Solar Cells," *Thin Solid Films,* 431 –432, 148–152, 2003.
57. T. Asperger, C. Miesner, K. Brunner, and G. Abstreiter; "Admittance Spectroscopy of Ge Quantum Dots in Si," Thin Solid Films, 380, 227–229, 2000.
58. V. G. Karpov, D. Shvydka, U. Jayamaha, and A. D. Compaan; "Admittance Spectroscopy Revisited: Single Defect Admittance and Displacement Current," *Journal of Applied Physics,* 94(9), 5809–5813, 2003.

59. Y. W. Hong, J. H. Kim; "Impedance and Admittance Spectroscopy of Mn_3O_4-Doped ZnO Incorporated with Sb_2O_3 and Bi_2O_3," *Ceramics International*, 30, 1307–1311, 2004.
60. M. Burgelman and P. Nollet; "Admittance Spectroscopy of Thin Film Solar Cells," *Solid State Ionics*, 176, 2171 - 2175 (2005).
61. Yasuo Koide, S. Koizumi, H. Kanda, M. Suzuki, H. Yoshida, N. Sakuma, T. Ono, and T. Sakai, "Admittance spectroscopy for Phosphorus-Doped n-Type Diamond Epilayer," *Applied Physics Letters*, 86, 232105 (2005).
62. D. Poplavskyya and F. So; "Bipolar Carrier Transport in a Conjugated Polymer by Complex Admittance Spectroscopy," *Journal of Applied Physics*, 99, 033707 (2006).
63. S. W. Tsang, S. K. So, and J. B. Xu; "Application of Admittance Spectroscopy to Evaluate Carrier Mobility in Organic Charge Transport Materials," *Journal of Applied Physics*, 99, 013706 (2006).
64. J. V. Li, S. W. Johnston, X. Li, D. S. Albin, T. A. Gessert, and Dean H. Levi; "Discussion of some 'Trap Signatures' Observed by Admittance Spectroscopy in CdTe Thin-Film Solar Cells." *Journal of Applied Physics*, 108, 064501 (2010).
65. H. J. de Bruin and S. P. S. Badwal; "Faradaic Reaction Kinetics in Solid Electrolytes by Impedance Dispersion Analysis," *Physica Status Solidi* (a), 49, K181-K184 (1978).
66. H. J. de Bruin and S. P. S. Badwal; "Faradaic Impedance Dispersion of the Noble-Metal/Ceramic Electrolyte Interface," *Journal of the Australian Ceramic Society*, 14(2), 20–28, 1978.
67. J. F. McCann and S. P. S. Badwal; "Equivalent Circuit Analysis of the Impedance Response of Semiconductor/Electrolyte/Counterelectride Cells," *Journal of Electrochemical Society*, 129, 551–559, 1982.
68. A. G. Milnes; "*Deep Impurities in Semiconductors*," John Wiley & Sons, Inc., N.Y. (1973).
69. J. R. Macdonald; "Linear Relaxation: Distributions, Thermal Activation, Structure, and Ambiguity," *Journal of Applied Physics*, 62, R51–R62, 1987.
70. G. E. Pike; "Electronic Properties of ZnO Varistors: A New Model," Materials Research Society Symposia Proceedings: *Grain Boundaries in Semiconductors*, Edited by H. J. Leamy, G. E. Pike, and C. H. Seager, Elsevier Scientific Publishing Company, New York, 5, 369–380, 1982.
71. H. W. Bode; *Network Analysis and Feedback Amplifier Design*, Van Nostrand, New York, 1945.
72. B. D. Cahan and C.- T. Chen; "The Nature of the Passive Film on Iron II: A-C Impedance studies," *Journal of the Electrochemical Society*, 129(3), 474–480, 1982.
73. B. D. Cahan and C.- T. Chen; "Questions on the Kinetics on Oxide-Covered Metals," *Journal of the Electrochemical Society*, 129(4), 700–705, 1982.

74. J.-H. Lee, S.-J. Park, K. Hirota; "Temperature Dependence of Electrical Conductivity in Polycrystalline Tin Oxide," *Journal of the American Ceramic Society*, 73(9), 2771–2774, 1990.
75. M. A. Alim; "Comment on 'Temperature Dependence of Electrical Conductivity in Polycrystalline Tin Oxide'," *Journal of the American Ceramic Society*, 74(7), 1757–1758, 1991.
76. J.- H. Lee, S.- J. Park, and K. Hirota; "Reply to 'Comment on 'Temperature Dependence of Electrical Conductivity in Polycrystalline Tin Oxide,''" *Journal of the American Ceramic Society*, *74*(7), 1759 (1991).
77. G. D. Mahan, L. M. Levinson, and H. R. Philipp; "Single Grain Junction Studies of ZnO Varistors – Theory and Experiment," *Appl. Phys. Lett.*, 33(9), 830–832, 1978.
78. M. A. Seitz, A. K. Verma, and R. W. Hirthe; "AC Electrical Behavior of Individual MOV Grain Boundaries," *Ceramic Transactions: Advances in Varistor technology*, L. M. Levinson edited, Published by The American Ceramic Society, Westerville, Ohio, 3, 135–144, 1989.
79. Y. –T. Tsai and D. W. Whitmore; "Nonlinear Least-Squares Analysis of Complex Impedance and Admittance Data," *Solid State Ionics,* 7, 129–139, 1982.
80. L. Yang; "Electrical Impedance Spectroscopy for Detection of Bacterial Cells in Suspensions Using Interdigitated Microelectrodes," *Talanta*, 74, 1621–1629, 2008.
81. L. Yang, C. Ruan, and Y. Li; "Detection of Viable Salmonella Typhimurium by Impedance Measurement of Electrode Capacitance and Medium Resistance," *Biosensors and Bioelectronics,* 19, 495–502, 2003.
82. L. Yang and R. Bashir; "Electrical/Electrochemical Impedance for Rapid Detection of Foodborne Pathogenic Bacteria," *Biotechnology Advances*, 26,135–150, 2008.
83. L. Yang, Y. Li, C. L. Griffis, and M. G. Johnson; "Interdigitated Microelectrode (IME) Impedance Sensor for the Detection of Viable Salmonella Typhimurium," *Biosensors and Bioelectronics*, 19, 1139–1147, 2004.
84. L. Yang, Y. Li, and G. F. Erf; "Interdigitated Array Microelectrode-based Electrochemical Impedance Immunosensor for Detection of *Escherichia Coli* O157:H7," *Analytical Chemistry*, 76, 1107–1113, 2004.
85. R. Gomez-Sjöberg and R. Bashir; "Impedance Microbiology-on-a-Chip: Microfluidic Bioprocessor for Rapid Detection of Bacterial Metabolism,: *Journal of Microelectromechanical Systems*, 14(4), 829–838, 2005.
86. R. Gomez, R. Bashir, and A. K. Bhunia; "Microscale Electronic Detection of Bacterial Metabolism," *Sensors and Actuators B*, 86, 198–208, 2002.
87. R. Gomez, R. Bashir, A. Sarikaya, M. R. Ladisch, J. Sturgis, J. P. Robinson, T. Geng, A. K. Bhunia, H. L. Apple, and S. Werely; "Microfluidic Biochip for Impedance Spectroscopy of Biological Species," Biomedical Microdevices, 3(3), 201–209, 2001.

88. Y.-S. Liu, P. P. Banada, S. Bhattacharya, A. K. Bhunia, and R. Bashir; "Electrical Characterization of DNS Molecules in Solution Using Impedance Measurements," *Applied Physics Letters*, 92, 143902 (2008).
89. Samples of *E. Coli* bacteria-suspended brain heart infusion (BHI) were prepared at the Department of Food and Animal Science at Alabama A & M University, and investigated at the Electrical Characterization Laboratory in the Department of Electrical Engineering and Computer Science.
90. F. A. Grant; "Use of Complex Conductivity in the Representation of Dielectric Phenomena," *Journal of Applied Physics*, 29(1), 76–80, 1958.
91. C. J. Felice, R. E. Madrid, J. M. Olivera, V. I. Rotger, and M. E. Valentinuzzi; "Impedance Microbiology: Quantification of Bacterial Content in Milk by Means of Capacitance Growth Curves," *Journal of Microbiological Methods*, 35(1), 37–42, 1999.
92. J. Wu, Y. Ben, H.-C. Chang; "Particle Detection by Electrical Impedance Spectroscopy with Asymmetric-Polarization AC Electroosmotic Trapping," *Microfluid Nanofluid*, 1, 161–167, 2005.
93. S, Bayoudha, A. Othmaneb, L. Ponsonnet, and H. B. Ouadaa; "Electrical Detection and Characterization of Bacterial Adhesion using Electrochemical Impedance Spectroscopy-based Flow Chamber," *Colloids and Surfaces A: Physicochemical Eng. Aspects*, 318, 291–300, 2008.
94. C. Xiao and J. H. T. Luong; "Assessment of Cytotoxicity by Emerging Impedance Spectroscopy," *Toxicology and Applied Pharmacology*, 206, 102–112, 2005.
95. C. Xiao, B. Lachance, G. Sunahara, and J. H. T. Luong; "Assessment of Cytotoxicity Using Electric Cell-Substrate Impedance Sensing: Concentration and Time Response Function Approach," *Analytical Chemistry*, 74#22, 5748–5753, 2002.

8

Liquid Systems

8.1 Non-Crystalline Systems: Liquids

Liquids consist of atoms and/or molecules that are connected by interatomic and/or intermolecular bonds. These are not crystalline as there is no specific bond length and systematic orientation of the atoms. It can be similar to the structure of glasses. In other words, particles in a liquid are clustered together and are free to flow. Thus, a liquid has a definite volume but it does not have a definite shape as it takes the shape of its container.

When conduction takes place in a liquid system then it is called an electrolyte because the charge carrying species are usually ions. An electrically conducting solid or liquid system may be referred to as an electrolyte when the conduction process takes place by ions. Alternately, the solute forming solution capable of conducting electricity is called electrolyte. Basically an electrolyte is an ionic conductor, solid or liquid. When electricity passes through an ionic compound or an electrolyte solution a chemical reaction takes place which is called electrolysis.

If common edible salt, sodium chloride, is added to pure distilled water, then the solution becomes electrolytic. There are many liquid systems that are of research interest because of their applications [1–6]. The liquid system can undergo physical changes through chemical reactions. Corrosion is a part of the process where the chemical reaction is dominant. Often biological materials are of interest where ions are conducting. The underlying processes via operative mechanisms in liquids are dependent on time which can be monitored via immittance measurements and thereby subsequent analysis.

Two types of processes allow current conduction across the electrode-solution interface. One process involves a direct transfer of electrons via oxidation reaction at one electrode and the other process involves a reduction reaction at the other electrode. Processes of this type are called Faradaic processes because they are governed by the Faraday's Laws of electrolysis [1–3]. This has two parts: (1) the first part states that the amount of chemical deposition via reaction at an electrode due to the flow of current through the electrolyte medium is directly proportional to the current or quantity of electricity, and (2) the second part states that when the same quantity of electricity passes through the electrolyte, then the mass of the deposited material (cation) is proportional to the chemical equivalent or equivalent weight of the deposited material. The chemical equivalent is given by:

$$\text{Chemical Equivalent} = \text{Equivalent weight} = \left(\frac{\text{Atomic weight}}{Valency}\right). \quad (8.1)$$

The resulting current passing through the electrolyte is called Faradaic current. In order to assess Faradaic processes, immittance spectroscopy can be used. Faraday's Law of electrolysis or Faradaic processes is also known as the kinetics of the electrode processes. Electrode processes are chemical transformations at the metal-to-electrolyte interface accompanied by the transfer of electrical charge between the phases in contact. Thus, ions are converted to neutral atoms or molecules, or vice versa [1].

Faraday's experiment on a copper sulfate ($CuSO_4$) solution having two electrodes connected externally to the dc power source or battery. Since $CuSO_4$ is ionic its molecules split into positive and negative ions. The positive ions (metallic Cu ions) move to the electrode connected with the negative terminal of the battery, and negative ions (sulfate ions or radical sulphions) move to the electrode connected with the positive terminal of the battery. The positive ions take electrons from the negative terminal (cathode) implying positive ions gain electrons at the cathode and become pure neutral Cu atom to get deposited on the electrode. The negative ions give up extra electrons, implying negative ions lose electrons at the anode and become neutral SO_4 radical. This transfer of electrons has the same effect as a flow of electrons, and the solution conducts electricity. Since the SO_4 radical cannot exist as-is in an electrically neutral state, it attacks the metallic positive electrode to form metallic (electrode terminal) sulfate. This metallic sulfate dissolves again in the solution. If the positive electrode is Cu then additional Cu will be deposited in the cathode (negative electrode) whereas for non Cu electrode the solution will be contaminated [1–6].

The container that handles electrolysis is sometimes referred to as the cell. This cell has three components to operate or use [1–6]:

1. an electrolyte solution ready to get reaction upon passing current;
2. two metallic electrodes across the electrolyte solution; and
3. an external power source or battery across the two electrodes.

The factors affecting the electrolysis reaction [1–6] of the liquid solution in the cell are the:

1. concentration or density of the liquid electrolyte,
2. type of electrolyte, and
3. type of electrodes.

The density or concentration of the liquid electrolyte dictates the rate of reaction. Usually higher concentration makes anion discharging constructive. Solid electrolytes react differently in dissolved liquid which is associated with the valency of the cation. Metallic electrodes as well as the semiconductor electrodes work differently than a specific metallic electrode. There are electrolyte-based semiconductors which can support as electrodes. Inert electrodes are always desired in electrolysis to prevent going into the solution. Nonreactive platinum or graphite is used as an inert electrode. Overall, a cell is used in making new products that employ anodizing and/or electroplating. The science and engineering behind achieving successful products are still challenging with the electrochemical aspect of the materials.

The behavior of the electrolytic cell under the ac signal-voltage is illustrated via an equivalent circuit shown in Figure 8.1(a) and Figure 8.1(b). Essentially this circuit comes from the impedance plane. The corresponding Bode plot is shown in Figure 8.1(c) where R_E or R_S is the electrolyte resistance or solution resistance, R_{CT} or R_P is the charge transfer resistance or Faraday impedance [3, 4], and C_{DL} is called the double layer capacitance [3, 4, 7–12]. The circuit displayed in Figure 8.1(a) or Figure 8.1(b)

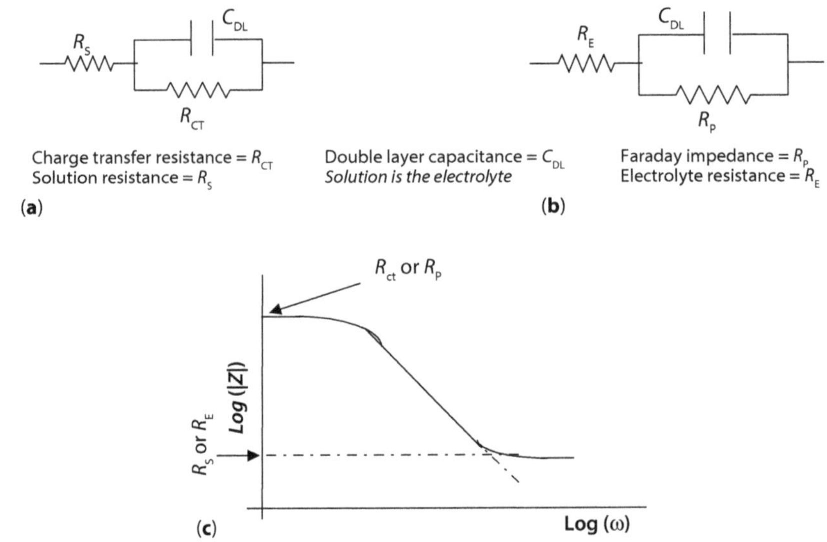

Figure 8.1 (a) and (b) Equivalent circuit for electrolysis with two notations, and (c) Bode plot for the circuit in (a) and (b).

is referred to as the Randles circuit as it is related to the electrolytic cell referred to as the Randles cell. The resistance denoted as R_{CT} or R_P is complicated as it is termed as impedance though sometimes referred to as the polarization resistance. This complicated term gives rise to Warburg and Faradaic impedances [3, 4, 8–10].

Do the data for the Randles cell in the impedance plane imply electrochemical impedance spectroscopy? How does the interpretation of the impedance plot differ from usual impedance spectroscopy for the polycrystalline material? Are the impedance plots of the Randles cell and the polycrystalline system distinguishable? What is the demarcation between these two impedance plots? Warburg and Faradaic impedances are also observed in the solid state systems, if rigorous research is conducted. In many cases the level of severity or degree of relentlessness is not as complicated for the solid systems as observed for the electrolytic cell. Thus, the solid state impedance spectroscopy and the electrochemical impedance spectroscopy belong to the same boat.

The solution resistance is important as a component of the impedance of the electrochemical cell. The resistance of an ionic solution depends on the ionic concentration, type of ions, temperature and the associated geometry through which the current is passed. Whenever the potential of an electrode is different from the value at the open circuit during electrolysis, then it is referred to as the polarizing electrode. The polarized electrode can cause current to flow via electrochemical reactions that occur on the electrode surface. The amount of current is controlled by the kinetics of the reactions, and the diffusion of the reactants both toward the electrode and away from the electrode.

An electrical double layer exists at the interface between an electrode and its surrounding electrolyte. This double layer is formed as ions from the solution accumulated on the electrode surface. Charges in the electrode are separated from the charges of these ions. This charge separation causes capacitance. The value of the double layer capacitance depends on electrode potential, temperature, concentration of ions, types of ions, oxide layers (if any), electrode roughness, electrode inertness, impurity adsorption, etc.

In the kinetically controlled electrochemical reaction, charge transfer resistance arises from the metal in contact with the electrolyte solution when electrons enter the metal and metal ions diffuse into the electrolyte. The charge transferred this way has a certain rate or speed which is dependent on the type of reaction taking place, ambient temperature, concentration of the reacting species, and potential. The diffusion process of ions taking place during the electrolysis can generate impedance known as the Warburg impedance. This impedance depends on the frequency of the potential perturbation. At high frequencies the Warburg impedance is small since diffusing reactants don't have to move very far. At low frequencies the reactants have to diffuse farther, thereby increasing the Warburg impedance [3, 4, 8–10, 13, 14].

8.2 Warburg and Faradaic Impedances

Usually Warburg and Faradaic impedances are identified and confirmed via Bode plot though a complex plane plot indicates these responses. Both impedances are related to the electrochemical reactions. Whenever the rate of an electrochemical reaction mechanism is strongly influenced by the diffusion then the impedance is termed as Warburg impedance. This means diffusion effects completely dominate the electrochemical reaction mechanism [13, 14]. It implies that the resistance to mass transfer under a diffusion-controlled situation gives Warburg impedance. Warburg postulated that the resistance and capacitance of an electrode-electrolyte interface vary inversely with the square root of frequency, where

$$R = R_O\, f^{-0.5} = R_O\, f^{-(1/2)} \tag{8.2}$$

and

$$C = C_O\, f^{-0.5} = C_O\, f^{-(1/2)} \tag{8.3}$$

for the equivalent R-C series circuit with R_O and C_O are equilibrium values dependent upon electrode material, electrolyte concentration, and temperature [13, 14]. This is shown in Figure 8.2. Z*-plane plot of the same response is shown in Figure 8.3 where 45° is the phase angle for the Warburg response. This means for the diffusion-controlled electrochemical reaction, the current is 45° out-of-phase with the imposed potential. With this phase relationship, the real and imaginary components of the impedance are equal at all frequencies [13, 14]. In terms of a simple equivalent circuit, the Warburg impedance is midway between that of a resistor having 0° phase shift, and a capacitor having 90° phase shift. There cannot be any further simple electrical equivalent for the Warburg impedance [13, 14].

The frequency dependence comes from the Warburg Impedance. Thus, the linear correlation of Z′ and Z″ in Figure 8.3 is the characteristic of a diffusion controlled electrode process. As the frequency increases the

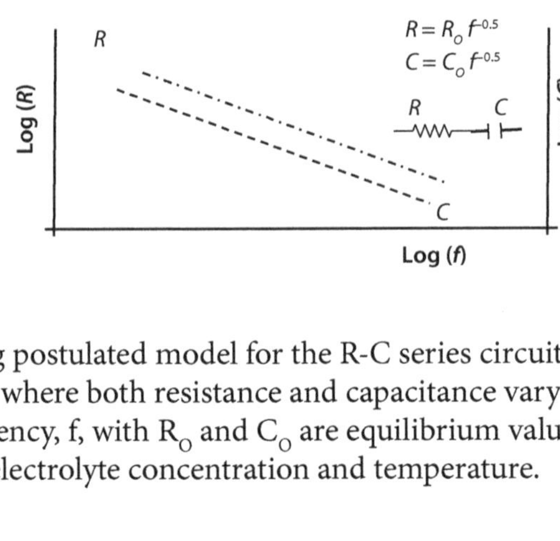

Figure 8.2 Warburg postulated model for the R-C series circuit of the electrode-electrolyte interface where both resistance and capacitance vary inversely with the square root of frequency, f, with R_O and C_O are equilibrium values dependent upon electrode material, electrolyte concentration and temperature.

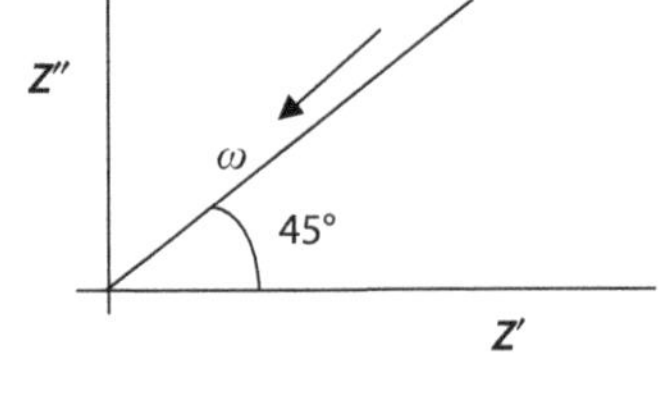

Figure 8.3 Electrode-electrolyte interface via 45° phase shift for the Warburg impedance.

charge transfer resistance and the double layer capacitance become more important. This impedance depends on the frequency of the potential perturbation. Warburg impedance represents resistance to mass transfer which is dependent on the frequency of the perturbation. At high frequencies the Warburg impedance is small since diffusing reactants don't have to move very far. At low frequencies the reactants have to diffuse farther, thereby increasing the Warburg impedance in an infinite manner:

$$\begin{aligned} Z_W &= Z_{Warburg} = R_{Warburg} + j\, X_{Warburg} \\ &= A\,\omega^{-0.5} + A\, j\,\omega^{-0.5} = A\,\omega^{-0.5}\left(1 + j\right) \end{aligned} \tag{8.4}$$

where A is the Warburg coefficient. In the complex plane the infinite Warburg impedance appears as a diagonal line with a slope of 0.5 as in Figure 8.3 while in Bode plot, the Warburg impedance exhibits a phase shift of 45° via slope of –0.5 as in Figure 8.4.

Fricke's polarization capacitance C_p varied as f^{-m}, and polarization resistance R_p as f^{m-1}, with the value of m dependent upon the type of interface [13, 15]. Electrode polarization impedance can also be characterized by the phase angle θ where tan θ is given by:

$$\tan\theta = \omega\, R_P\, C_P = \left(\frac{\text{real part}}{\text{reactive part}}\right) \tag{8.5}$$

For polarization resistance and capacitance as power functions of frequency, according to Fricke [13–15]:

$$\theta = \frac{\pi\,(m-1)}{2} \tag{8.6}$$

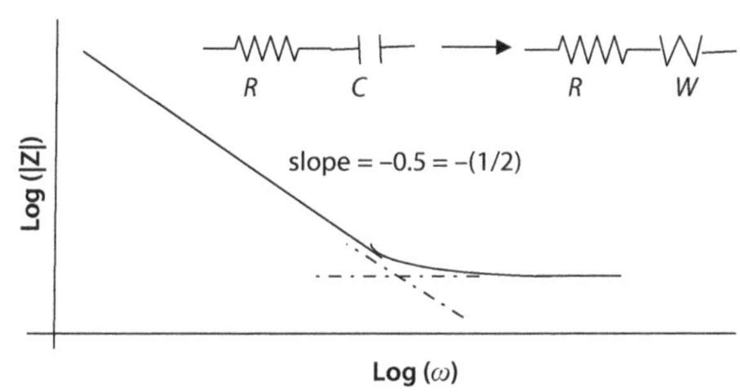

Figure 8.4 Bode plot of the Warburg impedance where the slope for the slant line is 0.5 and the capacitive element is referred to as *W*.

which is referred to as Fricke's Law. For Warburg case m = 0.5 and the corresponding phase angle is 45°. The polarization resistance and reactance are equal to each other at all frequencies [13, 14]. The Bode plot for the Warburg impedance is shown in Figure 8.4 where the slope for the slant line is 0.5 and the capacitive element is referred to as *W*. Although Fricke's Law does not apply in practical sense, it can be useful in many cases with approximations [13, 14].

For other values of m, Faradaic impedance is achieved. There are several situations for this impedance as observed in many systems. Two types of processes can conduct currents across the solution and the electrode-solution interface. The first kind involves direct transfer of carriers (i.e., electrons) via oxidation reaction at one electrode and reduction (opposite to oxidation) reaction at the other. Processes of this type are called Faradaic processes because these processes are believed to be governed by the Faraday's Law of electrolysis. Faraday's Law states that the amount of chemical reaction at an electrode is proportional to the current and the resulting current is called Faradaic Current. For this purpose measured impedance of the electrode-solution interface gives the value of m other than one-half. Ideal Faradaic impedance can be written as:

$$Z_{Faradaic} = R_{Faradaic} + X_{Faradaic} = A\,\omega^{-\beta} + A\,\omega^{-\beta}, \tag{8.7}$$

where $\beta \neq$ m but fraction making less than unity. The parameters $R_{Faradaic}$ and $C_{Faradaic}$ denote apparent Faradaic resistance and capacitance, respectively. An ideal Faradaic polarization recalls a reaction-related transport mechanism yielding two exponents of equal value as in equation (8.7). Deviation from the nature of equation (8.7) causes non-Faradaic polarization giving rise to the impedance of the form:

$$Z_{Non-Faradaic} = R_{Non-Faradaic} + X_{Non-Faradaic} = A\,\omega^{-\beta_1} + A\,\omega^{-\beta_2}, \tag{8.8}$$

where β_1 and β_2 are unequal. Overall, equation (8.8) indicates complicated type non-Faradaic impedance because of the inequalities in the exponents of β_1 and β_2. This equation becomes more complicated when the coefficient of $\omega^{-\beta_1}$ and $\omega^{-\beta_2}$ are unequal which is expressed as:

$$Z_{Non-Faradaic} = R_{Non-Faradaic} + X_{Non-Faradaic} = A\,\omega^{-\beta_1} + B\,\omega^{-\beta_2} \tag{8.9}$$

implying A and B are unequal. Possible theoretical treatment to delineate the coefficients and the exponents are not available till to date.

8.3 Constant Phase Element (CPE)

As of today the constant phase element (CPE) is found in the world of electrochemical sciences. It constitutes a separate domain within the immittance spectroscopy but popularly under the name of electrochemical impedance spectroscopy, bearing an acronym EIS. Is CPE or EIS complex plane specific? Is CPE or EIS a generalized acronym that satisfies every complex plane analysis? A most recent text [11] on electrochemical impedance spectroscopy has demonstrated the importance of a single complex plane (SCP) or simply Z*-plane analysis and another [8] advocated for the same.

The instrument manufacturers serving the community also adopted the same nomenclatures regardless of the window of capabilities. Sometimes CPE or EIS serves as synonym, and at certain times it is a misnomer or misleading nomenclature or acronym. However, it exists strongly for at least three decades or so presumably since 1984 [16] certifying Fricke's pioneering work of 1932 [15]. The existence of CPE is somehow drawn high volume relationship with the capacitance or double layer capacitance [11]. The first two texts on impedance spectroscopy [9, 10] noted the Cole-Cole classic work on the identification of the frequency-dependent impedance having phase angle constant independent of frequency as the CPE. This frequency-dependent impedance having phase angle constant independent of frequency it not designated as the CPE by the interpretation of the Cole-Cole [17] classic work. The first two texts [9, 10] noted the following equation as the Cole-Cole CPE:

$$Z_{Colnstant-Phase-Element} = Z_{CPE} = A\left(j\,\omega\right)^{-\alpha} \tag{8.10}$$

where α lies between 0 and 1, $j=\sqrt{-1}$, A is a constant parameter, and ω (=2πf) is the angular frequency. However, equation (8.10) refers to the presence of the depression angle or the depression parameter in the Z*-plane instead of the Cole-Cole plot. The Z_{CPE} becomes a resistor for $\alpha = 0$ having A = R; the same Z_{CPE} becomes a capacitor for $\alpha = 0$ having A = C^{-1} (= elastance = reciprocal of capacitance); and Z_{CPE} becomes an inductor for $\alpha = -1$ having A = L. Thus, for a resistor R, a capacitor C, and an inductor L equation (8.10) can be expressed as:

$$Z = A\left(j\,\omega\right)^{-0} = A = R \tag{8.11}$$

where equation (8.11) gives the in-phase resistance with A = R,

$$Z = A\left(j\,\omega\right)^{-1} = A\left(\frac{1}{j\,\omega}\right) = -\,A\left(\frac{j}{\omega}\right) = -\,j\left(\frac{1}{\omega\,C}\right), \tag{8.12}$$

where equation (8.12) gives the out-of-phase capacitive reactance with A = C^{-1}, and

$$Z = A\left(j\,\omega\right)^{-(-1)} = A\left(j\,\omega\right) = j\,\omega\,A = j\omega\,L, \tag{8.13}$$

where equation (8.13) gives the out-of-phase inductive reactance with A = L.

At this context classic work of Cole-Cole [17] must be mentioned for the frequency-independent component obtained from the complex permittivity plane as:

$$Z = \frac{\tau_{peak}\left(j\,\omega_{peak}\,\tau_{peak}\right)^{-a}}{\varepsilon_0 - \varepsilon_\infty} \tag{8.14}$$

where there is no mention about the CPE but there is an indication of demonstrating depression angle $\theta = \alpha\pi/2$ or depression parameter α. This frequency-independent term came from the Cole-Cole proposed equation:

$$\varepsilon^{\star} = \varepsilon_\infty + \frac{(\varepsilon_0 - \varepsilon_\infty)}{1 + (j\omega\tau)^{1-a}}. \tag{8.15}$$

The admittance from equation (8.10) gives:

$$Y_{Colnstant-Phase-Element} = Y_{CPE} = \left(\frac{1}{A\left(j\,\omega\right)^{-a}}\right). \tag{8.16}$$

This is a popular expression for the CPE in conjunction with equation (8.10). For capacitance using $\alpha = 1$, then equation (8.16) becomes:

$$\begin{aligned} Y_{Constant-Phase-Element} &= Y_{CPE} = \left(\frac{1}{A\left(j\,\omega\right)^{-a}}\right) \\ &= \left(\frac{1}{C^{-1}\left(j\,\omega\right)^{-1}}\right) = j\,\omega\,C \end{aligned} \tag{8.17}$$

The double layer capacitance often behaves as the CPE. That is why a leaky capacitor is considered for this capacitor.

8.4 Biological Liquid: *E. Coli* Bacteria

For the sake of developing a conducting equivalent circuit model for the biological liquid system instead of the dc blocking R-C series [18–26] circuit, samples of the *E. Coli* bacteria-suspended brain heart infusion (BHI) were prepared [27]. While presenting the R-C series circuit for the liquid samples, there was no backing of the complex plane plot except the Bode plot. It appeared from the Bode plot [18–26] that the window of judgment was narrow in the frequency domain because the low frequency end was not sufficient enough to view the shunt resistance to make the dc blocking R-C series circuit conductive.

The brain heart infusion (BHI) broth based material system used as a medium for the *Escherichia Coli* (hereafter *E. Coli*) bacteria growth has been investigated via impedance or admittance (immittance) measurements to determine underlying equivalent circuit analog. The bacteria used were from a non-pathogenic *E. Coli* culture obtained from the American Type Culture Collection (ATCC 8739) in Manassas, Virginia. The cultures were activated in a manner similar to Mason and Powelson [31] by first transferring the bacterial inoculum from a refrigerated slant to tryptic soy agar (TSA - Difco, Sparks, MD) plates and incubated at 37 °C for 24 h. A well-isolated single colony forming unit (CFU) was inoculated into 100 mL of brain heart infusion (BHI) broth (Difco, Sparks, MD) and incubated aerobically at 37 °C.

The bacteria were grown in BHI at 37 °C and transferred every 24 h for at least 3 days prior to use [28]. A 24 h culture of 3 μL was inoculated into 30 mL of BHI in a culture tube equipped with stainless steel probes for electrical impedance measurements. Commercially dehydrated or concentrated media were used and reconstituted according to the manufacturer's directions.

Two electrodes inserted in the test tube were connected to the impedance analyzer (HP 4192A). The ac small-signal electrical data were obtained in the frequency range 5 Hz $\leq$ f $\leq$ 10 MHz. These data were acquired in the admittance form ($Y^* = G_p + j\,\omega\,C_p$) where G_p is the conductance, C_p is the capacitance ($\omega \times C_p$ is the susceptance), ω (= 2π f) is the angular frequency, and $j = \sqrt{-1}$. The measurements were performed at room temperature in the vicinity of about 25 °C. These data were analyzed via complex plane formalisms and used in the Bode plane representation.

The ac small-signal data when displayed in the complex plane formalisms [29–34] the underlying conduction processes between the two electrodes across the material system is delineated for the operative phenomena. This display supports traditional Bode plot and, thus, does not violate the general requirements of the elements constituting the equivalent circuit model. The advantage of the complex plane plot is that an equivalent circuit model can easily be extracted unless dual representation of the complex plane plot of the same data is observed. Semicircular relaxation achieved in more than one complex plane provides at least one reasonable operative equivalent circuit. This has been demonstrated for a number of solid state material systems [35–44]. Figure 8.5 shows the measured data in the Y*-plane for the sterile BHI medium, the *E. Coli* after 3 hours, after 16.5 hours, and after 24 hours. This plot shows relaxation, resonance, and the existence of the small intercept on the left side of the circle. This small intercept on the left side of the semicircle in the Y*-plane is the shunt for the R-C series circuit extracted from the relaxation. Overall, the circle has a resemblance with that of the ZnO varistor [40, 45]. The same data are displayed in the C*-plane in Figure 8.6 where the semicircle is evident but the resonance is very much compressed. This semicircle has distortion at the low frequency end indicating the influence of the dc conductance [43, 46]. The equivalent circuit containing R_1-C_1 series combination with R_2 as the shunt is shown in Figure 8.7. The compressed portion is stretched in Figure 8.8 where the positive and negative domains are visible for the capacitance values. In reality this part of the resonance is forming a very small circle with respect to the rest of the curve.

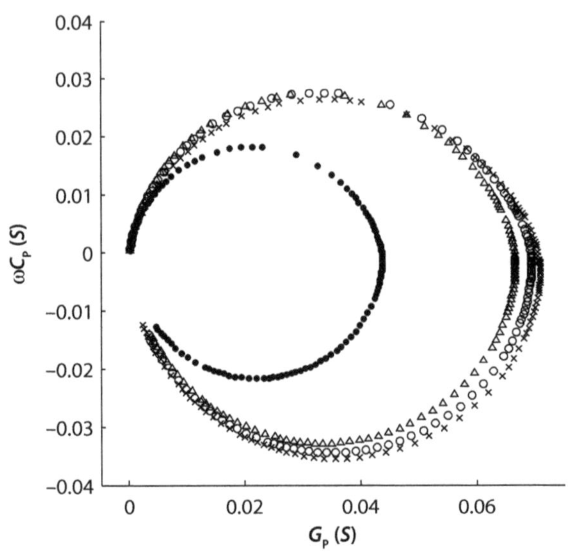

Figure 8.5 Measured data of the sterile BHI medium and *E. Coli* bacteria suspended in BHI medium in the Y*-plane for: the sterile BHI medium (•), the *E. Coli* after 3 hours (Δ), after 16.5 hours (o), and after 24 hours (×).

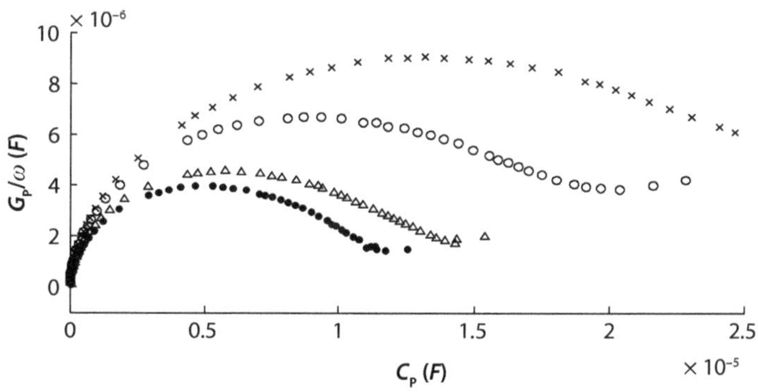

Figure 8.6 Measured data of the sterile BHI medium and *E. Coli* bacteria suspended in BHI medium in the C*-plane for: the sterile BHI medium (•), the *E. Coli* after 3 hours (Δ), after 16.5 hours (o), and after 24 hours (×).

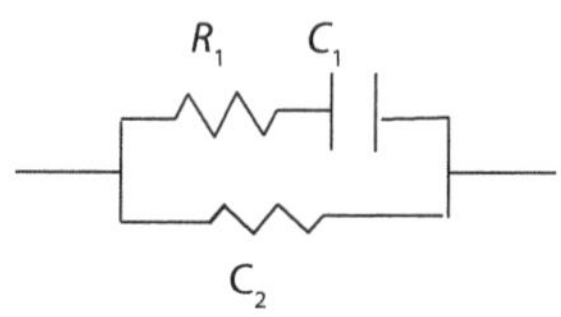

Figure 8.7 Equivalent circuit obtained from the Y*-plane of Figure 6.5.

The Bode plot for the same data is shown in Figure 8.9. There is no way to visualize that the shunt resistor exists in this plot. Therefore, the R_1-C_1 series circuit becomes obvious from the Bode plot whereas the Y*-plot reveals R_1–C_1 series circuit with a shunt resistor R_2 shown in Figure 8.7. Only one curve of *E. Coli* is shown in Figure 8.9 for the data taken after 16.5 hours. Since the other two are not significant, they are eliminated.

The sterile and bacterial media give rise to the equivalent circuit model comprising of the simplistic series R-C combination [18–26]. This type of equivalent circuit model is blocking in nature, prohibiting the flow of direct current (dc). The immittance data generation for a variety of peak-to-peak voltages indicates that there is no existence of the blocking element prohibiting the flow of dc through the material system. Invariably the dc condition can be extrapolated from the convenient Bode plot [29–31].

The ac electrical measurements and complex plane plotting techniques are used to establish operative mechanisms via representative equivalent circuit model for the material system regardless of the state. Such an approach has proved to be a useful tool/technique in characterizing the electrical nature of a number of heterogeneous systems [32–44]. This

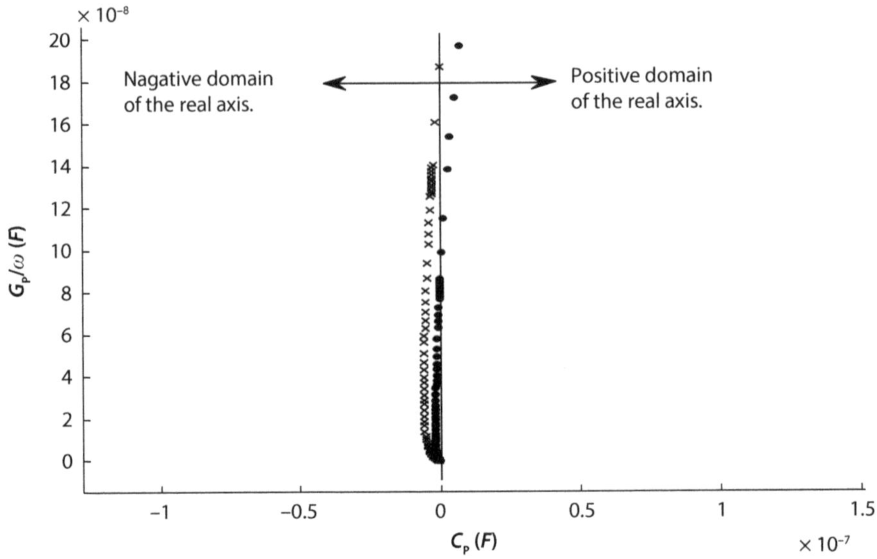

Figure 8.8 Measured data of the sterile BHI medium and *E. Coli* bacteria suspended in BHI medium in the C*-plane showing negative and positive domain of the data for: the sterile BHI medium (•), and the *E. Coli* after 16.5 hours (o).

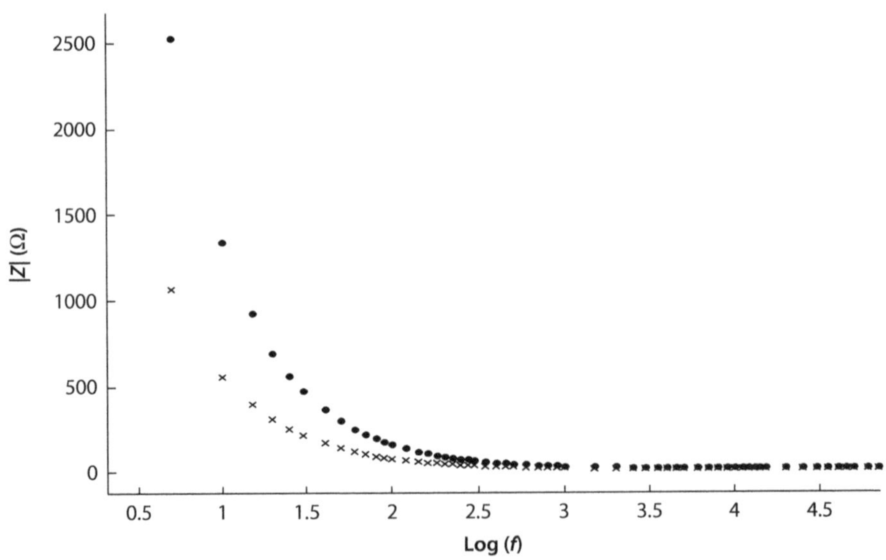

Figure 8.9 Bode plot representing absolute impedance versus frequency of the sterile BHI medium (•), the *E. Coli* after 16.5 hours (o).

technique unravels the underlying competing phenomena via lumped parameter/complex plane analysis (LP/CPA). The total ac response of the material system can be modeled in terms of equivalent circuit elements which identifies conduction mechanisms in an operating electrical path between the electrode terminals. Based on the slope of the straight line of the Bode plot and the semicircular relaxation of the complex plane plot, Debye and non-Debye responses are distinguished.

The data obtained for the *E. Coli* are analyzed via complex plane formalisms as well as conventional Bode plots. From these plots it is revealed that the R-C series combination is shunted while it is supporting the dc non-blocking nature of the operative path of the material system. Thus, the existence of the shunt resistor is verified for the *E. Coli.*

In this way the equivalent circuit presented in this work is not the same as seen elsewhere [18–26]. Often a series R-C circuit is extracted from the Bode plot as a final equivalent circuit model of this kind of liquid material system. That is why the correct or appropriate representation of the equivalent circuit model may need support from the complex plane plot for the same measured data.

Problems

1. What is the chemical equivalent of Cu? What is the chemical equivalent of Zn? Will the deposition of Cu and Zn be the same amount for a $CuSO_4$ solution and a $ZnSO_4$ solution, each having the same concentration and the same amount of current flowing through each for the same duration?
2. What is the expression for the absolute value of Warburg impedance? Determine the value of A if Z_W is 100 kΩ at 100 kHz.
3. What is Faradaic impedance? What is non-Faradaic impedance? For impedance, $Z_{Non\text{-}Faradaic}$, A and B are unequal but the exponents β_1 and β_2 are equal. Classify the category of this impedance with explanation.
4. The equation $Z_{Colnstant-Phase-Element} = Z_{CPE} = A\left(j\,\omega\right)^{-\alpha}$ satisfies the Z*-plane feature. Write the expression for the Y*-, C*-, and M*-plane features. Here, both features apply to the relaxation possessing depression angle or depression parameter α.
5. Which complex plane emphasizes the constant phase element? Why is the constant phase element tied to the capacitance type element? Explain.

References

1. D. A. MacInnes; "*The Principles of Electrochemistry*," Dover Publications, New York, 1961.

2. T. Erdey-Gruz; "*Kinetics of Electrode Processes,*" John Wiley & Sons, Inc. New York, 1972.
3. K. J. Vetter; "*Electrochemical Kinetics: Theoretical and Experimental Aspects,*" English Translation, Academic Press, N.Y., 1967.
4. G. Kortum; "*Treatise on Electrochemistry,*" 2nd English Edition, Elsevier Publishing Co., Amsterdam, Holland, 1965.
5. E. Gileadi, E. Kirowa-Eisner and J. Penciner; "*Interfacial Electrochemistry: An Experimental Approach,*" Addison-Wesley Publishing Co., Reading, Massachusetts, 1975.
6. D. T. Sawyer and J. L. Roberts; "*Experimental Electrochemistry for Chemists,*" John Wiley & Sons, Inc., New York, 1974.
7. J. E. B. Randles; "Kinetics of Rapid Electrode Reactions," *Faraday Transactions Society of Discussion* (*London*), 1, 11–19, 1947.
8. V. F. Lvovich; *Impedance Spectroscopy: An Application to Electrochemical and Dielectric Phenomena*, John Wiley & Sons, Inc., Hoboken, New Jersey, 2012.
9. E. Barsoukov and J. R. MacDonald; *Impedance Spectroscopy: Theory, Experiment, and Applications,* Edited, John Wiley & Sons, Inc., Hoboken, New Jersey, 2005.
10. J. R. MacDonald; *Impedance Spectroscopy: Emphasizing Solid materials and Systems*, Edited, John Wiley & Sons, Inc., Hoboken, New Jersey, 1987.
11. M. E. Orazem and B. Tribollet; *Electrochemical Impedance Spectroscopy*, John Wiley & Sons, Inc., Hoboken, New Jersey, 2008.
12. C. Gabrielle; "Identification of Electrochemical Processes by Frequency Response Analysis," Solartron Analytical Report Number 004/83, Issue 3, 1998.
13. J. J. Ackmann and M. A. Seitz; 'Methods of Complex Impedance Measurements in Biologic Tissue," *CRC Critical Reviews in Biomedical Engineering,* 11(4), 281–311, 1984.
14. M. A. Seitz; "Study of Heterogeneous Composite Materials via Lumped Parameter/Complex Plane Analysis," *International Journal of Hybrid Microelectronics*, (I.S.H.M.), 3(1), 1–7, 1980.
15. H. Fricke; "The Theory of Electrolytic Polarization," *Philosophical Magazine*, Series 7, Volume 14, Issue 90, 310–318, 1932.
16. G. J. Brug, A. L. G. van den Eeden, M. Sluters-Rehbach, and J. H. Sluyters; "The Analysis of Electrode Impedances Complicated by the Presence of a Constant Phase Element," *Journal of Electroanalytical Chemistry,* 176, 275–295, 1984.
17. K. S. Cole and R. H. Cole; "Dispersion and Absorption in Dielectrics I. Alternating Current Characteristics," *Journal of Chemical Physics*, 9, 341–351, 1941.
18. L. Yang; "Electrical Impedance Spectroscopy for Detection of Bacterial Cells in Suspensions Using Interdigitated Microelectrodes," *Talanta*, 74, 1621–1629, 2008.

19. L. Yang, C. Ruan, and Y. Li; "Detection of Viable Salmonella Typhimurium by Impedance Measurement of Electrode Capacitance and Medium Resistance," *Biosensors and Bioelectronics*, 19, 495–502, 2003.
20. L. Yang and R. Bashir; "Electrical/Electrochemical Impedance for Rapid Detection of Foodborne Pathogenic Bacteria," *Biotechnology Advances*, 26,135–150, 2008.
21. L. Yang, Y. Li, C. L. Griffis, and M. G. Johnson; "Interdigitated Microelectrode (IME) Impedance Sensor for the Detection of Viable Salmonella Typhimurium," *Biosensors and Bioelectronics*, 19, 1139–1147, 2004.
22. L. Yang, Y. Li, and G. F. Erf; "Interdigitated Array Microelectrode-based Electrochemical Impedance Immunosensor for Detection of *Escherichia Coli* O157:H7," *Analytical Chemistry*, 76, 1107–1113, 2004.
23. R. Gomez-Sjöberg and R. Bashir; "Impedance Microbiology-on-a-Chip: Microfluidic Bioprocessor for Rapid Detection of Bacterial Metabolism,: *Journal of Microelectromechanical Systems,* 14(4), 829–838, 2005.
24. R. Gomez, R. Bashir, and A. K. Bhunia; "Microscale Electronic Detection of Bacterial Metabolism," *Sensors and Actuators B*, 86, 198–208, 2002.
25. R. Gomez, R. Bashir, A. Sarikaya, M. R. Ladisch, J. Sturgis, J. P. Robinson, T. Geng, A. K. Bhunia, H. L. Apple, and S. Werely; "Microfluidic Biochip for Impedance Spectroscopy of Biological Species," *Biomedical Microdevices,* 3(3), 201–209, 2001.
26. Y.-S. Liu, P. P. Banada, S. Bhattacharya, A. K. Bhunia, and R. Bashir; "Electrical Characterization of DNS Molecules in Solution Using Impedance Measurements," *Applied Physics Letters,* 92, 143902, 2008.
27. Samples of *E. Coli* bacteria-suspended brain heart infusion (BHI) were prepared at the Department of Food and Animal Science at Alabama A & M University, and investigated at the Electrical Characterization Laboratory in the Department of Electrical Engineering and Computer Science.
28. H. M. Al-Qadiri, N. I. Al-Alami, M. Lin, M. Al-Holy, A. G. Cavinato, and B. A. Rasco; "Studying of the Bacterial Growth Phases using Fourier Transform Infrared Spectroscopy and Multivariate Analysis," *Journal Rapid Methods Automation Microbiology*, 6, 73–89, 2008.
29. M. A. Alim; "Electrical Characterization of Engineering Materials," *Active and Passive Electronic Components*, 19(3), 139–169, 1996.
30. M. A. Alim; "Immittance Data-Handling/Analyzing Criteria for Heterogeneous Systems," in: Materials Research Society Symposium Proceedings: *Electrically Based Microstructural Characterization*, R. A. Gerhardt, S. R. Taylor, E. J. Garboczi edited, 411, 13–23, 1996.
31. M. A. Alim; "Phenomena Related to Concurrent Multiple Complex Plane Representations of the Immittance Data," Materials Research Society Symposium Proceedings: *Electrically-Based Microstructural Characterization*, R. A. Gerhardt, S. R. Taylor, E. J. Garboczi edited, 411, 113–22, 1996.

32. M. A. Alim, A. K. Batra, M. D. Aggarwal, J. R. Currie; "Immittance Response of the SnO_2–Bi_2O_3 Based Thick-Films," *Physica* B, 406 1445–1452, 2011.
33. M. A. Alim, A. K. Batra, S. Bhattacharjee, M. D. Aggarwal; "Complex Capacitance in the Representation of Modulus of the Lithium Niobate Crystals," *Physica* B, 406 1088–1095, 2011.
34. A. K. Batra, J. R. Currie, M. A. Alim, and M. D. Aggarwal; "Impedance Response of Polycrystalline Tungsten Oxide," *Journal of Physics and Chemistry of Solids*, 70, 1142–1145, 2009.
35. M. A. Alim, S. R. Bissell, A. A. Mobasher; "Analysis of the AC Electrical Data in the Davidson–Cole Dielectric Representation," *Physica* B, 403, 3040–3053, 2008.
36. A-. M. Azad, L. L. W. Shyan, and M. A. Alim; "Immittance Response of $CaSnO_3$ Prepared by Self-Heat-Sustained Reaction," *Journal of Materials Science*, 34, 1175–1187, 1999.
37. A-. M. Azad, L. L. W. Shyan, and M. A. Alim; "The AC Electrical Characterization of the Solid-State Reaction Derived $CaSnO_3$," *Journal of Materials Science*, 34, 3375–3396, 1999.
38. C. C. Wang, V. D. Patton, S. A. Akbar, and M. A. Alim; "Effect of Zirconia Doping on the Electrical Behavior of Yttria," *Journal of Materials Research*, 11(2), 422–429, 1996.
39. C. C. Wang, W. H. Chen, S. A. Akbar, and M. A. Alim; "High-Temperature A.C. Electrical Behaviour of Polycrystalline Calcium Zirconate," *Journal of Materials Science*, 32, 2305–2312, 1997.
40. M. A. Alim, M. A. Seitz, and R. W. Hirthe; Complex Plane Analysis of Trapping Phenomena in Zinc Oxide Varistors," *Journal of Applied Physics*, 63(7), 2337–2345, 1988.
41. I. M. Hodge, M. D. Ingram, and A. R. West; "Impedance and Modulus Spectroscopy of Polycrystalline solid Electrolytes," *Journal of Electroanalytical Chemistry*, 74(2), 125–143, 1976.
42. I. M. Hodge, M. D. Ingram, and A. R. West; "New Method for Analyzing the ac Behavior of Polycrystalline Solid Electrolytes," *Journal of Electroanalytical Chemistry and Interfacial Electrochemistry*, 58(2), 429–432, 1975.
43. F. A. Grant; "Use of Complex Conductivity in the Representation of Dielectric Phenomena," *Journal of Applied Physics*, 29(1), 76–80, 1958.
44. M. A. Seitz; "Study of Heterogeneous Composite Materials via Lumped Parameter/Complex Plane Analysis," *International Journal of Hybrid Microelectronics* (I.S.H.M.), 3, 1–7, 1980.
45. M. A. Alim; "High-Frequency Terminal Resonance in ZnO-Bi_2O_3-Based Varistors," *Journal of Applied Physics*, 74(9), 5850–5853, 1993.
46. R. Coelho; *Physics of Dielectrics for the Engineer*, Elsevier Scientific Publishing Company, Amsterdam, 1979.

9

Case Studies

9.1 Analysis of the Measured Data: Aspects of Data-Handling/ Analyzing Criteria

In order to fabricate solid state devices it is necessary to get familiar with the solid materials such as metals, semiconductors, and insulators. These materials are classified in terms of the concept of band gap or work function used in contact purposes to assure ohmic response. Upon conducting measurements interpretation of the identified mechanisms operative within the DUT can be done from the analytical results. These results can seriously be influenced by the predetermined nomenclature of the electrical parameters selected with or without incorporating physical geometrical factors [1]. This step may cause confusion in portraying the electrical parameters associated with the mechanism. Therefore, it is emphasized as how the nomenclature containing the contribution of the microstructural features excluding inductive effect [1–4] originating from the DUT, electrode material, DUT-electrode interface, contact leads, etc. can mislead the physical occurrence or location or region of the phenomena due to the physical geometrical factors. The frequently used nomenclatures of the derived circuit elements in series or parallel equivalent are as follows: (1) resistivity/conductivity or resistance/conductance, and (2) permittivity or dielectric constant or capacitance. These include some case examples and their implications on the MPCHPH systems considered below to illustrate the underlying drawbacks associated with the nomenclatures of the electrical parameters with or without incorporating physical geometrical factors. Also, the extraction of the equivalent circuit elements from the displayed data in the complex planes is a serious issue regarding how to determine each of them accurately or precisely following legitimate rules. A lack of proper understanding about the extraction procedures of the equivalent circuit elements via complex plane analysis leads to confusion within the investigators. To resolve all issues, a few cases are noted.

9.2 Case 1: Proper Physical Geometrical Factors

The lumped bulk phenomenon of the MPCHPH system must display frequency-independent responses. Usually bulk phenomenon corresponds to the SCSJ-like dielectric behavior at high frequencies. The bulk resistance results from a homogeneous-like conduction process throughout the series-parallel microstructural network of conducting paths. It is reasonable to assume a uniform distribution of electrical field stress as all junctions implying grain-boundaries or intergranular regions are shorted out at high frequencies so that lumped bulk response occur. This means that the electrical barriers via the presence of interfacial polarization or space charge layer or barrier layer are no longer operative at high frequencies and, thus, leaving only bulk phenomenon. Under this circumstance, the bulk resistance can easily be transformed to bulk resistivity using physical geometrical factors [1].

The main criterion for this transformation or conversion is that the electrical conduction across the junctions in the microstructural network behaves like a single-crystal. A similar situation is experienced at extremely high-voltages or high-currents where the electrical barriers across the grain-boundary junctions are completely in the breakdown process yielding only single-crystal like lumped grain behavior. This is once again analogous to the single-crystal like dielectric response. These two experimental situations at high-frequencies and high voltages or high currents neglect any influence of the disorder in the grain orientation as it may have no impact whatsoever on the total electrical behavior. In the case of a sensitive DUT this issue should be considered. Thus, bulk conductance or bulk conductivity would bear a meaning identical to the bulk resistance or bulk resistivity, respectively. Any of these four nomenclatures indicates the same mechanism and must yield the average grain carrier density, if doped or added deliberately with cations or additives or imperfections. The average grain carrier density assumes even distribution of carriers in each grain. In the same way, utilizing foregoing arguments bulk phenomenon resolves the other four nomenclatures derived from the capacitance when series and parallel configuration are considered with or without incorporating physical geometrical factors.

Zinc oxide based varistor material is a good example of the MPCHPH system [1]. Discussion on ZnO varistor is useful in understanding other complicated polycrystalline materials. Schematic illustrations displaying immittance spectra of the ZnO-Bi_2O_3-based varistor system [2] are used in describing the advantage and/or purpose of the multiple complex plane representations. Thus, a systematic approach reveals development of an equivalent circuit for the ZnO-based varistor. Each operative phenomenon is isolated from the total behavior, and the lumped response is further delineated via several simultaneously operative mechanisms. These are identified as trapping, de-trapping, barrier layer effect, lumped grain behavior, inductive response, etc. If popular template Z*-plane is used in the first place before exploiting C*- and Y*-planes, the scenario with or without the resonating behavior at high frequencies, it is easy to provide a confusing interpretation. This is because Z*-plane exhibits a severe lumped behavior than the delineated features of C*- and Y*-planes [2–4]. The advantage of presenting three complex planes in a simultaneous fashion is that they display the same dielectric behavior confirming resonating parameters and the lumped single-crystal like device-related parameters [2–5].Bode plane analysis does not reveal such depth of information as the curve represents heavy lumped behavior.

The bulk phenomenon is extracted in the high-frequency range that provided tiny intercept on the left side of the semicircular relaxation in the Z*-plane. This left-intercept is identified as a bulk resistance R_{ZnO} on the real-axis attributing to the lumped resistance of the ZnO grains between the electrodes. This situation has also been recognized elsewhere [6, 7]. However, Y*-plane [3] clarifies bulk phenomenon when the chord of the circle on real-axis obtained as conductance coincides with the resistance value extracted as R_{ZnO} in the Z*-plane. This is not only a confirmation of the value of R_{ZnO} but also inductance parameters are also derived from the resonance behavior [3]. The resonant frequency f_r is obtained from demarcation between the positive and the negative domain of the terminal capacitance [2, 3] which is used in determining inductance and corresponding capacitance responsible for the R-L-C resonance. The bulk capacitance obtained as C_{ZnO} represents the lumped ZnO grains [2]. The bulk parameters obtained from Z*-, C*-, and Y*-planes can be converted to meaningful parameters such as bulk resistivity (R_{ZnO}) and bulk permittivity (ε_{ZnO}) using physical geometrical factors implying electrode area and thickness of the DUT.

From the C*-plane it is evident that the semicircular relaxation in the Z*-plane cannot simply be designated as the grain-boundary phenomenon [2–4]. The Z*-plane relaxation yielded several other relaxations in the C*-plane. Although the left-intercept in the Z*-plane is identified as the bulk resistance, two consecutive semicircles were not obvious as the resonance phenomenon intercepted the bulk dielectric behavior at f_r. The bulk dielectric quantity denoted as C_{ZnO} is obtained as a parallel parameter which effectively is also in parallel with the bulk resistance, R_{ZnO}. The bulk phenomenon, thus, identified can be normalized to obtain the average carrier density in the grains for the ZnO-Bi_2O_3-based varistors. This is analogous to the SCSJ-like situation and, therefore, a carrier density can be obtained [2]. In fact the carrier density can be obtained via conventional Mott-Schottky plot using lumped barrier layer capacitance and lumped ZnO capacitance [2, 5]. Essentially these two geometrical capacitances yield device related parameters from the Mott-Schottky slope [2, 5]. This process is identical to what is observed in the single crystal based metal-semiconductor junction or Schottky diode [8–11].

9.3 Case 2: Improper Normalization

The identification of the grain-boundary resistance becomes obvious at a frequency range lower than that of the bulk resistance phenomenon. The grain-boundary resistance is evident in the Z*-plane for the ZnO-Bi_2O_3-based varistor [1] by the chord of the semicircular relaxation. This resistance can easily be transformed to the grain-boundary conductance. It cannot give meaningful value of resistivity or conductivity if converted using electrode area and thickness of the DUT. This is because each grain boundary region does not have direct physical connection with the physical geometrical factors [1]. Upon understanding that the chord of the Z*-plane is resistance [2], it is definitely not the situation to convert it to resistivity because it lacks direct connection or association with the physical electrode area and/or thickness of the DUT.

The grain boundary phenomenon is associated with the intergranular region where two consecutive grains are in intimate contact, possibly with the presence of occasional second phase at the interfaces. This intimate contact can cause a potential barrier or discontinuity in the conducting path due to the charge storage and/or trapping at the defect sites, primarily originating from the dangling bonds. Each potential barrier constitutes a certain electrical thickness, corresponding to a transition or depletion region for a Schottky-like or similar barrier. Since these regions in the microstructure are more resistive than the surrounding grains, a major portion of the applied electric field is dropped across these regions. Thus, the microstructural geometrical factors constitute contact area between the successive grains (grain-to-grain) and electrical thickness of the grain-boundaries. These factors should be used in converting the grain-boundary resistance for obtaining the grain-boundary resistivity. Applying the same argument, grain-boundary capacitance can be converted to obtain the grain-boundary permittivity. It is obvious that such a conversion is meaningful only when the grain boundary phenomenon is properly identified in the frequency-domain, and these two microstructural geometrical factors (grain-to-grain contact area and electrical thickness) have nothing to do with the physical geometrical factors of the DUT. In reality single grain junction for two successive grains must be investigated in order to determine contact area and relevant electrical thickness [1].This kind of investigation is definitely tedious. However, first single grain junction for two successive grains was investigated initially at several laboratories [12–14].

The grain-boundary relaxation via Z^*-plane yields resistance [2] from the chord resulting from the electrically non-shorting grain-boundary junctions. However, the depletion region thickness and the corresponding grain-to-grain contact area are not known. Therefore, the grain boundary resistance cannot be normalized to the grain boundary resistivity [2–4]. If each single grain-boundary junction is investigated systematically [12–14] and thoroughly only then individual thickness of the depletion region and corresponding grain-to-grain contact area will be known.

In the event when arbitrarily selected single-frequency (spot-frequency) is used in the measurement of the immitance components in investigation, then the nature of the dispersion is not identified. In that case the operative phenomena can only be assigned by powerful guessing. This is one of the reasons that the single-frequency measurement, preferably considered at so-called arbitrarily selected high frequency, for the dielectric evaluation incorporating physical geometrical factors can cause confusion. Extending this concept further, until the grain-boundary resistivity is ascertained, the term grain-boundary conductivity will not be appropriate to use. If a fraction of the contact area contributes to the total conduction process due to the non-planer contact configuration across the grain-boundary interfaces, this contact area should be better termed as the effective contact area for the grain-to-grain contact. In order to understand the role of either contact area or effective contact area, a grain-to-grain single-junction study is certainly helpful [12–14]. The effective contact area can be determined using classical Mott-Schottky equation [8–11]if the carrier concentration is spatially uniform in each grain, as demonstrated in the literature [2, 7].

9.4 Case 3: Effect of Electrode and Lead Wire

The electrode-related contribution to the immittance components can be observed at frequencies much lower than the frequency-range for the grain-boundary phenomenon. The contributions associated with each electrode terminal may not be identical, although their physical nature or origin is likely to be the same considering the degree of non-uniformity and relevant concentration of the dangling bonds at the DUT's surface. A mismatch in the work-function between the electrode material and the DUT is often responsible for the electrode-related contribution in the measured immittance. Another source of electrode-related contribution, commonly observed in ionic or electrolytic materials, is ascribed to the charge accumulation or blocking of the charge carriers. For reasons similar to that of the grain-boundary phenomenon the electrical parameters associated with the electrode, once identified cannot be normalized using physical geometrical factors. Although the physical electrode area may be applied, the barrier layer thickness (i.e., electrical thickness) should be determined and used to replace the physical thickness of the DUT.

Another perspective of the electrode effect is the evolution of the resonance phenomena in certain material systems [3]. In general, resonance effect is observed at very high frequencies which is not necessarily related to the piezoelectric resonance phenomena for the ZnO-based varistor [3] but may involve electrode, lead-wire, and possible bulk ZnO grains. Combined lumped electrode-lead-material effect may contribute to the resonance effect. When a material exhibits piezoelectric resonance, possible contribution from the electrode may not be separated. Therefore, electrode effect can be observed depending on the type of the material and the nature of investigation. It is, therefore, important to have sufficient knowledge about the type of materials investigated when a resonance phenomenon is observed. For many liquid systems resonance is observed via Z^*-plane instead of Y^*- or C^*-plane as found for the solid state materials [2, 3]. The associated circuit requires thorough understanding to absorb the underlying operative mechanisms in the DUT.

9.5 Case 4: Identification of Contributions to the Terminal Immittance

The trapping contribution to the terminal immittance is another important issue. This contribution causes dispersion of immittance with frequency. For ZnO-based varistor dispersion with frequency is observed in the C^*-plane [2]. For a lumped grain-boundary phenomenon the identity of trapping is completely masked when the same data are observed as a single semicircular relaxation in the Z^*-plane [2]. However, C^*-plane unravels the lumped Z^*-plane response and more than one trapping contribution becomes evident in the frequency domain. Invariably, the traps are physically responsible for constituting the electrical barriers at the grain-boundaries. They sustain these electrical barriers under the applications of the ac small-signal amplitude in conjunction with the superimposed dc electric field when applied externally. These traps may vary in their contribution to the series R-C elements [2]. Thus, the relaxation time (time constant) is sensitive to the frequency domain measurements, and may vary or remain the same. Often these traps give rise to a lossy system [2]. Depending on their nature of contribution to the total immittance components, the entire dispersion of immittance as a function of frequency may vary.

The origin of the traps may be so different that the traps can respond at several frequency-domain steps for a wide range of measurement frequency. This response can be masked by other simultaneously occurring phenomena (such as grain-boundary resistance and capacitance), especially in an MPCHPH system. Therefore, a full exploitation of the LP/CPA, BPA, and SA techniques (i.e., involving wide frequency-domain measurements and analyses) are essential to evaluate the trapping contributions [1, 2]. Although the traps are distributed all over the MPCHPH based DUT, their frequency-domain response may become distinct when they actively respond from the regions where the electrical field drops significantly. The relaxation time is not affected when the physical geometrical factors are incorporated to its constituting elements. This is because the associated factors cancel each other from the reciprocal relationship.

Thus, normalization of the constituting elements of the relaxation time for the traps given by:

$$\tau = \mathrm{RC} = \rho\varepsilon = \frac{\varepsilon}{\sigma} \tag{9.1}$$

is not essential. In other words, although trapping phenomenon is physically associated with certain physical or geometrical configuration, it does not require exact knowledge of the physical area and thickness of the DUT. However, for zero left-side intercept of the single semicircular relaxation in the Z^*-plane the normalization of R and C to obtain ρ and ε, respectively, will not affect the purpose. This type of normalization is not productive effort or does not add anything to the extracted information. Of course circuit elements obtained in the Z^*-plane do not correspond to or reflect the trapping behavior.

Each trapping relaxation for the ZnO-based varistor provides R-C series elements in the C^*-plane. None of these parameters can be normalized though their physical location is within the electrical field falling depletion region of the grain-boundary electrical barrier. The associated physical area and thickness including location within the electrical field falling region for each trap is completely unknown. Therefore, using physical electrode area and thickness of the DUT will never work logically, if normalized. The trapping response is attributed to the lossy nature of the ZnO-based varistors [1, 15–17]. Extending the concept further, more lossy varistor will provide better visibility, possibly without sub-Hz measurement domain, in shaping the low-frequency arc of the semicircle in the Z^*-plane. None of these traps can be ascribed to the physical geometrical factors. Since the relaxation time, τ, is a geometry-independent parameter, their constituting elements can be evaluated using non-equilibrium conditions such as temperature, voltage, composition, processing variables, ambient gases, etc. Exercising foregoing argument, there is no need of the state of normalization if τ and the constituting elements are systematically evaluated [15–17].

The demonstration of the multi-plane analytical approach via Z^*- and C^*-plane provided a complete picture of the trap-controlled grain-boundaries. However, associated analysis indicate that the state of normalization of the as-measured immittance data is a confusing approach in the first place when extraneous contributing elements related to the traps exist in addition to the regular geometric contributions. Thus, the state of normalization cannot be used as a universal practice for the trap-related parameters. The other non-trap parameters such as bulk capacitance and barrier layer capacitance may be normalized under certain situations when exact geometrical parameters associated with the grains and grain boundaries are known.

9.6 Case 5: Use of Proper Unit

The choice of plotting electrical quantities and their units is usually left to the investigator. A series of arbitrary units can be derived from a basic root unit of a parameter. In the LP/CPA technique, Z^*- or ρ^*-plane plot utilizes usual unit of resistance Ω or resistivity Ω-cm, respectively. Sometimes Ω-cm^2 is also used while discussing Z^*- or ρ^*-plane by many investigators. Existing books [18–21] have introduced Ω-cm^2 while demonstrating template impedance plot. What specialty caused the usage of the unit Ω-cm^2? What is the name of the parameter that has the unit Ω-cm^2? What is the physical significance or scientific basis of the unit Ω-cm^2? Is there a fifth complex plane that requires this unit?

The root or basic unit Ω for resistance or normalized unit Ω-cm for resistivity (specific resistance) using physical geometrical factors has a physical significance, and also both are based on fundamental physical parameters. Ascertaining the same fundamental cause, the physical significance of Ω-cm^2 is vague or unclear or does not have real cause or does not have scientific meaning. It is not a unit of a scientific origin parameter or a physical quantity. This type of usage of the unit has no connection with the existing concept of sheet resistivity for the SCSJ-like system. In this case either resistivity or conductivity is exploited through a dc conduction process describing exclusively the real or in-phase quantity. This real quantity is invariant whether series or parallel configuration is considered. Nevertheless, Ω-cm^2 appears to be an arbitrary meaningless unit used in many scientific and technical journals for the so-called satisfaction of the new complex plane parameters.

A derived aimless unit has no physical meaning or sense when it does not reflect the true nature of a physical quantity. Often the physical basis of such derived unit is not checked or adjudged. Depending on the material system, a derived unit will have an impact on the interpretation for the series (impedance form) or parallel (admittance form) configuration of the circuit in the data-analysis. The reactance of the impedance:

$$Z^* = Rs - j/(\omega Cs) \tag{9.2}$$

can be converted to the equivalence of the unit Ω-cm^2 incorporating the physical electrode area in the series capacitance C_S representing capacitance per unit area. The physical electrode area incorporated to the reactance of the impedance is not involved in the real world interpretation of the data. This is because the electrode area is essentially associated with the parallel capacitance as observed for the SCSJ system. Moreover, measured capacitance is usually in the parallel form as in the term admittance unless specifically mentioned for impedance measurement. This parallel capacitance is used in portraying dielectric constant or permittivity of a material system. Therefore, the logic behind propagating the unit Ω-cm^2 has no physical basis [1, 22].

After measuring the impedance components incorporating electrode area for the series capacitance in the reactance term is misnomer and confusing. By manipulating the reactance term through importing the unit Ω-cm^2 is out of significance. Since the unit of the axes is not impedance Z^* or resistivity ρ^* by definition, practicing such a plot is either not worthwhile or meaningful. To extend the argument further, Ω-cm^2 can be used as a unit of the so-called hypothetical parameter which is not similar to that of considering sheet resistance only under dc ($f \rightarrow 0$ Hz) condition. At this conjecture, the significance of the series capacitance C_S is totally diluted in comparison to that of the parallel C_P which is invariably used in determining the permittivity. Use of capacitance per unit area in the series capacitance is out of place and absolutely a misleading concept or term.

To validate a derived unit such as Ω-cm^2, it is necessary to understand the role of the series capacitance C_S dealt with the physical electrode area. The reactance component of the impedance $[1/(\omega\ C_S)]$ is in series with the series resistance R_s yielding a dc blocking configuration of the DUT as $f \rightarrow 0$ Hz. This situation can be explained with zero left-side intercept i.e., no left-intercept on the real-axis of the single semicircular Debye-like or non-Debye relaxation, which is represented by an equivalent circuit consisting of a parallel R-C combination obtained in the usual Z^*-plot. This R-C parallel circuit is open to conduction at $f \rightarrow 0$ Hz because of the operative resistance branch.

When capacitance per unit area is considered then this capacitance is not series capacitance. Essentially capacitance per unit area is meaningful for the parallel capacitance C_P which is used in calculating relative permittivity or dielectric constant. For example, the capacitance used in the Mott-Schottky plot [2, 5] is the parallel capacitance C_P, not series capacitance C_S, unless $C_P \approx C_S$ under certain conditions like extremely large resistance of the DUT. In the Mott-Schottky plot capacitance per unit area is used in the equation [8–11] that deals with it. Despite such an explanation of the terminal capacitance or circuit capacitance (series or parallel form), there is no meaning of the real component for resistance or imaginary component for reactance displaying the unit Ω-cm^2.

In reality, the unit Ω-cm^2 has no physical significance regardless of the number of semicircular relaxations observed in the Z*-plane. This is because the real component on the x-axis is not a representative of the sheet resistivity or sheet resistance. In semiconductor engineering few logical units are used in handling the Si wafers such as Ω/□, where "□" is referred to as the "square" which is a dimensionless quantity originated from the physical dimensions of the Si wafer. Therefore, in a true sense "□" is not a unit but a dimensionless quantity or a factor associated with the unit Ω.

In fact, the situation becomes slightly more complicated if multiple semicircular relaxations are observed. This type of behavior is very common in the polycrystalline materials [1, 2, 23]. Further complication is expected if each relaxation or any of the relaxations is non-Debye. Although an equivalent circuit can be constructed for the non-Debye relaxation, the elements therein will have no meaning with respect to the microstructural features when Ω-cm^2 is used. The same argument applies to the analysis using BPA or SA technique with this kind of confusing unit [18–21]. Thus, a derived unit obtained through improper usage of the geometry is not the generalized representation of the unknown DUT, as it cannot explain the purpose and meaning of each equivalent circuit element concerning the network involving microstructures or media of electrical paths.

The unit Ω-cm^2 can alternately be viewed as the reciprocal of the unit of the conductance per unit area. This means $(\Omega\text{-cm}^2)^{-1}$ is much better to use for conductance than Ω-cm^2 for resistance or resistivity. The parameter susceptance can have the same unit as the conductance per unit area. Thus, the admittance components have a better appearance using the reciprocal of this unit. The unit $(\Omega\text{-cm}^2)^{-1}$ can be used as S/cm^2 or mho/cm^2 which is a standard unit for conductance per unit area or susceptance per unit area.

It is useful to add that the complex electrical quantities Y* or C* per unit area is more meaningful than that of Z* or M* per unit area. This is because the parallel configuration parameters are directly related to the physical situations of the DUT. However, achieving R-C parallel circuit in a semicircular relaxation obtained from the Z*- or M*-plane plot may enlighten the argument. Again, this achievement does not justify the concept of capacitance per unit area on a series configuration for the as-measured data.

In many cases the ac small-signal electrical data can provide semicircular relaxation simultaneously in two or three complex planes. Of course, there will be only one valid or reasonable equivalent circuit to satisfy multiple complex planes at the same time. Several material systems are demonstrated with the ac small-signal electrical data in multiple complex planes [24–28]. The complex plane representations provide congruity within two or three complex planes. However, only one equivalent circuit may be emphasized for each material system corresponding to specific experimental condition for any of the complex plane plots. More than one equivalent circuit provides a better resolution when other dominating mechanisms cannot be delineated by one complex plane representation. Thus, multi-plane analytical approaches unravel simultaneously operative mechanisms and provide a reasonable concept on the derivation of the equivalent circuit modeling.

The multiple relaxation phenomena are not disturbed by the type of unit of the electrical parameter used during the data-acquisition process. Modification or derivation of a pseudo unit from a basic unit does not add anything to the underlying operative phenomena. Nevertheless, BPA and SA techniques are exploited to assure the same operative mechanisms along with the equivalent circuit model as obtained via the LP/CPA for these material systems. Since normalized unit inherently possesses limitations in interpreting simultaneously operative phenomena representing multiple semicircular relaxations for the current or voltage dividing electrical paths, Ω-cm or similar unit of other electrical quantities or parameters constituting immittance function should be used with ultra caution for the MPCHPH system unless the DUT is SCSJ type. In the event when the as-measured data are used freedom exists to choose whether multiple operative phenomena really require the state of normalization. Exploitation of the as-measured data is usually more meaningful than that of the normalized data in extracting and establishing universal characteristics or meaningful interpretation for a DUT. It is definitely worthy when investigation allows uniform geometry for the DUTs. Only then as-measured data can be used and compared internally, if needed.

9.7 Case 6: Demonstration of the Invalid Plot

A plot conveys messages to the scientists and engineers as does the traffic signal to the motor vehicle drivers. A demonstration of an invalid plot in the book [20] spreads an erroneous message to the investigators and learners. The incongruent explanation for the C*-plane representation of the relaxation with respect to the interpretation of Grant [29] and Coelho [30] is misleading for the interested audience. For the C*-plane representation of a perfect semicircle, with the presence of a left-intercept or not, the outcome represents a perfect dc blocking circuit.

Distortion from the semicircular loci (or skewness) at low frequencies for the same C*-plane indicates the influence of the dc conductivity (resistivity) of the DUT. This C*-plane feature showing a vertical line, parallel to the y-axis, originating at the point of distortion is never observed for any material system. In the real-world scenario, the nature of distortion takes a smooth transition from the particular point on the semicircular arc, near the low-frequency end, and climbs upward in such a way that the curve marches with the dc limit. This smooth transition never has similarity with the parallel nature or perpendicular-like feature in the C*-plane, where vertical line originated about one-third of the loci from the dc value of the real part of the capacitance.

Grant [29] has successfully demonstrated with the real data for two complex planes, C*- and Y*-, in a concurrent manner where the influence of the dc conductivity was highlighted in the C*-plane. Coelho [30] endorsed gradual increment of conductivity values of the DUT, and offered the smooth shape of the transition from the departing point on the semicircular arc toward the dc limit. Thus, multiple curves reflecting various levels of conductivity values incorporated to the C*-plane representation show the influence in conductivity toward the dc limit. The dc limit has a significance in the C*-plane that offers complete blocking when $f \rightarrow 0$ Hz. At this conjecture, it is worthy to note that the unit Ω-cm^2 presented some vague parameter to obtain the C*-plane plot for the book. Although the unit on the abscissa and ordinate for the so-called complex capacitance plane representation emerged as reasonable, the origin is tied up with an invalid scientific basis [20].

9.8 Case 7: Obscuring Frequency Dependence

In a recent book [20] the format of the three complex planes: Z*-, Y*-, and C*- are considered to obscure the frequency dependence. It is true that the frequency dependence is obscure in each complex plane and cannot be adjudged spontaneously by a glance of the semicircle. There is no way to comprehend each datum on the semicircular loci corresponding to each frequency. This is not the case for the Bode plot because the frequency is comprehendible from the frequency on the x-axis with respect to the y-axis parameter. The spectroscopic approach is not possible with the single measurement presumably at a fixed temperature. Upon evaluation, the Bode plot does not delineate the underlying operative mechanisms of the DUT.

By knowing the range of the measurement frequency and initial inspection at the Z*-plot, an experienced investigator can apprehend several aspects of the observed behavior obtained in the complex plane. This inspection does not mandate or require selected characteristic frequencies on the semicircular arc. Assessing semicircular behavior is totally based on the handling experience of the investigator. A sharp eye inspection can resolve the ideal or non-ideal relaxation. Frequent exercise with the DUTs and handling frequency ranges give a strong understanding on the nature of the semicircular relaxation. With technology advancing, computer-controlled data and subsequent plotting can bear identification of the data points for every frequency. Handling the same data in more than one complex plane gives a strong feeling for the complex plane. First-time or initial-stage investigators may face a great challenge on the format for the obscure frequency dependence of the DUT. That is why dealing with impedance spectroscopic analysis requires massive experience concerning handling the data and the range of frequency for the DUT. It is a sharp repeated practice that allows an investigator on the driver's seat at a shorter time. This kind of practice also allows an investigator to assess the quality-check of a product for qualifying the entire batch in a matter of a few minutes, by choosing a handful of random products from the product-line for a few selected spot-frequencies. In this case, no computer is necessary as the hand measurement shortens the time for quick checking. Thus, regular or frequent handling of a semicircular response for any DUT for a certain number of observations in a specific range of frequency allows an investigator to become a champion of the data users.

In 1941, Cole-Cole [31] introduced the complex permittivity (ε^*) plot which is the normalized parameter of the complex capacitance (C^*). In 1969 Bauerle [32] introduced admittance (Y^*) for which complex conductivity (σ^*) is the normalized parameter. Grant [29] demonstrated duality of the measured data in 1958 by extracting a single circuit satisfying both complex conductivity (σ^*) and complex permittivity (ε^*) plots in a simultaneous fashion that are basically admittance (Y^*) and complex capacitance (C^*) parameters, respectively. In order to maintain uniformity, each of these complex planes are normalized. This track record of three pioneers indicates that Z*-plot or resistivity (ρ^*) plot had no popular impact until during that time among the complex plane users. If a book [20] campaigns to use the Z*-plane plot and note so-called popularity justifying the shape of the locus of points yields insight into possible mechanism or governing phenomena, then it presents terrible damage to the learners and investigators. This is because the window of research for a DUT gets limited by this kind of promotion.

Does research or investigation go on with the popularity? Does research get privilege on the basis of popular demand of any of the complex planes? How does a researcher or an investigator know in advance which complex plane to get the opinion poll? It is true that only a limited number of investigators or users employ concurrent multiple complex plane analysis. This is because most users do not comprehend the purpose or impact of using multiple complex plane analysis in a concurrent manner to get maximum benefit from the knowledge of impedance spectroscopy.

After the demonstration of Grant [29], concurrent Z^*- and M^*-planes were used [33–36] during the mid-1970s and later again in the 1980s. In the early 1980s the same two complex planes were used for biological materials [37, 38]. By 1988, C^*- and Z^*-planes were simultaneously [2] employed for the same data; and again in 1993, C^*- and Y^*-planes were concurrently used with the same 1988 electrical data [3]. Thus, three concurrent complex planes (C^*-, Z^*-, and Y^*-) have been used for the same data of the ZnO varistor to obtain a complete meaningful set of information on the ZnO varistor. The M^*-plane plot was averted in 1988 because of the dc blocking feature, which is unrealistic for the ZnO varistor as it is applied in surge protection since 1972. However, in 1997, Z^*- and M^*-planes were simultaneously presented [42] for the same data. Nevertheless, for a single device all four complex planes have been explored to allow a much better understanding for the ZnO varistor. While a number of global investigators [2, 3, 29, 33–39] find a deep-root advantage in delineating underlying operative mechanisms by exploring all four complex planes in a concurrent manner, particular encouragement for the Z^*-plane format surfaces in a book [20] as the so-called popular complex plane for the potential users. This sort of state-of-thinking allows a narrow window for future learners and investigators. It may be noted that sometimes ignorance may lead to the so-called popularity of a tool or technique.

9.9 Case 8: Misnomer Nomenclature for the Complex Plane Plot

It is observed that the name of the complex plane plot is gradually shifting toward adopting a confused state of nomenclature. In reality, there is nothing called the Nyquist plot or Nyquist diagram that emphasizes any of the four complex plane plots. Macdonald repeatedly advocated [40, 41] in 1990 against the use of the name Nyquist plot referring to any of the four complex plane plots. By that time, this name propagated rapidly as the newest investigators somehow adopted it. It is very difficult to detect from where it got indication to advance among the users. Basically Macdonald [40, 41] had discouraged such usage as a legitimate explanation had been given about the origin of the complex quantity of Nyquist's work [42], which is totally different than the origin of impedance or of admittance, and thereby no similarity with the phasor forms of the data. When a book [20] on the subject refers to a misleading nomenclature for the complex plane plotting, then improper content propagates among the serious learners or within the dedicated investigators in the area. If an erroneous or incorrect term is introduced by mistake then somehow that should impede instead of finding avenue to accelerate.

Once, there was a reckless propagation on the usage of the modified Mott-Schottky equation[43] after it appeared satisfying a physical basis in 1979, as if polycrystalline semiconductors received a golden avenue for the solution of the multiple frequency-dependent Mott-Schottky straight lines. No one noticed that the underlying problem of choosing arbitrary measurement frequency never got away from the core problem. The solution was given in 1988, but it seemed difficult to square away the frequency dependence of the terminal capacitance as the resonance phenomenon in ZnO varistor was not very obvious for each device. An explanation of the resonance phenomenon was forwarded [3] in 1993, and in 1995 a discussion on the Mott-Schottky analysis [5] provided additional information. Combining stepwise information between 1988 and 1995, a complete answer for the underlying frequency-dependent problem of the polycrystalline system was erected that has physical basis. In this way, the academic corner can take the lead to provide positive guidance to the learners and the new investigators in this area.

9.10 Case 9: Extraction of Equivalent Circuit from the Straight Line or the Non-Relaxation Curve

Another analytical perspective of the complex plane analysis is that when a relaxation process is not straightforward or evident in the complex plane. This means the straight line feature may not be ignored in the complex plane as the same data may appear as semicircle in another complex plane. Some form of straight-line or segmented curvature response of the data in the complex plane is not adequate to receive backup from a chosen or selected Bode plot, and then develop the equivalent circuit model by extrapolating the straight line or the segmented curvature on the real-axis. In this case, exploring other complex planes can lead to the viable meaning having semicircular relaxation.

In this straight line or segmented curvature situation, extraction of an equivalent circuit becomes a critical issue only on the basis of the conventional Bode plane analysis. If all operative phenomena are not fully explored, then the meaning of such an extracted circuit may be diluted. The possibility of masking behavior of a mechanism by other dominant mechanisms extraneous to the system or not in a single complex plane analysis cannot be ruled out. Measurements under non-equilibrium conditions can delineate the masking behavior or distortion of the relaxation process via prominence of the emerging semicircular behavior. Systematic evaluation will aid in the interpretation of the data. Furthermore, a spectroscopic approach can delineate the underlying meaning of the unresolved straight line response observed in the complex plane.

On many occasions the immittance data may yield a straight line in the form of a vertical or slant line on the real-axis in a complex plane. As an example, if a semicircular relaxation is observed in the C^*-plane, the high-frequency intercept can be obtained through the semicircular curve-fitting using CNLS procedures. The fitting procedures include extrapolation on the left and right side of the semicircular relaxation if experimental points are either missing or not possible to acquire. This happens because no relaxation is perfect or ideal for exhibiting a sharp boundary on the high-frequency limit for a single relaxation or a demarcation between the relaxations for multiple relaxations in the C^*-plane. Assuming that the masking behavior in the frequency domain is explored, this left-side intercept can be used in representing the bulk phenomenon implying bulk dielectric response.

In the situation where only a straight line, vertical or slant, is evident in the complex plane in lieu of a semicircular relaxation, no extrapolation either at the low frequency end or at the high frequency end may be valid or justified to obtain the so-called bulk dielectric response or any other potential response without establishing a sound scientific basis for doing so. In fact, attempting approximate extrapolation from the straight-line feature is absolutely unethical or unreasonable to obtain some parameter from the real-axis attributing to some possible mechanism [1, 22] and then using it as a valid parameter for the equivalent circuit. Under certain circumstances this procedure may follow an occasional coincidental result with the expected bulk phenomenon, in particular, when a masking behavior is not resolved. This coincidental result does not bear scientific validity. A systematic evaluation of the DUT as a function of non-equilibrium experimental condition such as voltage, temperature, composition, fabrication or processing methods, etc. may be required to validate the straight-line response in a complex plane, especially when it is used in extracting the bulk phenomenon. It is worthwhile noting that a straight line, vertical or slant, in a complex plane is likely to exhibit semicircular behavior in another complex plane which is necessary to examine or exploit before conducting unethical efforts to corrupt the scientific basis [1, 22].

In order to delineate complexity due to the extrapolation of the straight line at the high-frequency end having frequency direction anti-clockwise to achieve an intercept for the left side on the real-axis of the C^*-plane, the immittance data must be displayed in another complex plane such as Z^*- or Y^*- or M^*-plane. In general, a straight line, perpendicular to the x-axis or not, in one complex plane yields one or more semicircular arc in another complex plane. Upon proper identification of each of these relaxations the bulk dielectric response can easily be identified utilizing the high-frequency end. Multiple relaxations in one complex plane i.e., Z^*- or Y^*- or M^*-plane may yield a severe lumped behavior in the C^*-plane in the form of a straight line in conjunction with the features of the Bode planes. Therefore, the use of the straight line in extracting a parameter, bulk or not, is somewhat misleading when a specific phenomenon is intentionally emphasized. Straightforward extrapolation, using the high frequency end of the straight line in the C^*-plane, is not equivalent to the process of obtaining a parameter such as bulk dielectric behavior in a curve-fitting procedure for the semicircle without additional exploitation of the immittance data assuming availability of wide range of frequency measurements.

Some form of straight-line or linear response of the data in the complex plane is not adequate to receive backup from a chosen or selected Bode plot, and then develop the equivalent circuit model. In this case, exploring other complex planes can lead to the viable meaning having semicircular relaxation. The meaning of the extracted equivalent circuit elements becomes important for the delineated relaxation obtained in each of the complex plane formalisms. Arbitrary usage of the straight line or curvature is not the valid approach to establish the scientific findings.

Problems

1. What should be done with the measured ac small-signal electrical data in the frequency range 10^{-2} Hz $\leq f \leq 10^9$ Hz if no semicircular shape or arc is observed in any of the four complex planes? Assume that in each complex plane, a straight line or near straight line curve is obtained.
2. A material system exhibited no semicircular arc in any of the four complex planes for the ac small-signal electrical data obtained in the frequency range 10^{-2} Hz $\leq f \leq 10^9$ Hz. Assume that in each complex plane, a straight line or near straight line curve is obtained. What type of material is this? What should be the action items for investigation of this type of material? What are the dominant factors causing this material to behave in this way? Classify destructive and non-destructive approaches for investigation of this material.
3. An end-product procured from the market exhibited no semicircular arc in any of the four complex planes for the ac small-signal electrical data obtained in the frequency range 10^{-2} Hz $\leq f \leq 10^9$ Hz. Assume that in each complex plane, a straight line or near straight line curve is obtained. Since reverse engineering is necessary to understand as how this end-product was manufactured at a mass-scale, describe the sequence of destructive and non-destructive experiments on the basis of the taste or view point of each professional investigator assuming the background in: (1) material science, (2) electrical engineering, (3) mechanical engineering, (4) chemical engineering, (5) civil engineering, (6) metallurgy, (7) physics, (8) chemistry, and (9) thirty years experience as reverse engineer possessing no academic degree.
4. One n-type semiconductor (polycrystal or single-crystal) surface requires insulating passivating material to prevent surface conduction at high voltages. The investigator decided to use ion bombardment to change the morphology of the surface of the semiconductor so that the surface becomes more resistive to achieve the purpose. Assume that the resistance of the surface of the semiconductor is increased at least one order or a minimum of ten times. The ac small-signal electrical data are obtained for this ion-bombarded

semiconductor in the frequency range 10^{-2} Hz ≤ f ≤ 10^{9} Hz. What kind of equivalent circuit may be expected when measured between the passivating surface and the other side of the semiconductor? What kind of equivalent circuit may be expected for the passivating surface? Consider the n-type semiconductor had the R-C parallel equivalent circuit before the ion bombardment.

5. One n-type semiconductor (polycrystal or single-crystal) surface requires insulating passivating material to prevent surface conduction at high voltages. The investigator decided to use ion bombardment to change the morphology of the surface of the semiconductor so that the surface becomes more resistive to achieve the purpose. The semiconductor is annealed at about 400 °C and cooled down to room temperature. Assume that the resistance of the surface of the semiconductor is increased at least one order or a minimum of ten times. The ac small-signal electrical data are obtained for this ion-bombarded semiconductor in the frequency range 10^{-2} Hz ≤ f ≤ 10^{9} Hz. What kind of equivalent circuit may be expected when measured between the passivating surface and the other side of the semiconductor? What kind of equivalent circuit may be expected for the passivating surface? Discuss the effect of annealing of the semiconductor on each equivalent circuit. Consider the n-type semiconductor had the R-C parallel equivalent circuit before the ion bombardment.

References

1. M. A. Alim; "Electrical Characterization of Engineering Materials," *Active and Passive Electronic Components*, 19, 139–169, 1996.
2. M. A. Alim, M. A. Seitz and R. W. Hirthe; "Complex Plane Analysis of Trapping Phenomena in Zinc Oxide Varistors," *Journal of Applied Physics*, 63(7), 2337–2345, 1988.
3. M. A. Alim; "High-Frequency Terminal Resonance in ZnO-Bi_2O_3-Based Varistors," *Journal of Applied Physics*, 74(9), 5850–5853, 1993.
4. M. A. Alim; "Admittance-Frequency Response in Zinc Oxide Varistor Ceramics," *Journal of the American Ceramic Society*, 72(1), 28–32, 1989.
5. M. A. Alim; "An Analysis of the Mott-Schottky Behavior in ZnO-Bi_2O_3 Based Varistors," *Journal of Applied Physics*, 78(7), 4776–4779, 1995.

6. K. A. Abdullah, A. Bui, and A. Loubiere; "Low Frequency and Low Temperature Behavior of ZnO-Based Varistor by ac Impedance Measurements," *Journal of Applied Physics*, 69, 4046–4052, 1991.
7. A. Bui, K. A. Abdallah, A. Loubiere, M. Tao, and Q. C. Nguyen; "Impulse-Degradation Analysis of ZnO-Based Varistors by ac Impedance Measurements," *Journal of Physics D: Applied Physics*, 24, 757–762, 1991.
8. E. H. Rhoderick; "*Metal-Semiconductor Contacts*," Clarendon Press, Oxford, U.K., 1978.
9. R. S. Muller and T. I. Kamins; "*Device Electronics for Integrated Circuits*," J. Wiley & Sons, Inc., N.Y., 1977.
10. S. M. Sze; "*Physics of Semiconductor Devices*," 2nd Edition, Wiley Interscience Publication, N.Y., 1981.
11. E. S. Yang; "*Fundamentals of Semiconductor Devices*," McGraw-Hill Book Co., N.Y., 1978.
12. G. D. Mahan, L. M. Levinson, and H. R. Philipp; "Single Grain Junction Studies of ZnO Varistors – Theory and Experiment," *Applied Physics Letters*, 33(9), 830–832, 1978.
13. M. A. Seitz, A. K. Verma, and R. W. Hirthe; "AC Electrical Behavior of Individual MOV Grain Boundaries," *Ceramic Transactions: Advances in Varistor technology*, 3, 135–144, 1989, L. M. Levinson edited, Published by The American Ceramic Society, Westerville, Ohio.
14. H. Wang, W. Li, and J. F. Cordaro; "Single Junction in ZnO Varistors Studied by Current-Voltage Characteristics and Deep Level Transient Spectroscopy," *Japanese Journal of Applied Physics*, 34 Part 1#4A, 1765–1771, 1995.
15. M. A. Alim; "Influence of Intrinsic Trapping on the Performance Characteristics of ZnO-Bi_2O_3 Based Varistors," *Active and Passive Electronic Components*, 17, 99–118, 1994.
16. M. A. Alim; "Influence of Multiple Trapping Phenomena on the Applications Criteria of ZnO-Bi_2O_3 Based Varistors," *Active and Passive Electronic Components*, 17, 57–65, 1994.
17. L. C. Sletson, M. E. Potter and M. A. Alim; "Influence of Sintering Temperature on Intrinsic Trapping in Zinc Oxide-Based Varistors," *Journal of the American Ceramic Society*, 71, 909–913, 1988.
18. E. Barsoukov and J. R. MacDonald; *Impedance Spectroscopy: Theory, Experiment, and Applications,* Edited, John Wiley & Sons, Inc., Hoboken, New Jersey, 2005.
19. J. R. MacDonald; *Impedance Spectroscopy: Emphasizing Solid materials and Systems*, Edited, John Wiley & Sons, Inc., Hoboken, New Jersey, 1987.
20. M. E. Orazem and B. Tribollet; *Electrochemical Impedance Spectroscopy*, John Wiley & Sons, Inc., Hoboken, New Jersey, 2008.
21. V. F. Lvovich; *Impedance Spectroscopy: An Application to Electrochemical and Dielectric Phenomena*, John Wiley & Sons, Inc., Hoboken, New Jersey, 2012.

22. M. A. Alim; "Immittance Data-Handling/Analyzing Criteria for Heterogeneous Systems," Materials Research Society Symposium Proceedings: *Electrically Based Microstructural Characterization*, R. A. Gerhardt, S. R. Taylor, and E. J. Garboczi edited, 411, 13–23, 1996.
23. D. C. Sinclair and A. R. West; "Impedance and Modulus Spectroscopy of Semiconducting $BaTiO_3$ Showing Positive Temperature Coefficient of Resistance," *Journal of Applied Physics*, 66, 3850–3856, 1989.
24. C. C. Wang, V. D. Patton, S. A. Akbar, and M. A. Alim; "Effect of Zirconia Doping on the Electrical Behavior of Yttria," *Journal of Materials Research*, 11, 422–429, 1996.
25. V. D. Patton, C. C. Wang, S. A. Akbar, and M. A. Alim; "The ac Electrical Behavior of Polycrystalline Yttria," *Journal of Applied Physics*, 78, 1757–1762, 1995.
26. C. C. Wang, W. H. Chen, S. A. Akbar, and M. A. Alim; "High-Temperature a.c. Electrical Behaviour of Polycrystalline Calcium Zirconate," *Journal of Materials Science*, 32, 2305–2312, 1997.
27. A-. M. Azad, L. L. W. Shyan, and M. A. Alim; "Immittance Response of $CaSnO_3$ Prepared by Self-Heat-Sustained Reaction," *Journal of Materials Science*, 14, 1175–1187, 1999.
28. A-. M. Azad, L. L. W. Shyan, and M. A. Alim; "The AC Electrical Characterization of the Solid-State Reaction Derived $CaSnO_3$," *Journal of Materials Science*, 14, 3375–3396, 1999.
29. F. A. Grant; "Use of Complex Conductivity in the Representation of Dielectric Phenomena," *Journal of Applied Physics*, 29(1), 76–80, 1958.
30. R. Coelho; *Physics of Dielectrics for the Engineer*, Elsevier Scientific Publishing Company, Amsterdam, 1979.
31. K. S. Cole and R. H. Cole; "Dispersion and Absorption in Dielectrics I. Alternating Current Characteristics," *Journal of Chemical Physics*, 9, 341–351, 1941.
32. J. E. Bauerle; "Study of Solid Electrolyte Polarization by a Complex Admittance Method," *Journal of Physics and Chemistry of Solids*, 30, 2657–2670, 1969.
33. I. M. Hodge, M. D. Ingram, and A. R. West; "A New Method for Analyzing the A. C. Behavior of Polycrystalline Solid Electrolytes," *J. Electroanal. Chem.*, 58, 429–432, 1975.
34. I. M. Hodge, M. D. Ingram, and A. R. West; "Impedance and Modulus Spectroscopy of Polycrystalline Solid Electrolytes," *J. Electroanal. Chem.*, 74, 125–143, 1976.
35. D. P. Almond and A. R. West; "Impedance and Modulus Spectroscopy of 'Real' Dispersive Conductors," Solid State Ionics, 11(1), 57–64, 1983.
36. D. C. Sinclair and A. R. West; "Impedance and Modulus Spectroscopy of Semiconducting $BaTiO_3$ Showing Positive Temperature Coefficient of Resistance," *Journal of Applied Physics*, 66, 3850–3856, 1989.

37. M. A. Seitz; "Study of Heterogeneous Composite Materials via Lumped Parameter/Complex Plane Analysis," *Inter. J. of Hybr. Microelec.* (I.S.H.M.), 3(1), 1–7, 1980.
38. J. J. Ackmann and M. A. Seitz; 'Methods of Complex Impedance Measurements in Biologic Tissue," *CRC Critical Reviews in Biomedical Engineering,* 11(4), 281–311, 1984.
39. M. Andres-Verges and A. R. West; "Impedance and Modulus Spectroscopy of ZnO Varistors," *Journal of Electroceramics,* 1–2, 125–132, 1997.
40. J. R. MacDonald; "Impedance Spectroscopy," *Annals of Biomedical Engineering,* 20, 289–305, 1992.
41. J. R. Macdonald; "Impedance Spectroscopy: Old Problems and New Developments," *Elecrrochimica Acta*, 15(10), 1483–1492, 1990.
42. H. Nyquist, "Regeneration Theory," *Bell System Technical Journal,* 11, 126–147, 1932.
43. K. Mukae, K. Tsuda, and I. Nagasawa; "Capacitance-vs-Voltage Characteristics of ZnO Varistors," *J. Appl. Phys.*, 50(6), 4475–4476, 1979.

10

Analysis of the Complicated Mott-Schottky Behavior

10.1 Capacitance – Voltage (C-V) Measurement

The single-frequency capacitance-voltage (C-V) measurement is widely used in determining device related parameters for various junction devices [1–4]. Usually an arbitrary high-frequency is chosen for the measurement knowing confidently that the DUT is based on near defect-free single-crystal systems. This arbitrary high-frequency, such as 10^6 Hz, is reasonable where most single-crystal based systems are near trap-free. The same measurement frequency may not be an option for the complicated structured material systems, such as multi-component polycrystalline systems where various types of defects and traps are found. If the system comprises a multi-phase structure then the complexity of the measurement becomes severe resulting in frequency-dependent behavior. Invariably arbitrary measurement frequency may not be the choice in any case [5] without proper justification.

The C-V measurement is useful in a wide range of applications for new materials, processes, devices, and circuits. This measurement is important for improving processes and device performances. Reliability engineers use this measurement to qualify material suppliers, monitor process parameters, and analyze failure mechanisms. With appropriate methodologies, instrumentation, and software a multitude of semiconductor devices and material parameters can be derived. This information is used in conjunction with the evaluation of the layered epitaxial growth including parameters such as average doping concentration, doping profiles, trapping profiles, and carrier lifetime. In wafer processes, C-V measurement can reveal oxide thickness, oxide charges, mobile ions (i.e., contamination), and interface trap density, etc. These measurements may use lithography, etching, cleaning, dielectric and polysilicon depositions for MOSFETs, and electroding (metallization purposes). Sensitive issues are dealt with quality check of various semiconductor products [6].

After devices are fabricated on the wafer, C-V measurement is used in characterizing threshold voltages and other parameters during reliability and basic device testing besides modeling the performance. A junction device may be used in the circuits wherein this simple structure is easy to control in the fabrication steps. In a layered device the capacitance C is determined by [7]:

$$C = \frac{\varepsilon_0 \varepsilon_r A}{d}, \tag{10.1}$$

where

A	=	electrode area of the capacitor,
ε_O	=	permittivity of free-space = 8.85×10^{-14} F/cm,
ε_r	=	dielectric constant of the insulator, and
d	=	separation of the two opposite parallel plates.

Thus, the larger electrode area and the thinner insulator layer can cause higher capacitance. The procedure for the C-V measurement involves the application of the dc bias voltages across the capacitor using the ac small-signal amplitude. Commonly applied ac frequencies from arbitrary selection are used for the measurement. This convention originates from the non-dependence of the terminal capacitance on the applied measurement frequency. The dc bias is applied as a voltage sweep that drives the junction from physical region within the device. Thus, the ac small-signal in conjunction with the applied dc bias voltage actually experiences or drops across the junction. This voltage drop can be described as the electrical field-falling for the region or junction. This electrical field-falling region actually accounts for the capacitance identical to the separation of the two opposite parallel plates [7]. In this way the measured capacitance is geometric that employs equation (10.1). Different ac small-signal frequencies can reveal additional details.

Low frequencies reveal quasi-static characteristics, whereas high frequencies indicate dynamic performance. The C-V measurement at both range of frequencies are often required to ascertain performance characteristics. Since the C-V measurement is actually done at applied frequencies, the capacitance for the DUT is calculated by:

$$C_{DUT} = \frac{I_{DUT}}{2\pi f V_{ac}} = \frac{I_{DUT}}{\omega V_{ac}}, \tag{10.2}$$

where

I_{DUT} = magnitude of the alternating current passing through the DUT,
f = test or applied frequency,
ω = $2\pi f$ = angular frequency, and
V_{ac} = magnitude and phase angle of the measured ac voltage.

This is often done internally by the measuring instrument where the digital display is the result of the internal circuitry that provides necessary calculations. In other words, the test measures impedance of the DUT by applying ac voltage and measuring the resulting alternating current. Thus, using the ac voltage and the impedance phase angle between them yields the impedance. This measurement takes into account series and parallel resistance associated with the capacitance as well as the dissipation factor indicating leakage. Certain challenges are associated with the C-V measurement and, typically, the problems arise in the following cases:

1. measurement of low capacitance,
2. measurement of leaky capacitance possessing high dissipation factor D,
3. measurement of the frequency-dependent C-V data,
4. C-V instrument connections through probe or lead to the device,
5. use of hardware and software to acquire the data,
6. conversion of parallel mode capacitance to series mode capacitance or vice versa, and
7. understanding the difference between the parallel mode and series mode of capacitances.

Nevertheless, overcoming the challenges requires careful attention of the C-V measurement used in conjunction with appropriate hardware and software. If the measured C is small, the ac response current of the DUT is small and becomes difficult to measure. At higher frequencies, the impedance is reduced so that the current increases and becomes easier to measure. Often semiconductor capacitance is very low which is below the capabilities of monitoring by many instruments. Measuring small capacitance values may have confusing specifications that make difficult to determine the accuracy in the measurement. If accuracy of the instrument in the full measurement range is not explicitly stated then clarification from the manufacturer is useful.

In addition to the low value of C, the semiconductor capacitor may also be leaky. That is the case when the equivalent R is in parallel (or shunt) with C is too low. This results in resistive impedance overwhelming the capacitive impedance, and the C value gets lost in the noise. For ultra-thin layered devices the dissipation factor D can be smaller. In general, as D increases, the accuracy in the measurement of C rapidly degrades. Thus, high D is a limiting factor in the practical use of a capacitance meter. At high frequencies the capacitive impedance (i.e., reactance) is lower resulting in a capacitive current that is higher and more easily measured. The dissipation factor is the ratio of the real current versus the imaginary current flowing through the DUT. If G_p is the as-measured parallel conductance and C_p is the as-measured parallel capacitance for the DUT at a given voltage, then the dissipation factor becomes [8]:

$$D = \frac{G_P}{\omega C_P}. \tag{10.3}$$

In most cases the junction device is connected to the C-V instrument through a probe. At high frequencies, special corrections, and compensation may be necessary. Because of the complexity of the hardware, cabling, wiring, and compensation technique, it is good to note that the long cables can cause inductive response via negative terminal capacitance.

Besides accuracy issues, practical considerations in the C-V data collection include the instrumentation range of test variables, versatility of parameter extraction software, and hardware usage. Usually C-V testing has been limited to some value of dc voltage or corresponding dc current for the junction device. However, many applications require tests at higher voltage or current. In this case, a separate high voltage dc power supply and a capacitance meter are required.

The foregoing discussion belongs to the single-crystal based junction devices or simplistic features of the layered devices. In the case of polycrystalline systems the C-V measurement is completely different due to the usual frequency-dependent response. The grain boundaries play major role for such devices though as usual Mott-Schottky plot is always achieved using C_P^{-2} versus V_{dc} plot. Any arbitrary selected frequency cannot provide a single-frequency meaningful C-V data that can be used in extracting the device-related parameters. This is because arbitrarily selected single frequency for measurement is misleading as it has no scientific or technical basis. In this case frequency-dependent device-related parameters will be achieved having no physical basis or justification. Essentially the frequency-dependent response can be handled via complex plane plots to delineate the frequency-independent response of the same device. This approach is absolutely necessary to achieve meaningful frequency-independent device-related parameters.

10.2 The Mott-Schottky Plot

As a customary practice, the ideal Mott-Schottky behavior is displayed via C_P^{-2} versus V_{dc} plot as shown in Figure 10.1 for metal to n-type semiconductor junction (Schottky diode) and single-sided p^+-n diode. V_{dc} is the reverse bias which is plotted as positive on the x-axis. This is a usual practice to represent the Mott-Schottky plot. A single straight line may be obtained regardless of the measurement frequency implying at each frequency the same geometric capacitance is measured. Since high purity single-crystal semiconductor is used in making these junction devices, it

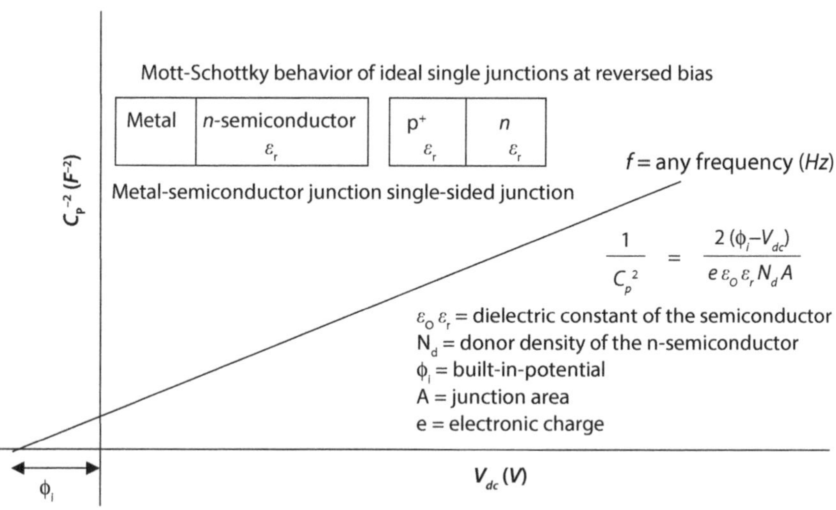

Figure 10.1 Frequency-independent Mott-Schottky straight line at any given frequency for the ideal near non-trap controlled metal-semiconductor and p^+-n junction devices.

is unlikely to observe the influence of traps. Thus, arbitrarily selected high frequency becomes effective for the single Mott-Schottky straight line. Several device-related parameters such as built-in-potential, carrier density, Fermi level (E_F) position, barrier height, etc. may be obtained from this plot. The intercept on the x-axis provides built-in-potential whereas the slope of the straight line provides carrier density. A relevant Mott-Schottky equation [1–4] is given by:

$$\frac{1}{C_P^2} = \frac{2(\phi - V_{dc})}{e\varepsilon_o\varepsilon_r N_d A}, \tag{10.4}$$

where $\varepsilon_o\varepsilon_r$ = dielectric constant of the semiconductor, N_d = donor density of the n-semiconductor, ϕ_i = built-in-potential, e = electronic charge, and A = junction area. The barrier height can be calculated as [1–4]:

$$\phi_B = \phi_i + (E_C - E_F) \tag{10.5}$$

for the n-semiconductor. The extracted parameters sometimes can be manipulated from various types of plots, such as Bode plot presented via terminal C versus Log (frequency) and similar plots. If the semiconductor is highly resistive approaching near insulating condition, then parallel capacitance C_P and series capacitance C_S become nearly equal, that is $C_P \approx C_S$. Thus, the Mott-Schottky plot for this kind of semiconductor junction doesn't make difference in the interpretation of the device-related parameters.

Two distinct situations may arise from the frequency-dependent C-V data. In one case the Mott-Schottky plot can give multiple straight lines corresponding to each measurement frequency converging at a fixed point on the x-axis when extrapolated. This feature is displayed in Figure 10.2 where a single valued built-in-potential [9–15] may be obtained, which is not frequency-dependent [9, 10]. The presence of traps causes the frequency-dependent straight lines. Thus, the slope of each straight line gives the effect of linear summation of carrier density and trap density in the form of $N_d + N_T$, where N_T is the density of traps [11, 14]. The parameter N_T causes the frequency-dependent Mott-Schottky straight lines as it gradually disappears with increase in frequency where the extrapolated lines meet at one single point on the x-axis yielding the built-in-potential [9–15]. Usually the associated trapping species is not complicated but it gives very straightforward device-related parameters once the frequency-dependent nature is eliminated at specific high frequency. Sometimes this

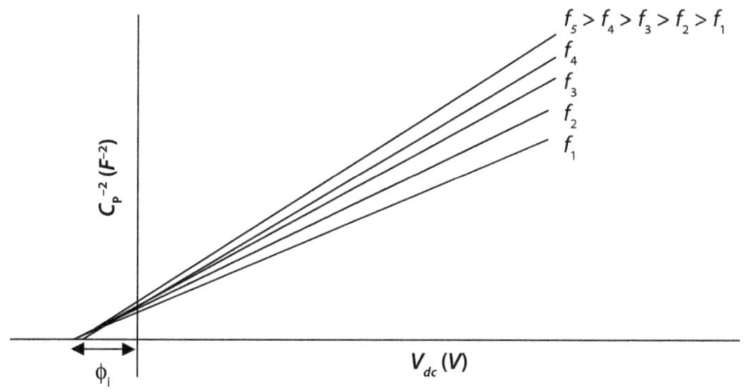

Figure 10.2 Frequency-dependent converging Mott-Schottky straight lines at various frequencies for trap-controlled junction device converging at a single built-in-potential ϕ_i.

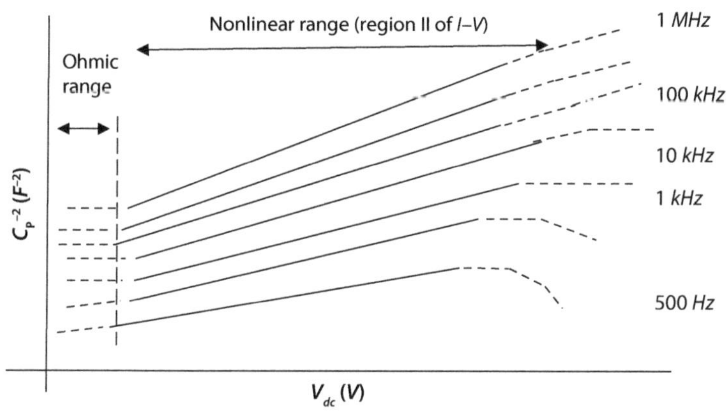

Figure 10.3 Frequency-dependent non-converging Mott-Schottky straight lines at various frequencies as always found in a multi-trap controlled ZnO-based varistor.

type of trapping may be identified as of the Debye type possessing single energy depth and singular time constant. The influence of trapping on the Mott-Schottky straight line causes $N_d + N_T$ for the slope, where N_T decreases as the frequency increases. Thus, the value of N_T can be estimated from the Mott-Schottky straight lines obtained between high and low frequencies. Once again, for the highly resistive semiconductor $C_P \approx C_S$ does not alter the underlying meaning of the extracted device-related parameters.

Another class of frequency-dependent Mott-Schottky straight lines arising from the polycrystalline devices [16–18] is seriously complicated. The Mott-Schottky plot of the ZnO-Bi_2O_3-based varistors exhibits a series of non-converging frequency-dependent [16] straight lines shown in the generic form in Figure 10.3. It is a surprise that the Mott-Schottky slope constitutes a straight line [17] when plotted against Log(frequency) as shown in Figure 10.4. This means that the Mott-Schottky straight lines do not converge at a single point upon extrapolation on the negative side of the voltage-axis. Thus, the extracted device-related parameters depend on the arbitrary selection of the measurement-frequency. Under these circumstances, the frequency-dependent Mott-Schottky representation can lead to significant errors if single-frequency C-V data at fixed temperature are used in extracting device-related parameters, i.e., built-in-potential, barrier height, carrier density, Fermi-level position, etc. Therefore, a meaningful insight satisfying a reasonable physical basis or scientific logic is lost.

The set of single–frequency C-V data implies that the terminal capacitance lies on the curve of the generic representation depicted in Figure 10.5. It is clear that the value of the terminal capacitance above C_5 (i.e., $\omega \rightarrow$ increasing) is going to yield frequency-dependent Mott-Schottky plot. Therefore, observation of resonance clearly provides C_5 and C_6. These two capacitances are related to barrier layer and bulk ZnO, respectively. All ZnO varistors do not exhibit this kind of straightforward resonance[19] behavior. There are three kinds of resonances, usually observed, in the ZnO varistors. The other two kinds do not provide geometric capacitances but overwhelms the resonating circle in the C*-plane. Thus, without observing straightforward resonance as in Figure 10.5 it is absolutely impossible to obtain geometric capacitances.

The error in extracting the device-related parameters can easily exceed 100% depending on the type of device, and an analysis using the classical or modified Mott-Schottky equation with single-frequency C-V data [20–22]. Figure 10.3 confirms non-parallel Mott-Schottky straight lines

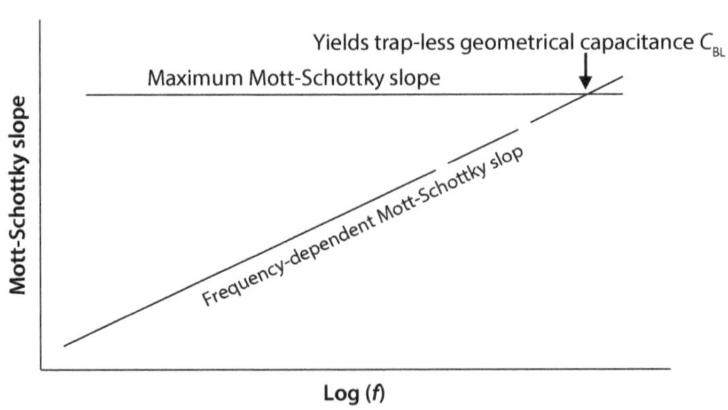

Figure 10.4 The Frequency-Dependent Mott-Schottky Slope for a series of non-converging straight lines in the C_P^{-2} versus V_{dc} plot observed in ZnO-based varistor.

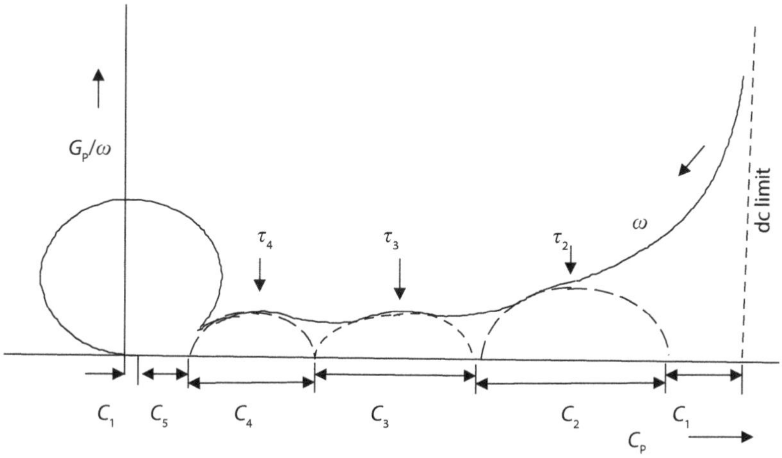

Figure 10.5 Generic zero-bias complex capacitance plot in the frequency range 10^{-2} Hz $\leq f \leq 10^9$ Hz showing the dc limit at f → 0 Hz, resonance behavior at f → 100+ MHz, non-Debye trapping relaxations τ_4, τ_3, and τ_2 along with the heavily distorted τ_1 relaxation associating C_1.

when the frequency-dependent measurement is performed. The series of non-converging frequency-dependent Mott-Schottky straight lines is universal [16–18] for the ZnO varistor.

The non-converging Mott-Schottky straight lines observed for ZnO varistor shown in Figure 10.3 appear as near parallel [16–18] regardless of their initial recipe or processing steps. Between 1976 and 1988, this unresolved Mott-Schottky plot was referred to unknown extraneous contribution to the terminal capacitance [18] although it confirmed the presence of the back-to-back Schottky barrier attributing to the biasing with both polarities. There is a correspondence between this observation and the current-voltage (I-V) behavior displayed in Figure 10.6. The Mott-Schottky feature exhibits distinct ohmic behavior via near flat region in Region I as frequency increases, and then straight line in Region II of the I-V response. Region II is distinguished from Region III having a demarcation at electric field $E_{1\ mA/cm^2}$ where 1 mA/cm^2 is the current density. No C-V measurement is possible in Region III and post-nonlinear ohmic response designated as Region IV, as there is a limitation in instrumentation from damage from the lacking protection via blocking dc biasing attributing to high current exposure.

Usually in the ohmic bias range the curve is flat at higher frequencies while this segment is a part of the straight line for the low frequencies. Though traps are contained in the terminal capacitance, such capacitance exhibits a geometric role via depletion width. This means that the ohmic

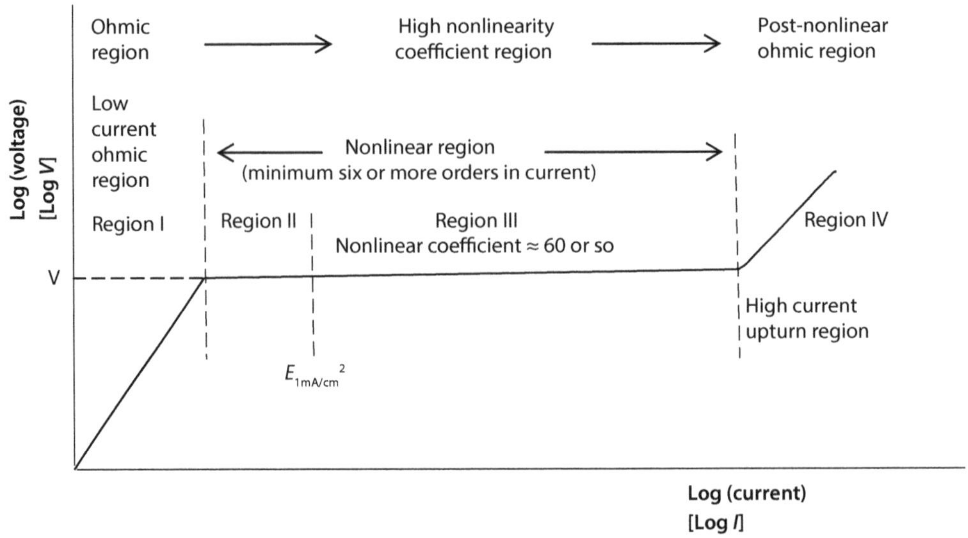

Figure 10.6 Nonlinear current-voltage behavior of the ZnO-based varistor.

voltage drops or falls evenly across the grain boundary electrical barrier making no net change in the depletion width. Somehow at low frequencies, terminal capacitance is not constant in the ohmic bias range but has a slight decrease with applied frequency like those of high frequencies of Region II, as if there is elimination of trapping or discharge of traps taking place. Indeed this is not the case at low frequencies. Furthermore, the electrical barrier behaves somewhat like a symmetric barrier comprising of forward bias side and reverse side.

When the dc bias enters in the nonlinear range Region II then the reverse bias side of the electrical barrier begins to accumulate the voltage implying band bending and the increased depletion width on the reverse bias side. The increased depletion width reduces the terminal capacitance, causing a change in slope for the curves [16, 17]. The device exhibits continued linearity for Region II. The device shows steady value of constant nonlinear coefficient of about 60 or so in Region III, and then again shows upturn or linearity in Region IV at extreme bias corresponding to the surge or lightning level.

Since the classical Mott-Schottky equation (10.4) could not resolve the issues with the series of straight lines for the ZnO-Bi_2O_3-based varistors, a modified form appeared [23–25], as if the underlying problem is resolved or removed. Upon examining the approach to the capacitance per unit area for the modified form of the Mott-Schottky equation is given by [23–25]:

$$\frac{C}{A}=\left(\frac{e\varepsilon N_d}{2\phi_i}\right)^{0.5} \tag{10.6}$$

For back-to-back Schottky barrier, one side behaves like forward biased and the other side behaves like reversed biased. This gives the split in the applied voltage V as V_1 for the forward side causing a decrease of the barrier, and V_2 for the reverse side causing an increase for the barrier such that $V = V_1 + V_2$ is maintained for the grain boundary electrical barrier. This gives [23–25]:

$$\frac{A}{C}=\left(\frac{2}{e\varepsilon Nd}\right)^{0.5}\left[\left(\phi_i - V_1\right)^{0.5}+\left(\phi_i + V_2\right)^{0.5}\right] \tag{10.7}$$

By assumption the forward biased decrease is negligible so that the entire voltage drops in the reverse bias side making $V \approx V_2$. Thus, equation (10.7) becomes [23–25]:

$$\left(\frac{1}{C_P}-\frac{1}{2C_o}\right)^2=\frac{2\left(\phi_i + V_{dc}\right)}{e\varepsilon_o\varepsilon_r N_d A} \tag{10.8}$$

where

$$\left(\frac{1}{C_o}\right)^2=4\left(\frac{2\phi_i}{e\varepsilon_o\varepsilon_r N_d A}\right). \tag{10.9}$$

Though the foregoing logic of the derivation of the modified Mott-Schottky equation (10.8) looks reasonable, this does not remove the frequency dependence. Multiple straight lines again appear as frequency remains open for selection. This convincing-like work [23–25] never mentions the measurement frequency as it is presumed to be considered around 1 MHz. Several papers appeared noting a so-called modified Mott-Schottky equation without prior thinking as to how this equation has resolved the underlying "frequency-dependent" problem. Basically taking arbitrary measurement frequency and then generating a Mott-Schottky plot using equation (10.4) or equation (10.8) never makes any difference as single straight line is displayed in each case. This is because it is not the issue to presume to get rid of the underlying problem, as it is a mere disguising or hiding process of the issue by taking shelter under a deceptive equation. In reality, both aforementioned equations have limitations in handling the measured single-frequency C-V data when each straight line does not converge on a single point extrapolated on the voltage-axis.

Several reasons can account for the limitation in using classical and modified forms of the Mott-Schottky equations for single-frequency measured C-V data at a given temperature. The measurement frequency is arbitrary even if a sharp transition capacitance, originating from the minimum in dispersion of the terminal admittance (especially terminal parallel capacitance) with ac frequency and the onset of the resonance event, is achieved. This transition capacitance is a potentially erroneous parameter in the C-V analysis. In general, it is greater than the net or total geometric capacitance and, thus, yields different results. The method of extracting the device-related parameters, using specific single-frequency C-V data and incorporating the original and/or modified Mott-Schottky equations, is incorrect at least for the following possible reasons [17]:

1. frequency-independent geometric capacitance as a function of dc voltage is not ensured;
2. the forward-bias barrier height is not strictly constant;
3. Fermi level pinning at the interface of the back-to-back electrical barrier;
4. the forward-bias barrier does not continuously provide an effect of the electrical field drop (in the forward-bias barrier region) and is thereby unlikely to achieve a flat-band situation;
5. trapping states within the depletion layers contribute to the terminal parallel capacitance (at quasi-equilibrium condition) so their contribution is not completely eliminated;
6. continuous/discontinuous trapping and de-trapping under non-equilibrium conditions affecting terminal parallel capacitance;
7. possible leakage of the trapped charges (i.e., destruction of the trap sites) at any instant (i.e., slow or ultra-slow time-dependent processes under ac/dc biasing at a given temperature) which contributes to the conduction processes affecting terminal parallel capacitance;
8. the effective cross-sectional area of conduction across the grain-boundary electrical barriers is not precisely known (this area is not necessarily the same as the physical contact area across each grain-boundary);

9. invariant nature of the net/total depletion layer thickness at any instant of a specific experimental situation due to the non-identical grain-boundary junctions (or electrical barriers);
10. non-symmetric distribution of grain-size and non-symmetric (or asymmetric) depletion layer thickness involving adjacent grains constituting a back-to-back grain-boundary configuration; and
11. inhomogeneous or random distribution of secondary and other phases at the grain-boundary interfaces resulting into the non-symmetric/asymmetric trapping contribution in conjunction with the variation in the effective cross-sectional area of conduction affecting terminal parallel capacitance.
12. Effect of the non-uniform network of conduction consisting of non-linear, ohmic, bad and/or poor junctions causing an overlapped behavior between the geometric and trapping capacitances.

The traps in varistors are not singular type but originated from multiple type traps where each is different from the other though lie within the same electrical field falling region. Moreover, each trap is non-Debye type and often exhibit overlapped behavior in the relaxation process among themselves. Thus, each trap corresponds to the frequency domain with respect to the time constant or relaxation time. Often within one single type trap there is a certain level of complications with the energy depth and distribution for the electrical field falling region [17, 26–29]. This frequency domain behavior gives a series of non-converging Mott-Schottky straight lines. The term non-converging means that upon extrapolation on the voltage axis frequency-dependent built-in-potential values are obtained. Therefore, extraction of the device-related parameters becomes frequency-dependent [16, 17]. In this case arbitrary selection of the measurement frequency can jeopardize the device-related parameters via frequency dependence. In reality the set of the device-related parameters should not be frequency-dependent.

It is necessary to discuss the aforementioned errors, and provide insight on data-handling and analyzing criteria for the extraction of useful device-related parameters. The data-handling and analyzing criteria emphasize proper identification of a terminal resonance event when the dc voltage is applied at a fixed temperature. Only one type of terminal resonance event (out of three terminal resonance events in these devices) allows the extraction of the device-related parameters employing the lumped parameter/complex plane analysis (LP/CPA) technique [16, 17, 26–29].This method identifies the nature of the dispersion of the terminal parallel capacitance with ac frequency, and thereby establishes a frequency-independent response satisfying the basic or classical Mott-Schottky equation for the devices. Thus, this approach becomes identical with those of the single-junction-like metal-semiconductor contact and/or single-sided p^+-n junction.

The complexity of the trap-controlled back-to-back electrical barriers across the grain-boundaries in polycrystalline varistors was first pointed [18] out in 1976. The construction of the classical Mott-Schottky type straight line with the C-V data presumably acquired at an arbitrarily selected single high-frequency. This single-spot selection of the high-frequency measurement value was possibly prompted when the terminal parallel capacitance exhibited a minimum value in the high-frequency domain. The classical Mott-Schottky equation is commonly used in analyzing single-junction devices made of single crystals [1–4] where complexity with traps is not an important issue. However, this analysis is not likely to be valid for multi-component and/or multi-phase heterogeneous devices such as polycrystalline varistors containing multiple trapping [16, 17, 26–29].

10.3 Arbitrary Measurement Frequency and Construction of the Deceiving Mott-Schottky Plot

Usually arbitrary selection of the high-end of the measurement frequency presumes elimination of the trapping contribution for a device system. This presumption does not provide frequency-independent nature of the C-V data when device-related parameters are extracted for almost all the polycrystalline systems. The weakness of the Mott-Schottky (C_P^{-2} versus V_{dc}) response appears at any arbitrary measurement frequency yielding a series of frequency-dependent straight lines. The extraction of the device-related parameters from any of these straight lines is incorrect and invariably erroneous. Therefore, these straight lines are deceptive in nature as a Mott-Schottky response and need to be fixed concerning the frequency-independent behavior.

The frequency-dependent Mott-Schottky straight line gives two types of errors for the built-in-potential ϕ_i or barrier height ϕ_b and N_d. The first type of error provides physically unrealistic magnitudes of ϕ_i or ϕ_b in conjunction with a reasonable value of N_d. Despite using modified [23–25] Mott-Schottky equation such an error could not be eliminated as the barrier height in polycrystalline varistors was obtained [20–22] between

1.45 eV and 5.12 eV. These barrier height values are outrageous and often exceed the band gap of the ZnO. The second type of error is very common, where the values of the device-related parameters exceed the values obtained in the first type of error, regardless of using classical [1–4] or modified [23–25] Mott-Schottky equation. In general, well-formed or commercial varistors possessing stable properties meeting applications criteria [16, 17, 26–29] exhibit barrier height of about 0.7 eV. This is a sort of average as observed typically between 0.45 eV and 0.85 eV for most devices possessing excellent application characteristics and carrier concentration $N_d \approx 10^{17}/cm^3$. Depending on the selection of the measurement frequency and type of varistor response, the value of terminal capacitance and its variation with dc voltage are critical for the slope of the Mott-Schottky straight line. Since the net or total geometric capacitance overlaps with the trapping response in the frequency-domain [16, 17, 26–29] measurement, a nonlinear [17, 26–29] voltage dependence of the terminal capacitance influences the Mott-Schottky slope. If contributions from the trapping relaxations [16, 17, 26–29] denoted as τ_4 and τ_3 are contained in the Mott-Schottky straight line. The error in extracting device-related parameters could exceed easily 100% or more when compared to the actual values obtained by other methods [16, 17, 30, 31]. This is true for both classical and modified Mott-Schottky equations. In addition to the foregoing situations, the process of extracting device-related parameters can be vitiated with the usage of the physical electrode area for either form of the Mott-Schottky equations [31]. The cross-sectional area termed as effective area involved in every operative conduction path in a series-parallel microstructural network configuration [26–29, 32, 33] is different than the physical electrode area of the polycrystalline varistor.

10.4 Frequency-Independent Representation

Several frequency-dependent converging Mott-Schottky straight lines for various single-sided junction devices containing traps were analyzed [9, 10, 34, 35]. The net or total geometric capacitance was achieved by eliminating the trapping capacitance from the terminal parallel capacitance using the high-frequency measurement. In these cases, the measured capacitance is likely to exhibit a very nearly constant value which is a flat-region parallel to the frequency axis if plotted with terminal parallel capacitance on the other axis in the Bode-plane [36] indicating that the extraneous contribution to the terminal parallel capacitance has been eliminated. A similar analysis involving slightly more complicated junction devices [37–42] did not provide an explanation that satisfied the feature of the Mott-Schottky behavior observed [16, 17, 26–29] in the polycrystalline varistors. Based on the early work [9, 10, 34, 35], LP/CPA technique [16, 17] has been devised for analyzing the ac immittance data for the polycrystalline materials exhibiting frequency-dependent C-V data.

The frequency-dependent C-V data analyzed via employing the complex plane plot (LP/CPA), and then a new Mott-Schottky plot has been generated that yielded a set of frequency-independent useful device-related parameters satisfying classical Mott-Schottky equation [16, 17]. The net or total geometric capacitance was identified in this analysis and used in constructing true or new frequency-independent Mott-Schottky plot from the C*-plane plot. This exercise included a rigorous exploitation of the classical [1–4] and other forms of Mott-Schottky equations [9, 10, 20–23, 34, 35, 37–42] in a systematic manner for a large variety of well-formed or good varistors. Thus, the modified Mott-Schottky equation used in a number of studies is generally not valid due to the frequency-dependent C-V characteristics.

The LP/CPA technique helps to obtain the net or total geometric capacitance including the simultaneous effect of the forward-biased and reverse-biased electrical barriers across a single grain-boundary interface. The nature of the resonance events in polycrystalline varistors suggest that each of these events has direct influence on the frequency-dependent problem concerning the Mott-Schottky behavior. If the reasonable event implying straightforward resonance [19] is not exactly identified and its implications for the data-handling/analyzing criteria are not understood for each device system [16, 17], it is difficult to obtain the frequency-independent Mott-Schottky plot. Thus, errors can also be incorporated in extracting a set of device-related parameters even when the LP/CPA technique is employed. Nevertheless, a straightforward resonance event is necessary [19] to obtain exact frequency-independent Mott-Schottky using the LP/CPA technique.

In relation to the foregoing context, it is useful to summarize the immittance spectra of the polycrystalline varistors which include a straightforward resonance event [25]. The small-signal ac immittance data of these devices acquired over a frequency range 10^{-2} Hz $\leq f \leq 10^9$ Hz reveal multiple relaxations [16, 17, 26–29] and a resonance phenomenon [16, 17, 26–29, 43, 44] in the complex capacitance plane:

$$C^* = C' - jC'' = C_p - j(G_p/\omega), \tag{10.10}$$

where $C' = C_p$ = terminal parallel capacitance, $C'' = (G_p/\omega)$, G_p = terminal parallel conductance, $\omega = 2\pi f$, and $j = \sqrt{-1}$. Three distinct relaxations (denoted as $\tau_2 = R_2 C_2$, $\tau_3 = R_3 C_3$, and $\tau_4 = R_4 C_4$), a low-frequency distortion (C_1, or corresponding τ_1 related relaxation), and a barrier-layer capacitance ($C_{BL} = C_6 + C_5$) including the dielectric contribution of the lumped grains ($C_6 = C_{lumped\text{-}grains} = C_{ZnO}$) are evident [16, 17, 26–29] in this plane (Figure 10.5). In general, this behavior is a characteristic of polycrystalline varistor composites exhibiting good performance [26–29] regardless of the recipe, history, and processing methods. The chemistry or composition or formulation and processing methods influence relative magnitudes and visibility of all these parameters but do not affect the response of this dielectric composite. At higher frequencies (usually 10^6 Hz $\leq f \leq 10^9$ Hz) the resonance phenomenon emerges as a circle [16, 17, 26–29] possessing negative values of the terminal parallel capacitance in the C*-plane. A detailed description concerning resonance events is available elsewhere [19].

10.5 Extraction of the Device-Related Parameters

The frequency-independent Mott-Schottky representation requires achieving the maximum slope of the straight line. This implies achieving the minimum value of the capacitance that is due only to the net/total geometric capacitance. The Mott-Schottky slope gives a nearly straight line when plotted against Log(f). This straight line provides maximum slope corresponding to trap-less geometrical capacitance.

The geometrical capacitance is illustrated by the frequency-dependent C-V data presented in the generic Figure 10.3. The data for this generic curve lies in the frequency range 10^{-2} Hz $\leq f \leq 10^9$ Hz. Three distinct semicircular arcs are shown as τ_2-, τ_3-, and τ_4-relaxations in Figure 10.5. Usually the capacitance at 1 MHz lies at the semicircular loci of the τ_4-relaxation. This capacitance provides an incorrect Mott-Schottky straight line. The dc limit of the curve asymptotically rises upward as the frequency reduces for which the capacitance C_1 is heavily distorted. For such a case, DLTS (deep-level transient spectroscopy) can unravel some aspects of C_1 but not all. Quantification concerning the circuit elements remains tedious. This capacitance C_1 is more than likely to be associated with the non-Faradaic response which is the most complicated behavior remaining unresolved. It was established when the rest of the circuit elements was subtracted from the total circuit and evaluated in the Z*-plane. The respective Bode plot is assured to behave like the complicated non-Faradaic response. The underlying mechanism or non-relaxed trapping behavior responsible for C_1 still demands further investigation.

The origin of the non-parallel Mott-Schottky response is attributed to the dc bias dependence of each contributing capacitive element corresponding to specific relaxation process in the device. These non-parallel straight lines are deceiving if at the highest measureable frequency is chosen for taking the single-frequency data for making the C_P^{-2} versus V_{dc} plot. In reality, the slope of the straight line at this situation lies toward the upper level of the Mott-Schottky (M-S) slope line near to the horizontal line indicating maximum value [17] of the slope in Figure 10.4. Yet still it is not the correct set of the C-V data.

The generic illustration of Figure 10.7 has been systematically evaluated for well-formed devices, using the trend of the previously published Mott-Schottky data [16–18]. This is the true Mott-Schottky plot that gives the device-related parameters. It further provides an indication of the frequency-independent capacitance, C_{BL} (= $C_5 + C_6$), as the measurement frequency approaches the high-frequency limit. In actuality, C_{BL} requires to be extracted from the behavior presented in Figure 10.5. Thus, C_{BL} is the minimum capacitance that accounts for the net/total depletion layer thickness which is geometric in nature. This capacitance yields a maximum value of the slope for the frequency-independent Mott-Schottky straight line [16, 17]. C_6 is the lumped ZnO (bulk) response.

The maximum Mott-Schottky slope implying true or frequency-independent behavior includes the effect of the forward-biased barrier.

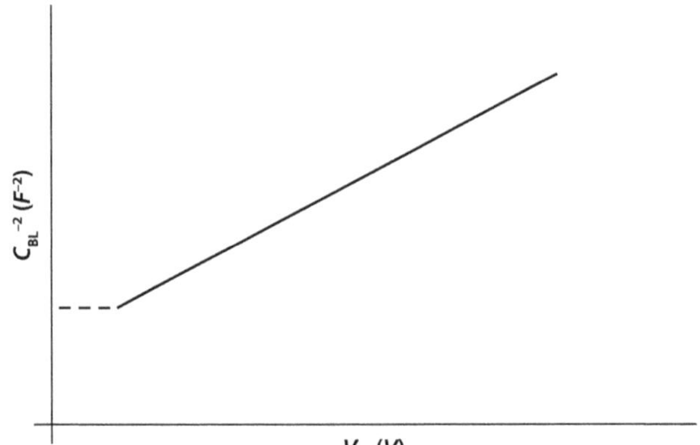

Figure 10.7 True Mott-Schottky behavior of the ZnO-based varistor reflecting frequency-independent straight line for the C_{BL}^{-2} versus V_{dc} plot, where $C_{BL} = C_5 + C_6$ per Figure 10.5.

Perhaps, the forward-biased barrier without flat-band situation has the smallest effect between the onset of nonlinearity and up to the current density J = 1 mA/cm^2 corresponding to the electric field $E = E_{1\ mA/cm^2}$ in determining the actual device-related parameters. Therefore, the net/total depletion layer should respond like a combined (lumped) effect of the forward- and reverse-biased electrical barriers across the grain-boundaries. Hence, in the J-E nonlinear range the C-V characteristics of ZnO-Bi_2O_3-based varistors are similar to those for reverse-biased single-junction devices. This concept is reasonable regardless of the actual status of the forward-biased barrier without reaching flat-band situation, and explains the as-measured C-V data.

In contrast with the aforementioned true or frequency-independent representation of the Mott-Schottky behavior, several erroneous Mott-Schottky straight lines can be constructed with or without using the LP/CPA technique. Perhaps three such cases are of primary interest. First, a series of Mott-Schottky straight lines [16–18] may be obtained from the arbitrary selection of the measurement-frequency for the as-measured parallel capacitance, essentially greater than C_{BL}, as a function of voltage. Second, a short-range nearly constant capacitance [16, 17, 26–29, 43, 44] with respect to increasing frequency identified as the third event in the resonance effect [19] yields a single, frequency-independent Mott-Schottky straight line. Third, the frequency-independent pseudo static capacitance value or values, consisting of the equivalent circuit elements [16, 17, 26–29] (i.e., $C_{BL}+ C_4$, $C_{BL}+ C_4+ C_3$, $C_{BL}+ C_4+ C_3 + C_2$, etc.), as a function of applied voltage, yields a finite number depending on the visibility of the number of traps of Mott-Schottky straight lines. The first two cases may be extracted from the Bode plots, and need not employ the LP/CPA technique. These are possibly demonstrated [18] in 1976 and by other investigators [45–47]. Despite using the LP/CPA technique, the third case can also create an erroneous C_{BL}-like capacitance resulting in a misleading Mott-Schottky plot. Thus, the erroneous C_{BL}-like parameter is essentially greater than true C_{BL}, and can yield a smaller Mott-Schottky slope value possibly lie on the curve of Figure 10.6.

The frequency dependence of the Mott-Schottky slope strongly suggests that the single-frequency C-V analysis can invariably lead to an erroneous set of device-related parameters. The masking behavior concerning overlapping mechanisms and dominance of the traps is efficiently resolved by the LP/CPA technique. The data-handling/analyzing criteria provided net/total geometric capacitance, wherein the depletion layer thickness includes the effect of the forward and reverse-biased barriers of the grain-boundary interfaces in a simultaneous fashion in the ohmic and nonlinear J-E regions. Only upon correct inspection of the resonance event (i.e., first type designated as straightforward), and subsequent identification of C_{BL} (or parallel combination of C_5 and C_6), one can generate a meaningful frequency-independent Mott-Schottky plot which in turn can be used in extracting a useful set of device-related parameters.

Problems

1. Choose an arbitrary straight line with a specific slope for the Mott-Schottky slope versus Log(f) plot. Take each Mott-Schottky straight line for the C_P^{-2} versus V_{dc} plot having capacitance value ranging between 100 pF and 200 pF with the low-frequency end at 100 Hz and the high-frequency end at 1 MHz. Develop a series of straight lines in the C_P^{-2} versus V_{dc} plot corresponding to each frequency having variation in the slope. Extend each straight line on the negative side of the x-axis and obtain the frequency-dependent value for the built-in-potential of the device. Use own choice of the values for the rest of the parameters that are needed for the classical Mott-Schottky equation.
2. For a space charge region, capacitance decreases by 10% for every 1 V increase in reverse bias. Take the zero-bias width of the space charge region as 100 μm to develop the data up to 6 V to for the C_P^{-2} versus V_{dc} plot. The contact area of the Si wafer with the metal is 10 μm^2. If it is a straight line then find the carrier density. If it is not a straight line then explain the reason for it. What is the nature of the space charge region and the corresponding electrical barrier in this case?
3. Develop six straight lines for a Si based junction Schottky diode using 0.1 V, 0.12 V, 0.14 V, 0.16 V, 0.18 V, and 0.2 V as the built-in-potential. Choose these straight lines in such a way that the reverse side dc biasing is ranging between 1 V and 6 V having 1 V interval. The Si wafer for the Schottky diode has doping density 10^{17}/cm^3.
4. Describe the circumstances for which the single-frequency (say 1 MHz or so) measurement may be useful for the Mott-Schottky plot? Under what circumstances, may the same

measurement not be useful? Describe the character of the device in each case.

5. In general, polycrystalline semiconductors exhibit a series of frequency-dependent straight lines. Explain the reasons. How can one resolve this problem and obtain a reasonable Mott-Schottky straight line?
6. Describe the differences between the two capacitances given below:

$$C = \frac{\varepsilon_o \varepsilon_r A}{d} \quad \text{and} \quad C_P = C_{Geometric} + C_{Frequency-Dependent}.$$

Which capacitance will give the Mott-Schottky straight line at a single-frequency measurement? Explain the reasons.

7. What is the geometric capacitance? Why a material system encounters geometric capacitance and frequency-dependent capacitance?
8. Why single-crystal based junction devices encounter almost no influence of traps? Why any arbitrary frequency can be used in measuring the terminal capacitance and treated as constant?

References

1. E. H. Rhoderick; "*Metal-Semiconductor Contacts,*" Clarendon Press, Oxford, U.K. 1978.
2. R. S. Muller and T. I. Kamins; "*Device Electronics for Integrated Circuits,*" John Wiley & Sons, Inc., New York, 1977.
3. S. M. Sze; "*Physics of Semiconductor Devices,*" 2nd Edition, Wiley Interscience Publication, New York, 1981.
4. E. S. Yang; "Fundamentals of Semiconductor Devices," McGraw-Hill Book Co., New York, 1978.
5. M. A. Alim; "Electrical Characterization of Engineering Materials," *Active and Passive Electronic Components*, 19, 139–169, 1996.
6. M. A. Alim; "Quality Assessment of Engineering Material Systems by Immittance Measurements," *Ceramic Transactions: Ceramic Manufacturing Practices and Technologies*, B. Hiremath, T. Gupta, and K. M. Nair edited, The American Ceramic Society, Westerville, Ohio, 70, 233–242, 1996.
7. R. Resnick and D. Halliday; *Physics I & II*, John Wiley & Sons, New York, 1966.

8. B. M. Oliver and J. M. Cage; *Electronic Measurements and Instrumentation*, McGraw-Hill Book Company, New York, 1971.
9. W. P. Gomes and D. Vanmaekelbergh; "Impedance Spectroscopy at Semiconductor Electrodes: Review and Recent Developments," *Electrochimica Acta*, 41(7–8), 967–973, 1996.
10. F. Fransent, M. J. Madou, W. H. Laflere, F. Cardon, and W. P. Gomes; "On the Dielectric Properties of Semiconducting Materials as Obtained from Impedance Measurements on Schottky Barriers," *Journal of Physics D: Applied Physics*, 16, 879–888, 1983.
11. Y. Zohta; "A New Method for Determination of Deep-Level Impurity Centers in Semiconductors," *Applied Physics Letters*, 17(7), 284–286, 1970.
12. Y. Zohta; "Frequency Dependence of C and $\Delta V/\Delta C^{-2}$ of Schottky Barriers Containing Deep Impurities," *Solid State Electronics*, 16, 1029–1035, 1973.
13. Y. Zohta; "Frequency Dependence of $\Delta V/\Delta C^{-2}$ of MOS Capacitors," *Solid State Electronics*, 17, 1299–1309, 1974.
14. G. L. Miller, D. V. Lang, and L. C. Kimmerling; "Capacitance Transient Spectroscopy," in *Annual Review of Materials Science*, R. A. Huggins edited, 7, 377–448, 1977.
15. F. Cardon and W. P. Gomes; "On the Determination of the Flat-Band Potential of a Semiconductor in Contact with a Metal or an Electrolyte from the Mott-Schottky Plot," *Journal of Physics D: Applied Physics*, 11, L63–L67, 1978.
16. M. A. Alim, M. A. Seitz and R. W. Hirthe; "Complex Plane Analysis of Trapping Phenomena in Zinc Oxide Varistors," *Journal of Applied Physics*, 63(7), 2337–2345, 1988.
17. M. A. Alim; "An Analysis of the Mott-Schottky Behavior in $ZnO\text{-}Bi_2O_3$ Based Varistors," *Journal of Applied Physics*, 78(7), 4776–4779, 1995.
18. W. G. Morris; "Physical Properties of the Electrical Barriers in Varistors," *Journal of Vacuum Science and Technology*, 13(4), 926–931, 1976.
19. M. A. Alim; "High-Frequency Terminal Resonance in $ZnO\text{-}Bi_2O_3$-Based Varistors," *Journal of Applied Physics*, 74#9, 5850–5853 (1993).
20. J. Fan and R. Freer; "The Roles Played by Ag and Al Dopants in Controlling the Electrical Properties of ZnO Varistors," *Journal of Applied Physics*, 77(9), 4795–4800, 1995.
21. S. N. Bai and T. Y. Tseng; "Influence of Sintering Temperature on Electrical Properties of ZnO Varistors," *Journal of Applied Physics*, 74(1), 695–703, 1993.
22. J. K. Tsai and T. B. Wu; Non-ohmic Characteristics of $ZnO\text{-}V_2O_5$ Ceramics," *Journal of Applied Physics*, 76(8), 4817–4822, 1994.
23. K. Mukae, K. Tsuda, and I. Nagasawa; "Capacitance-vs-Voltage Characteristics of ZnO Varistors," *Journal of Applied Physics*, 50(6), 4475–4476, 1979.

24. K. Mukae and K. Tsuda, "Interface States at Grain Boundaries and their Effects on the I-V Characterization in ZnO-Pr Varistors: Theoretical Calculation," *Ceramic Transactions: Grain Boundaries and Interfacial Phenomena in Electronic Ceramics,* edited by L. M. Levinson and S.- I. Hirano, The American Ceramic Society, Westerville, Ohio, 41, 195–205, 1994.
25. F. Schipani, C. M. Aldao, and M. A. Ponce; "Inadequacy of the Mott–Schottky Equation in Strongly Pinned Double Schottky Barriers with no Deep Donors, *Journal of Physics D: Applied Physics,* 45,495302, 2012.
26. M. A. Alim; "Influence of Intrinsic Trapping on the Performance Characteristics of ZnO-Bi_2O_3 Based Varistors," *Active and Passive Electronic Components,* 17(2), 99–118, 1994.
27. M. A. Alim; "Influence of Multiple Trapping Phenomena on the Applications Criteria of ZnO-Bi_2O_3 Based Varistors," *Active and passive Electronic Components,* 17(1), 57–65, 1994.
28. L. C. Sletson, M. E. Potter and M. A. Alim; "Influence of Sintering Temperature on Intrinsic Trapping in Zinc Oxide-Based Varistors," *Journal of the American Ceramic Society,* 71, 909–913, 1988.
29. M. A. Alim; "Admittance-Frequency Response in Zinc Oxide Varistor Ceramics," *Journal of the American Ceramic Society,* 72(1), 28–32, 1989.
30. M. A. Seitz, R. W. Hirthe, and M. E. Potter; "Thermal Runaway in Metal Oxide Varistors," Materials Research Society Symposia Proceedings: *Grain Boundaries in Semiconductors*, Edited by H. J. Leamy, G. E. Pike, and C. H. Seager, Elsevier Science Publishing Co., New York, 5, 405–410, 1982.
31. L. M. Levinson, and H. R. Philipp; "Optical Method for Determining the Grain Resistivity in ZnO-Based Ceramic Varistors," *Journal of Applied Physics*, 47(3), 1112–1116 (1976).
32. M. A. Alim and M. A. Seitz; "Singular Nature of Preferential Conducting Paths at High Electric Fields in ZnO-Based Varistors," *Journal of the American Ceramic Society*, 71, C246–C249, 1988.
33. M. A. Alim, M. A. Seitz and R. W. Hirthe; "High-Temperature/Field Alternating Current Behavior of ZnO-Based Varistors," *Journal of the American Ceramic Society,* 71, C52–C55, 1988.
34. W. H. Laflere, R. L. V. Meirhaeghe, F. Cardon, and W. P. Gomes; "On the Frequency-Dependence of the Impedance of n- and p-Type Gallium Arsenide Electrodes," *Surface Science*, 59, 401–412, 1976.
35. R. L. V. Meirhaeghe, W. H. Laflere, and F. Cardon; "On Interfacial Charges in the Oxide Layer of MIS-Type Ag- and Au-*n* GaAs Schottky Barriers," *Solid State Electronics*, 26(12), 11891192, 1983.
36. H. W. Bode; *Network Analysis and Feedback Amplifier Design*, Van Nostrand, New York, 1945.
37. R. D. Gryse, W. P. Gomes, F. Cardon, and J. Vennik; "On the Interpretation of Mott-Schottky Plots Determined at Semiconductor/Electrolyte Systems," *Journal of the Electrochemical Society*, 122(5), 711–712, 1975.

38. W. Smit; "Application of a Modified Colloid Chemical Site Binding Model to Mott-Schottky Plots Determined at Oxidic Semiconductor/Electrolyte Systems," *Journal of the Electrochemical Society*, 132(9), 2172–2175, 1985.
39. K. Uosaki and H. Kita; "Effects of the Helmholtz Layer Capacitance on the Potential Distribution at Semiconductor/Electrolyte Interface and the Linearity of the Mott-Schottky Plot," *Journal of the Electrochemical Society,* 130(4), 895–897, 1983.
40. G. Nogami; "Theory of Capacitance-Voltage Characteristics of Semiconductor Electrodes with Interface States," *Journal of the Electrochemical Society,* 133(3), 525–531, 1986.
41. G. Nogami; "Some Aspects of Large Frequency Dispersion of Mott-Schottky Plots in TiO_2 Electrodes, *Journal of the Electrochemical Society,* 132(1), 76–81, 1985.
42. G. Nogami; "Characterization of Semiconductor Electrodes with a Deep Impurity Level," *Journal of the Electrochemical Society*, 129(10), 2219–2223, 1982.
43. L M. Levinson and H. R. Philipp; "ac Properties of Metal-Oxide Varistors," *Journal of Applied Physics,* 47(3), 1117–1122, 1976.
44. L M. Levinson and H. R. Philipp; "High-Frequency and High-Current Studies of Metal-Oxide Varistors," *Journal of Applied Physics*, 47(7), 3116–3121, 1976.
45. M. Prudenziati, A. Masoero, and A. M. Rietto; "Conduction mechanism and flicker noise in ZnO varistors," *Journal of Applied Physics*, 58(1), 345–350, 1985.
46. A. Kusy and T. G. M. Kleinpenning; "Conduction mechanism and 1/f noise in ZnO varistors," *Journal of Applied Physics*, 54(6), 2900–2906, 1983.
47. L. K. J. Vanadamme and J. C. Brugman; "Conduction mechanisms in ZnO varistors," *Journal of Applied Physics*, 51(8), 4240–44, 1980.

11

Analysis of the Measured Data

11.1 Introduction and Background of the Immittance Data Analysis

Smart electronic materials and devices are often characterized by the popular analytical novel tool/technique immittance spectroscopy referring to either impedance or admittance (IS or AS) related data handling by a variety of investigators [1–32]. Immittance is a hybrid word in use presumably within the General Radio products since the 1960s. Published texts [33–36] are widely used for this purpose, and often restricted the display of the ac small-signal electrical data within the impedance (Z^*) plane only. The meaning of immittance is illustrated by Macdonald [37] as any of the four quantities used under the umbrella of impedance spectroscopy without the geometrical factors such as electrode area and thickness of the device under test (DUT): (1) impedance Z^*, admittance Y^*, complex capacitance C^*, and complex modulus M^*. These four quantities are complex parameters, and when used simultaneously to investigate unknown DUT may be referred to as the immittance spectroscopy (IS).

Concurrent use of multiple complex plane formalisms (multi-plane) for the same set of measured ac small-signal electrical data is very rare in the literature despite an explicit demonstration by Grant [24] in 1958. The current limitation of the same data analysis via only Z^*-plane provides a limited window of visibility or interpretation of the semicircular relaxation despite performing complex nonlinear least squares (CNLS) fitting [33–36] procedure. Thus, complete underlying operative phenomena or mechanisms may not be unraveled or delineated for any unknown device system.

The objective of this chapter is to provide guidelines for handling concurrent multi-plane analysis if the semicircular response is observed in more than one complex plane for the measured ac electrical data. The semicircular relaxation curve fitting is done with two separate methods termed as: (1) geometric and (2) CNLS. Each is dealt independently and then compared within each other for the extracted parameters. This is emphasized to provide information to the audience about the choice of using the method preferred for the semicircular fitting of the measured data. It is pointed out that fewer data points can be detrimental in the fitting process as the two methods may not provide identical results for the extracted parameters.

For a large quantity of data having a small interval between the successive points, it is beneficial to extract identical fitting parameters as these are very much similar in magnitude obtained for both methods. In the case of fewer data points either method loses accuracy and integrity of the results severely, despite knowing the equivalent circuit parameters. The references [1, 2, 9, 12, 15–18, 24] noted herein used multi-plane analysis for the complete scenario or interpretation of the equivalent circuit representation. Some case studies are illustrated and demonstrated with the real data as well as the simulated data.

The origin of the semicircular relaxation behavior observed for the measured data is described in the form of the equation of a circle in the 1955 text [38]. Available texts [33–36] did not provide adequate guidance of using concurrent multi-plane analysis though they mentioned about the existence of four complex plane formalisms for the same measured data. Among these available texts, one is completely dedicated to the liquid system [35] whereas others [33, 34, 36] demonstrated solid-state materials only via Z^*-plane analysis.

Cole-Cole [31] originally demonstrated complex plane analysis via complex permittivity (ε^*) in 1941 using ac small-signal electrical data. The parameter complex capacitance (C^*) can be converted to complex permittivity [10] when geometrical factors such as thickness and electrode area are used in normalizing it. Somehow the Z^*-plane analysis achieved popularity over the past few decades. It is worth noting that the direction of the measurement frequency for every measured datum point plotted in both Z^*- and C^*-planes is anti-clockwise. That is why many investigators possibly refer to the Z^*-plane plot as the "Cole-Cole" plot though there is no match for the equivalent circuit representation obtained in each of these two plots [39–41]. Precisely, the Z^*-plane yields an R-C (resistance-capacitance) parallel circuit [1–5] whereas the C^*-plane yields an R-C series circuit [9, 41] for the semicircular relaxation.

Sometimes the complex conductivity (σ^*) plot [24, 42] and C-C plot [24, 42] may be conducive in terms of extracting meaningful equivalent circuit model. The parameter complex conductivity is derived from the conversion of the terminal admittance ($Y^* = 1/Z^*$) when geometrical factors such as electrode area and thickness are used in normalizing it [10]. Although both σ^* (or Y^*) and ε^* (or C^*) plots yield a R-C series circuit [9, 12, 23, 24] for the semicircular relaxation, that does not imply both parameters are synonyms.

Some confusing issues found in the literature [43–50] are consistently misnomers though of minor concern. Worthy terminologies are often shadowed by these misnomers. Macdonald [37] pointed out that it is worthwhile to dispose of those misnomers and use the language as: "In recent years, it has become relatively common for writers in the IS field to refer to a Nyquist diagram, taken to mean a plot of the values of the real and imaginary parts of a complex quantity, such as impedance, in the complex plane. Such usage should be discouraged." A quantity such as impedance is basically derived from measurements of current and potential at a single (input) port. But Nyquist's work [51] dealt with two-port measurements of feedback in amplifiers and involved input and output voltage determinations. Thus, a complex-plane Nyquist plot is intrinsically quite different from an impedance or admittance plot in this plane. Instead of using a Nyquist diagram of impedance response, a better description would be a complex-plane impedance plot.

From Macdonald's standpoint [37] and despite noting elsewhere [10] beside Nyquist's original [50] paper, it appears that "Nyquist plot" is simply a misnomer when referred to any of the four complex plane plots. Specifically describing impedance plot having nomenclature of "Cole-Cole plot" [39–41] or "Nyquist plot" [43–50] is absolutely misleading due to the loss of justification of the physical basis. Emphasizing or using the term impedance (or admittance) as "ac impedance (or ac admittance)" or "complex impedance (or complex admittance)" is a redundant situation. There is no need of a "qualifier" as ac or complex for the term or parameter [10, 37]. This is because inherently impedance and admittance are associated with the ac small-signal measurements representing a complex quantity.

Cole-Cole (C-C) [31], Davidson-Cole (D-C) [32], and Havriliak-Negami (H-N) [25] provided complicated empirical non-Debye equations that have been distinguished diligently over the past few decades for the material systems. In the real world scenario, the Cole-Cole empirical equation [31] possesses a complicated depression parameter (expressed as α or conversely $1-\alpha$) handled on many occasions by the investigators [1–4, 6–9, 31] for which the impedance has the striking property that the resistance and reactance components are the functions of frequency, the phase angle is constant independent of the frequency and equal to $\alpha\pi/2$ [or conversely $(1-\alpha)\pi/2$]. The presumed "impedance element" causing the phase angle to be constant [31] is mostly referred to as the constant phase element (CPE) [33–36].

The empirical equations for both D-C [13] and H-N [38] are more complicated as both handle D-C skew parameter α and H-N skew parameters α and β. Each empirical equation portrays the generalized expression leading to the Debye condition when α and β do not exist. Recently D-C [13] and H-N [52]relaxations are demonstrated for the same set of data obtained for the Ni-Cd battery as α and β had small variation or difference between them. These two skew parameters are more complicated than the C-C depression parameter. The CNLS curve fitting of the distorted semicircular loci in the complex plane, designated as D-C and H-N relaxations becomes useful when more than one CPE is considered [53] in the equivalent circuit representation. The physical basis for such consideration, regardless of the underlying operative mechanisms within the material system, is not clear though curve fitting procedure appears suitable or satisfactory.

The advantage of using multiple complex planes in a concurrent manner yields maximum information about the DUT. As a device, an electronic system is very complicated as sometimes "polynary electronic materials" may be practical for applications as seen for the ZnO-based varistors [9] where more than half a dozen oxides are used in forming the sintered composite. In general, transport properties are strongly affected by the microstructure comprising of grains, grain boundaries, other secondary or tertiary phases, point defects, voids, etc. A reliable tool/technique is useful in delineating processing-microstructure-property relationships.

Often electrically active grain boundary junctions constitute complicated operative conducting path(s) between the two opposite electrodes (terminals) of the DUT. A combined "series-parallel network" of the grain boundary junctions is the effective representation between the terminals via a "single lumped" electrical barrier. The term series-parallel network implies that a significant microstructural influence on the electrical response of these materials is the total behavior, representing "m" junctions in parallel with "n" junctions in series, between the electrodes.

Immittance spectroscopy proves to be a useful tool/technique of characterizing the electrical nature of a number of heterogeneous composite systems [1–32]. It unravels the underlying competing phenomena via lumped parameter/complex plane analysis (LP/CPA) [8, 9]. It is advantageous to examine the measured ac electrical data in the form of terminal immittance in four complex planes. A semicircular fit of the data in any of these complex planes suggests an appropriate equivalent circuit representation of the observed dispersion.

The LP/CPA approach [8, 9] reveals the degree of structural uniformity, variation in electrical conduction path(s), stability, degradation, time-dependent processes, etc., in all kinds of inhomogeneous and heterogeneous composite-based devices. It elucidates intergranular electrical barriers, the origin of resistance or capacitance, their dispersion with ac small-signal frequency, and the role of defects within the regime of the electric field. This powerful analytical tool/technique is a reliable source of investigating complex material and device systems, and their design related to component and device applications. Finally, this approach allows the development of an equivalent circuit representation of the total ac response which incorporates the underlying mechanism(s) in a systematic manner for a given set of experimental conditions.

After the four complex plane formalisms were established [3, 4, 31, 54, 55], the usage of the immittance data accelerated investigating variety material systems in order to achieve smart materials and potential device applications. The immittance data were in use extensively for a while but no extensive curve fitting procedure existed for the semicircular relaxation for a long time until the concept of precision evolved gradually [56]. Mainframe-based semicircular curve fitting procedure was used in the interpretation of the transitions in the ripening fruits [57]. Almost concurrently the same approach used Levenberg-Marquardt [58, 59] Algorithm with the HP86 desktop computer for analyzing the immittance data of the ZnO-based varistors [8, 9].

It is noticed that the numerous users of the immittance data presumably exhibit reluctance in developing necessary algorithm and/or software for the curve fitting procedure, though published literature continued massive volumes on the usage of this tool/technique. At the same time it may be apprehended that there may be a lack of concept about the data-handling procedures but there is a desire for using the immittance data. In recent days it is also noticed that there are serious deficiencies [60–68] in the appropriate requirements for the data-handling procedures. These deficiencies include proper use of an equal grid on the axes, extraction of the equivalent circuit elements, relaxation time, identification of Debye or non-Debye response, etc. An unequal grid on the axes does not allow semicircular fitting accurately. For example, (1) equivalent circuit elements do not get extracted unless both ends of the semicircle intersect the real axis, (2) proper peak position for the frequency corresponding to the maximum of the imaginary part cannot be identified, (3) presence of depression angle does not become evident, etc. These losses of accuracy dilute the interpretation of the underlying operative phenomena for the DUT. That is why an equal grid is necessary for the extraction of the parameters.

This chapter illustrates with case studies how to handle the immittance data when semicircular relaxation is obtained in the complex plane formalisms. The case studies are provided with the real data and subsequent fitting using the necessary algorithms. Both geometric and CNLS approaches are highlighted with the results. Interpretation of the semicircular relaxation or other responses in the complex plane follows standard approaches [10].

11.2 Measurement of the Immittance Data and Complex Plane Analysis

A variety of commercial Impedance Analyzers having certain frequency range is often used in the acquisition of the ac small-signal electrical data. These data are used in the eventual characterization of the material or device system. Various response characteristics are observed depending on the type of the material. Thus, the processing-microstructure property of the material system can be ascertained through the analytical results. Some responses indicate perfect Debye-like behavior resulting from the perfect or the near perfect crystal structure [38]. Many responses indicate non-Debye type behavior originating from the distributed nature of the relaxation time resulting from small or large defects in the crystal structure of the materials [9, 10, 13, 19].

The actual characteristics of the materials become more evident when the acquired immittance data are separated in real and imaginary parts, and then plotted in a complex plane. While a Debye-like behavior exhibits semicircular loci having the center on the real axis, a non-Debye behavior is characterized by semicircular loci having the center located below the real axis [38]. A nonlinear regression algorithm is used to fit the plotted data and extract equivalent circuit parameters [9–13, 19, 24, 33–36]. An equivalent circuit representation of the data portrays the intrinsic dielectric behavior of the materials. Basics of nonlinear regression algorithms are presented in this paper followed by the fitting procedure of the measured data and case studies.

Immittance data can be acquired using either series or parallel mode of measurement beside phasor determination [10]. Two parameters are measured in each form of measurements as a function of applied frequency (f). For example, parallel mode of measurement having two parameters will be used in this paper for demonstration.

In the parallel mode C_p and G_p referred to the parallel capacitance and the parallel conductance, respectively whereas in the series mode C_s and R_s referred to the series capacitance and the series resistance, respectively. The data can be acquired in the frequency range, as an example, 0.1 Hz – 13 MHz using small voltage amplitude as the ac signal for the familiar impedance analyzers. The measured immittance, $I^*(\omega)$, is a complex quantity expressed as:

$$I^*(\omega) = I'(\omega) \pm jI''(\omega), \tag{11.1}$$

with $j = \sqrt{-1}$ and ± in front of the imaginary part depends upon the type of complex plane formalism selected. The measured data are inter-convertible to four complex planes: Z*-, Y*-, C*-, and M*- planes [10]. The same data can be presented in each complex plane by plotting the real part of $I^*(\omega)$ on x-axis and the imaginary part on the y-axis. The actual form of $I^*(\omega)$ and the corresponding real and imaginary parts in the four planes are shown in Figure 11.1. The relationships between two complex quantities are also shown therein. It is to be noted that the same measured data can be represented in all four complex planes using the relationships provided in the illustration.

On many occasions real versus imaginary plot of the immittance data results in semicircular loci in the complex plane. If the center of the semicircle lies on the real axis, the behavior of the material is termed as the Debye response. The non-Debye behavior is characterized by a depressed semicircle with center located below the real axis. The peak of the semicircle provides the relaxation time (or time constant) associated with the

Complex plane	Complex function	Relationships
Z*	$\frac{G_P}{G_P^2 + \omega C_P^2} - j\frac{\omega C_P}{G_P^2 + \omega C_P^2}$	$Z^* = 1/Y^* = M^*/j\omega$ $= 1/j\omega C^*$
C*	$C_P - j\frac{G_P}{\omega}$	$C^* = 1/j\omega Z^* = Y^*/j\omega$ $= 1/M^*$
Y*	$G_P + j\omega C_P$	$Y^* = 1/Z^* = j\omega C^*$ $= j\omega/M^*$
M*	$\frac{\omega^2 C_P}{G_P^2 + \omega^2 C_P^2} + j\frac{\omega G_P}{G_P^2 + \omega^2 C_P^2}$	$M^* = j\omega Z^* = 1/C^*$ $= j\omega/Y^*$

Figure 11.1 Complex plane representation of the measured immittance data in the parallel mode comprising of C_p and G_p as the function of frequency f.

underlying mechanism of the material. The non-Debye behavior can be attributed to the presence of distributed relaxation time for the material. Often more than one semicircle is observed in a single complex plane indicating the presence of multiple relaxation times corresponding to multiple elements in the circuit configuration for the DUT. Essentially relaxation response yields an equivalent circuit that requires a fitting procedure of the semicircle observed in a complex plane for the measured data points. The semicircle is fitted using the method of nonlinear regression.

11.3 Nonlinear Least Squares Estimation

The basics of nonlinear regression are presented by actual implementation in a complex plane. Nonlinear least squares (NLS) estimation, also known as nonlinear regression analysis, involves estimation of the parameters of a nonlinear function by minimizing the errors between the fitted and the observed data. Examples of nonlinear functions include the complex quantity of immittance response functions such as Cole-Cole [31], Davidson-Cole [32], and Havriliak-Negami [25]. The Cole-Cole immittance, I, is a function of the angular frequency ω and a set of parameters, β:

$$I^*(\omega;\beta) = I_{LI} + \frac{I_{RI} - I_{LI}}{1+(j\omega\tau)^a} = I' - jI'', \tag{11.2}$$

where I′ and I″ are the real and imaginary parts of the complex function, I. The parameters I_{LI} and I_{RI} denote the left-intercept and the right-intercept of the same semicircle in the complex plane, respectively. The non-negative parameter α is the Cole-Cole depression parameter which ranges between 0 and 1 for which $\alpha \rightarrow 1$ represents the ideal or Debye response, and $\alpha \rightarrow 0$ represents the extreme non-ideal or non-Debye behavior.

In reality, equation (11.2) represents the relaxation behavior in the Z*- and C*-plane with $I_{RI} > I_{LI}$, $I_{LI} = I_{HF}$, the high-frequency intercept on the left side of the semicircle, and $I_{RI} = I_{LF}$, the extremely low-frequency intercept. For Y*- and M*-plane, the negative sign in front of the imaginary part is replaced by the positive sign, and $I_{LI} = I_{LF}$ and $I_{RI} = I_{HF}$. The measurement frequency, ω (in radian for $\omega = 2\pi f$ with f in Hz) increases in the anti-clockwise direction along the semicircular arc for the Z*- and C*-plane, and in the clockwise direction for the Y*- and M*-plane. The relaxation time (in sec) is denoted by τ.

There exists one-to-one relationship between the parameters of two complex planes which are reciprocal to each other, for both Debye and non-Debye responses. The reciprocal relationship exists between Z*- and Y*-plane, and between C*- and M*-plane as presented in Figure 11.1. Using equation (11.2), which corresponds to Z*- or C*-plane, the equation for the reciprocal plane (Y* or M*) can be expressed as:

$$\left[I^{*}(\omega)\right]^{-1} = \frac{1+(j\omega\tau)^{a}}{I_{LI}(j\omega\tau)^{a} + I_{RI}} \tag{11.3}$$

Using the method of partial fractions, equations (11.3) can be expressed as:

$$\left[I^{*}(\omega)\right]^{-1} = I_{LI}^{-1} - \frac{I_{LI}^{-1} - I_{RI}^{-1}}{1+(j\omega\tau_1)^{a}}, \tag{11.4}$$

where

$$\tau_1 = \tau\left(I_{LI} I_{RI}^{-1}\right)^{a^{-1}}. \tag{11.5}$$

Equation (11.4) has the same form as in equation (11.2), but with different intercepts on the real axis and the relaxation time. The relaxation times of the two planes are related as shown in equation (11.5). When I_{LI}

is non-zero, Equation (11.4) is valid, i.e. a small non-zero left-intercept is obtained in Z*- or C*-plane. If the left-intercept in Z*- and C*-planes is zero, no semicircle is obtained in the reciprocal plane (Y*- and M*-planes). Setting $\alpha = 1$ in equation (11.4) and (11.5) provides the corresponding Debye version of the equation.

As shown in equation (11.2), $I^*(\omega)$ is linear in I_{LI} and I_{RI}, and nonlinear in α and τ. Thus, the nonlinear regression analysis involves the determination of $\beta = [I_{LI}, I_{RI}, \alpha, \tau]^T$ such that the errors between the model and observed responses are minimized. The general algorithms and the applications of NLS for the fitting of the complex function in equation (11.2) are discussed. Matrix notations are used wherever possible for computer implementations. Bold letters indicate matrices or vectors.

In general, the nonlinear least square estimation procedures minimize a residual sum of squares objective function. The objective function to be minimized is a scalar valued weighted residual sum of squares [69]:

$$\Phi = \sum_i w_i R_i^2 = R^T W R, \quad i = 1,2,..., \tag{11.6a}$$

in which

$$R = y - f(x;\beta), \tag{11.6b}$$

with $\beta = [\beta_1, \beta_1, \ldots, \beta_p]^T$ are p number of parameters to be determined such that the residual Φ is minimized having n (> p) number of observations $x = [x_1, x_2, \ldots, x_n]^T$. In equation (11.6) R is a residual column vector of size n and function of β, and y is a column vector of the observed data of size n, having $f(x; \beta)$ is a column vector of predicted (fitted) values of size n. Assuming that the observations are uncorrelated, the matrix W is a diagonal square matrix of size n with diagonal element w_i representing the weight for each measurement. Usually the weights are inversely proportional the measurement errors or the variance of each observation, $\sigma_i^2 = \mathrm{Var}(y_i)$; thus larger weights are used for smaller variance of y_i. When the weights are exactly equal to $1/\sigma_i^2$, the sum of squares Φ becomes χ^2 (Chi-Square) error criterion. The choices for the weights are included.

There are several algorithms of nonlinear least squares estimations used for the analysis of the immittance data [33–36, 21, 53]. There are two common algorithms that require an iterative solution procedure: (a) Gauss-Newton method [33–36, 69] and (b) Levenberg-Marquardt method [58, 59]. The fitted function is linearized in the neighborhood of a value of the parameter β, a set of linear equations are solved to obtain an estimate of the correction to the parameter, and the residual Φ is updated. In the next iteration, the new corrected values of the parameters are used. The iteration continues until convergence is achieved.

11.3.1 Gauss-Newton Method (Algorithm) of Least Squares Estimation

The estimate of β at kth iteration (k = 1, 2, ...; k = 0 corresponds to the initial guess) is given by:

$$\beta \approx \beta^{(k)} = \beta^{(k-1)} + \Delta\beta \tag{11.7}$$

where $\Delta\beta$ is the increment (correction, or shift vector) to the estimated value of β obtained in the previous iteration, and the superscripts indicate the iteration number. The first order Taylor series approximation of f(x; $\beta^{(k)}$) in the neighborhood of $\beta^{(k-1)}$ is:

$$f(x_i;\beta^{(k-1)} + \Delta\beta) \approx f\left(x_i;\beta^{(k-1)}\right) + \sum_j \frac{\partial f\left(x_i,\beta^{(k-1)}\right)}{\partial \beta_j}\Delta\beta_j \tag{11.8}$$
$$\text{with } i = 1,\ldots,n;\;\; j = 1,\ldots,p$$

From equation (11.6b), the derivatives of the residuals are:

$$\frac{\partial R_i}{\partial \beta_j} = -\frac{\partial f\left(x_i,\beta\right)}{\partial \beta_j}. \tag{11.9}$$

Then equation (11.8) becomes:

$$f(x_i;\beta^{(k)}) \approx f\left(x_i;\beta^{(k-1)}\right) - \sum_j \frac{\partial R_i^{(k-1)}}{\partial \beta_j}\Delta\beta_j. \tag{11.10}$$

The derivatives of the sum of squares in equation (11.6a) are:

$$\frac{\partial \Phi}{\partial \beta_j} = 2\sum_i w_i R_i \frac{\partial R_i}{\partial \beta_j} = 2\sum_i w_i\left(y_i - f\left(x_i;\beta\right)\right)\frac{\partial R_i}{\partial \beta_j}. \tag{11.11}$$

With equation (11.10), equation (11.11) becomes:

$$\frac{\partial \Phi^{(k)}}{\partial \beta_j} = 2\sum_i w_i\left(\Delta y_i^{(k-1)} + \sum_j \frac{\partial R_i^{(k-1)}}{\partial \beta_j}\Delta\beta_j\right)\frac{\partial R_i^{(k-1)}}{\partial \beta_j}. \tag{11.12}$$

where

$$\Delta y_i^{(k-1)} = y_i - f\left(x_i; \beta^{(k-1)}\right), \tag{11.13}$$

i.e. the difference between the observed and the predicted values. At the kth iteration, all quantities are calculated using their current values (i.e. the values obtained in the previous iteration). With

$$\frac{\partial R_i}{\partial \beta_j} = J_{ij}, \tag{11.14}$$

the elements of the Jacobian matrix of the residual function, equation (11.12) becomes:

$$\frac{\partial \Phi^{(k)}}{\partial \beta_j} = 2\sum_i w_i \left(\Delta y_i^{(k-1)} + \sum_j J_{ij}^{(k-1)} \Delta \beta_j \right) J_{ij}^{(k-1)}. \tag{11.15}$$

The parameter set β that minimizes Φ is obtained as:

$$\frac{\partial \Phi}{\partial \beta_j} = 0 \tag{11.16}$$

which gives

$$\sum_i w_i \left(\Delta y_i^{k-1} + \sum_j J_{ij}^{k-1} \Delta \beta_j \right) J_{ij}^{k-1} = 0 \tag{11.17a}$$

and

$$\sum_i J_{ij}^{k-1} w_i \sum_j J_{ij}^{k-1} \Delta \beta_j + \sum_i J_{ij}^{k-1} w_i \Delta y_i^{k-1} = 0 \tag{11.17b}$$

In matrix notations, equation (11.17b) becomes:

$$J^T W J \Delta \beta + J^T W \Delta y = 0. \tag{11.18}$$

which is evaluated using the values of the parameters obtained in the (k – 1)[th] iteration. The matrix J is of size n × p, W is of size n × n and 0 is the null vector. Here superscript "T" denotes the transpose of the matrix. Equation (11.18) is a set of p linear equations from which the correction to the parameter set is obtained as:

$$\Delta\beta = -\left(J^T W J\right)^{-1} J^T W \Delta y, \tag{11.19}$$

and the new estimates of the parameters for the next iteration are obtained from equation (11.7). The iteration starts with an initial estimate of the parameter β. Once the convergence is achieved, several statistics are calculated to obtain estimates of various errors. Some important issues related to this procedure and the general nonlinear least square procedure are noted below.

1. The convergence may be specified as

$$\Phi^{(k)} - \Phi^{(k-1)} \leq tol, \tag{11.20}$$

   where tol is a specified constant (e.g. 10^{-5}). The tolerance can also be specified for each parameter. The objective function is quadratic in terms of the parameters in the neighborhood of its minimum value where the truncated Taylor series expression may be a good approximation of the model function. The trial values of the parameters should be close to the final values for which $\Phi = \Phi_{min}$, so that Φ is quadratic function of the parameters and Taylor series expansion of the model is reasonably valid around the neighborhood of the trial values. The method is convergent when the objective function is approximately quadratic in parameters. The convergence of the problem may be achieved within reasonable number of iterations if the initial estimates are near the optimal values. A good estimate of the initial value of β may be obtained by observing the visual display of the data on a graph paper and fitting an approximate curve to the data.
2. If divergence occurs, a possible solution could be to multiply the shift vector $\Delta\beta$ by a fraction α,

$$\beta^{(k)} = \beta^{(k-1)} + a\Delta\beta, \tag{11.21}$$

   α may be optimized by a line-search algorithm which minimizes the objective function within $0 < \alpha < 1$. If divergence occurs even with this method if the direction of the shift vector is far from the ideal direction, or the value of α is very close to zero, Levenberg-Marquardt algorithm may be used in such cases.

3. The normal equations can be solved by Cholesky decomposition [1] or QR factorization [1]. Standard computing software packages such as MATLAB [2] provide efficient built-in functions for solving system of linear equations. If the parameters are not exactly correlated, the matrix J^TWJ can be inverted if enough data points are available. It is necessary that n > p and the matrix J^TWJ is positive definite (being a symmetric real positive definite, it can be decomposed into a unique lower triangular matrix and a unique upper triangular transpose matrix), otherwise it is not invertible. In cases where the matrix is close to singular large truncation and rounding errors may be induced in the solution if direct inversion is attempted.

11.3.2 Levenberg-Marquardt Method (Algorithm) of Least Squares Estimation

If divergence occurs because of the direction of the shift vector is very far from the ideal direction, Levenberg-Marquardt method [58, 59] may be used. The algorithm is a combination of Gauss-Newton method [69] and Gradient-Descent method. It is a robust algorithm which can handle an initial guess far from the optimal solution. To improve convergence, Levenberg [55] included a nonnegative scalar damping factor, λ, in equation (11.8):

$$\Delta\beta = -\left(J^TWJ + \lambda I\right)^{-1} J^TW\Delta y, \tag{11.22}$$

where I is an identity matrix of size p × p. To ensure large movement in the direction of small gradient, Marquardt [59] replaced the identity matrix with the diagonal matrix,diag(J^TWJ), the jth element of which is the gradient of Φ in the jth direction,

$$\Delta\beta = -\left(J^TWJ + \lambda\,\mathrm{diag}\left(J^TWJ\right)\right)^{-1} J^TW\Delta y. \tag{11.23}$$

The value of the damping factor is updated in each iterative procedure. If a choice of the damping factor reduces Φ rapidly, a reduced value is used in the next iteration, bringing the method close to Gauss-Newton method. Otherwise, if the reduction is small or insignificant, the damping factor may be increased in the next iteration, essentially bringing the method close to the Gradient-Descent method. Once convergence is achieved, the fitting statistics are calculated.

When the initial guess is far from the optimal value, Taylor series is not a good approximation; instead, a larger value of λ is used, essentially turning the method to Gradient-Descent. Near Φ_{min}, Taylor series is a good approximation and a smaller value of λ is needed which is equivalent to Gauss-Newton method. For $\lambda >> 1$, Levenberg-Marquardt method is similar to the Gradient-Descent method and for $\lambda = 0$, it is identical to the Gauss-Newton method.

11.3.3 Numerical Procedure to Calculate Jacobian Matrix

In the situation where closed form expressions of the derivatives of the residual function are not available, numerical calculations must be carried out to obtain the Jacobian matrix. This can be achieved by using the finite difference method. The second order central difference scheme for the first derivative may be used:

$$\frac{\partial R}{\partial \beta} \approx \frac{R(\beta + \delta\beta) - R(\beta - \delta\beta)}{2\delta\beta}, \tag{11.24}$$

in which $R(\beta + \delta\beta)$ and $R(\beta - \delta\beta)$ are the residual function values evaluated at $\beta + \delta\beta$ and $\beta - \delta\beta$, respectively, with $\delta\beta$ being a small increment of the parameter. The perturbation $\delta\beta$ should not be too large to avoid large approximation error and should not be too small to avoid round-off error. The closed form expressions for the derivatives of the residual function are always preferred for accuracy of the results.

11.3.4 Error Analysis: Analysis of Errors in Regression

Once the convergence is achieved, various asymptotic standard errors [33–36, 69, 70] are calculated. The estimation of the error variance, σ^2, is given by:

$$S^2 = \Phi_{min}/(n - p). \tag{11.25}$$

The variance-covariance matrix of β is given by:

$$C = \left(J^T W J\right)^{-1}. \tag{11.26}$$

The standard error estimates of the parameters are given by the diagonal elements of C:

$$S_{\dagger_j} = \sqrt{C_{jj}}. \tag{11.27}$$

If the diagonal elements of W are equal to $1/\sigma^2$ (equal variance measurements), then $W = \sigma^{-2} I,\ C = \sigma^2 \left(J^T J\right)^{-1}$, and the variance of the jth parameter is simply equal to the jth diagonal element in $(J^T J)^{-1}$ multiplied by S^2. The asymptotic standard error is a measure of how unexplained variation in the data affects the solution. The confidence interval (the lower confidence limit, LCL, and the upper confidence limit, UCL) of the parameter is then obtained as:

$$LCL = \beta_j - t_{\alpha/2,\, n-p} S_{\beta_j}, \quad UCL = \beta_j + t_{\alpha/2,\, n-p} S_{\beta_j} \tag{11.28}$$

in which $t_{\alpha/2,\, n-p}$ is the t-statistic with α as the level of significance (usually $\alpha = 0.05$) and $n - p$ degrees of freedom. The standard error of the fitted values of the function is:

$$S_f = \sqrt{J^T \left(J^T W J\right)^{-1} J}. \tag{11.29}$$

The standard error of the fitted value is a measure of how variations in the parameters affect the fitted values. The quality of the fit is determined by visual inspection of the data and calculating the R^2 statistic given by:

$$R^2 = 1 - \frac{\Phi_{\text{res}}}{\Phi_{\text{total}}} \tag{11.30}$$

in which Φ_{res} and Φ_{total} are the residual and total (with respect to the mean) sums of squares respectively. The R^2 statistic compares the unexplained variance with the total variance; a higher R^2 value indicates better fit. The Analysis of Variance (ANOVA) table is prepared by separating the total sum of squares into regression and error sums of squares. The F-statistic is also calculated to check overall model validity. It is to be noted that these statistics do not have the same meaning in nonlinear regression as they have in linear regression; however they are quite useful to check the general validity of the model.

11.3.5 Selection of the Weights

The weights [33–36] may be estimated by repeated observations at each frequency. As mentioned earlier, if $w_i = 1/\sigma_i^2$, the sum of squares becomes χ^2 (Chi-Square) error criterion. To obtain an independent estimate of σ_i, multiple observations at each frequency should be performed. In this ideal case, the minimum value of Φ becomes the estimate of χ^2 statistic and the χ^2 goodness of fit test [54] may be used to assess the quality of the regression in addition to the R^2 statistic.

Because of time brevity and practical reasons, single measurement is often performed at each frequency. Thus, independent estimate of σ_i for each observation is not available and Chi-Square goodness of fit criterion cannot be used since no such estimate is available for this case. Therefore, some assumption regarding σ_i must be taken. If all measurement standard deviations are assumed to be equal (equal uncertainty of each data point, fitted curve passes through the mean values of the observations, i.e. the fit is ideal), $\sigma_i = \sigma$, then $W = \sigma^{-2}I$ with σ being estimated by the standard error S. This is equivalent to setting all weights equal to 1.0 in equation (11.8), (11.10), and (11.11). Another strategy requires assuming $\sigma_i^2 = y_i^2$ known as the proportional weighting [33–36, 53]. In the presented examples, single observation at each frequency has used $\sigma_i = \sigma$.

11.4 Complex Nonlinear Least Squares (CNLS) Fitting of the Data

The two algorithms presented above may be used for fitting immittance data in complex planes. This can be achieved in two ways [7]:

1. geometrical fitting of the data in complex plane involving fitting of a nonlinear function to the plotted data in the complex plane, and
2. simultaneous fitting of real and imaginary parts of the data over the entire frequency domain.

These two procedures provide estimations of the parameters in equation (11.2) and the equivalent circuit parameters.

11.4.1 Procedure 1: Geometrical Fitting in the Complex Plane

In the geometrical fitting procedure, the nonlinear regression analysis is performed by directly fitting a nonlinear real valued function, $y = f(x; \beta)$, which regresses the imaginary values (y) to the real values (x) in the complex plane, provided such a function is available and can simulate the behavior. The real and imaginary parts of the complex function in equation (11.2) are given by:

$$I' = I_{LI} + \left(I_{RI} - I_{LI}\right)\frac{\left[1+\left(\omega\tau\right)^{a}\cos\left(a\pi/2\right)\right]}{\left[1+\left(\omega\tau\right)^{2a}+2\left(\omega\tau\right)^{a}\cos\left(a\pi/2\right)\right]} \tag{11.31a}$$

$$I'' = \left(I_{RI} - I_{LI}\right)\frac{\left(\omega\tau\right)^{a}\sin\left(a\pi/2\right)}{\left[1+\left(\omega\tau\right)^{2a}+2\left(\omega\tau\right)^{a}\cos\left(a\pi/2\right)\right]} \tag{11.31b}$$

Eliminating $\omega\tau$ from these equations and rearranging,

$$\left(I' - \frac{I_{RI}+I_{LI}}{2}\right)^{2} + \left(I'' + \frac{I_{RI}-I_{LI}}{2}\cot\left(a\pi/2\right)\right)^{2} = \left(\frac{I_{RI}-I_{LI}}{2\sin\left(a\pi/2\right)}\right)^{2}. \tag{11.32}$$

Equation (11.32) represents the equation of a circle with the center located at the coordinates $\left(\frac{I_{RI}+I_{LI}}{2}, -\frac{I_{RI}-I_{LI}}{2}\cot\left(a\pi/2\right)\right)$ and radius equal to $\frac{I_{RI}-I_{LI}}{2\sin\left(a\pi/2\right)}$. Therefore, an ideal Cole-Cole (i.e., ideal non-Debye) behavior is represented by the data as a semicircle in the complex plane with the center depressed below the real axis as shown in the schematic diagram in Figure 11.2(a). In case of the ideal Debye behavior, the center is situated on the real axis. The geometrical fitting procedure involves fitting a semicircle to the data. The purpose of the NLS is to obtain the coordinates of the center and radius of a semicircle which minimizes the residual sum of squares between the fitted semicircle and the observed data. The parameters of the Cole-Cole immittance function and the equivalent circuit are then determined from the fitted circle parameters.

As shown in equation (32) and Figure 11.2(a), the relaxation behavior of Cole-Cole model in the complex plane indicates semicircular behavior. Thus, the goal of NLS is to fit a semicircle to the acquired data to check if the relaxation behavior observed in the measurement is actually due to

Cole-Cole relaxation. The residual to be minimized may be constructed from the equation of a circle:

$$R_i = \sqrt{\left(x_i - x_C\right)^2 + \left(y_i - y_C\right)^2} - r, \quad \text{with } i = 1, \ldots, n. \tag{11.33}$$

The parameters to be optimized are the coordinates of the center, x_C and y_C, and the radius of the circle, r. The derivatives of the residuals in the Jacobian matrix are therefore:

$$\begin{aligned} J_{i1} &= \frac{\partial R_i}{\partial x_C} = \frac{-\left(x_i - x_C\right)}{\sqrt{\left(x_i - x_C\right)^2 + \left(y_i - y_C\right)^2}} \\ J_{i2} &= \frac{\partial R_i}{\partial y_C} = \frac{-\left(x_i - y_C\right)}{\sqrt{\left(x_i - x_C\right)^2 + \left(y_i - y_C\right)^2}} \\ J_{i3} &= \frac{\partial R_i}{\partial r} = -1 \end{aligned} \tag{11.34}$$

To obtain the complete description of Cole-Cole behavior, six parameters are to be estimated: I_{LI}, I_{RI}, x_C, y_C, r and τ. The coordinates of the center and the radius are estimated from the regression analysis using Gauss-Newton or Levenberg-Marquardt algorithm. The other parameters are estimated from the geometry of the fitted circle.

The two intercepts, I_{LI} and I_{RI}, and the coordinates of the center in equation (11.32) are indicated in Figure 11.2(a). Two chords are drawn by joining the points of the left- and right-intercepts to a point on the circle and the lengths are denoted as U and V. The ratio between U and V can be obtained from equation (11.31) as:

$$\ln\left(U / V\right) = a \ln \omega + a \ln \tau. \tag{11.35}$$

Thus, if ln(U/V) is plotted against ln ω, the resulting plot will be a straight line, as shown in Figure 11.2(b). At the peak frequency, U = V. Therefore the peak frequency may be obtained by fitting a straight line to the data ln(U/V) versus ln ω, setting ln(U/V) = 0 and solving for $\omega = \omega_{peak}$ from the equation of the straight line. The slope is negative for Z*- and C*-plane and positive for Y*- and M*-plane. The parameter α is usually obtained from the relation:

$$\theta = \left(1 - a\right)\pi / 2, \tag{11.36}$$

where θ is the depression angle shown in Figure 11.2(a). Once ω_{peak} (or f_{peak}) is determined, the relaxation time is obtained as:

$$\tau = 1/\omega_{peak} = 1/\left(2\pi f_{peak}\right) \tag{11.37}$$

The left- and right-intercepts are obtained from the circle equation:

$$I_{HF} = x_C - \sqrt{r^2 - y_C^2}, \quad I_{LF} = x_C + \sqrt{r^2 - y_C^2}. \tag{11.38}$$

The circuit parameters (resistance and capacitance, in parallel or series) are obtained from the following relations:

$$\text{for } Z^*\text{-plane:} \quad R = \text{chord} \quad \text{and} \quad C = \tau/R, \tag{11.39a}$$

$$\text{for } C^*\text{-plane:} \quad C = \text{chord} \quad \text{and} \quad R = \tau/C, \tag{11.39b}$$

$$\text{for } Y^*\text{-plane:} \quad R = 1/\text{chord} \quad \text{and} \quad C = \tau/R, \tag{11.39c}$$

and

$$\text{for } M^*\text{-plane:} \quad C = 1/\text{chord} \quad \text{and} \quad R = \tau/C. \tag{11.39d}$$

In these expressions:

$$\text{chord} = | I_{LI} - I_{RI} |. \tag{11.39e}$$

Another approach of fitting circle equation is to use a linear model[8] of the residual function. The residual may be defined as:

$$R_i^2 = \left(x_i - x_C\right)^2 + \left(y_i - y_C\right)^2 - r^2 = Ax_i^2 + By_i^2 + Cx_i + Dy_i + E \tag{11.40}$$

where the coefficients A, B, C, D, and E are related to the circle parameters x_C, y_C and r. Equation (11.40) is linear with respect to the coefficients which may be estimated by linear least square regression. Once they are obtained, circle parameters are estimated from them. The advantage of this method is that no initial guesses of the parameters are needed and the solution is obtained without any iteration. The asymptotic parameter statistics are calculated as explained in the procedure for NLS.

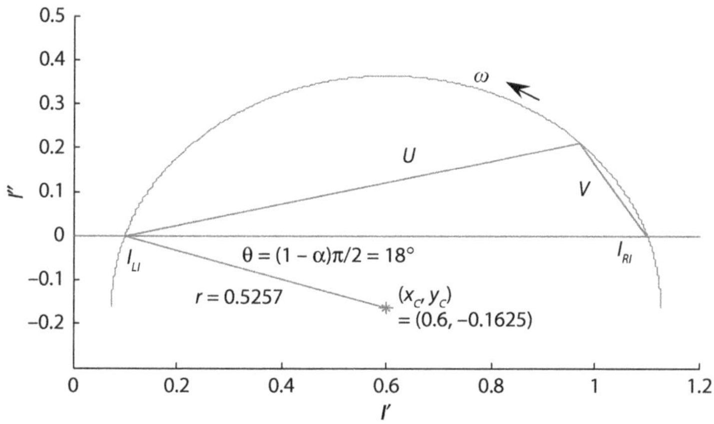

Figure 11.2 (a) Cole-Cole immittance response in complex plane where frequencies increase anti-clockwise along the arc of the semicircle for $\alpha = 0.8$, $I_{LI} = 0.1$, $I_{RI} = 1.1$

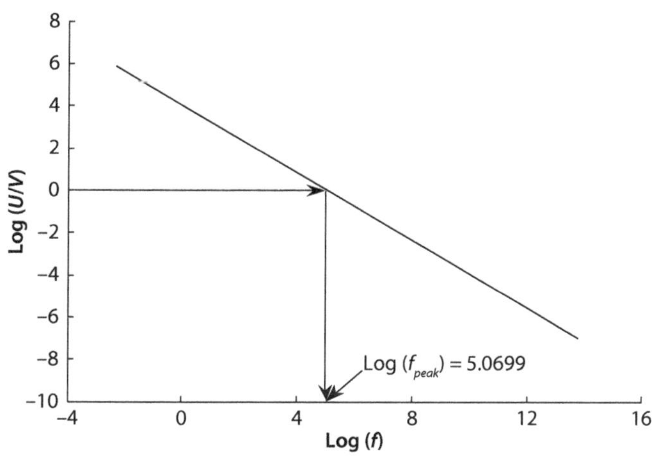

Figure 11.2 (b) Determination of the peak frequency, f_{peak}, for $\alpha = 0.8$ and $\tau = 10^{-3}$ sec.

The advantages of geometrical fitting are the following:

- The geometric fitting is intuitive and based on the shape of the relaxation curve;
- For this procedure a simple circle equation is needed; and
- Cole-Cole equation parameters are obtained directly from the fitted parameters from simple manipulations.

The disadvantages of this procedure are the following:

- The geometric fitting procedure is not applicable to immittance spectra of geometrical shape other than semicircle, such

as Cole-Davidson and Havriliak-Negami spectra, since these shapes are not readily describable in a functional form such as equation (11.32);
- The curve does not explicitly fit the real and imaginary parts since the two parts are not fitted simultaneously; and
- The errors associated with the Cole-Cole parameters are not readily available from the geometric fitting since the regression is not performed to estimate these parameters directly.

Nevertheless, because of its simplicity, the geometrical fitting procedure is very common, widely used by investigators and researchers to understand materials behavior. To overcome these disadvantages, especially if the error bounds of Cole-Cole parameters are needed, simultaneous fitting of the data in real and imaginary axes is needed. However, in many cases geometrical fitting is the only way to obtain meaningful estimations of the parameters compared to the simultaneous fitting procedure.

11.4.2 Procedure 2: Simultaneous Fitting of Real and Imaginary Parts

In this procedure, the real and imaginary parts of the complex immittance function are used simultaneously and the total error is minimized [6, 7]. The residual to be minimized is similar to equation (11.6), but computed separately over the real and imaginary components of the observations:

$$\begin{aligned}\Phi = \Phi' + \Phi'' &= \sum_i \left(w_i' R_i'^2 + w_i'' R_i''^2\right)\\ &= \sum_i \left[w_i'\left(I_i' - I'(\omega_i;\beta)\right)^2 + w_i''\left(I_i'' - I''(\omega_i;\beta)\right)^2 \right]\end{aligned} \tag{11.41}$$

where $\beta = [I_{LI}, I_{RI}, \alpha, \tau]^T$ are the four Cole-Cole parameters to be determined such that the residual Φ is minimized with n number of observations $\omega = [\omega_1, \omega_2, \ldots, \omega_n]^T$. In equation (41), I'_i and I''_i are the ith observed values of the real and imaginary parts respectively, $I'_i(\cdot)$ and $I''_i(\cdot)$ are the corresponding fitted values. The residuals involving the real and imaginary parts of the data are R′ and R″, respectively. The corresponding sum of squares are Φ' and Φ'', and the weights are w'_i and w''_i. Following the same procedure outlined in the Gauss-Newton algorithm, setting $\partial\Phi/\partial\beta = 0$, we have the following normal equations:

$$\left(J'^T W' J' + J''^T W'' J''\right)\Delta\beta + \left(J'^T W' J' + J''^T W'' J''\right) = 0 \tag{11.42}$$

in which

$$\begin{aligned} J'_{ij} &= \frac{\partial R'_i}{\partial \beta_j}, \quad J''_{ij} = \frac{\partial R''_i}{\partial \beta_j}, \\ \Delta y'_i &= I'_i - I'\left(\omega_i;\beta\right), \quad \Delta y''_i = I''_i - I''\left(\omega_i;\beta\right). \end{aligned} \tag{11.43}$$

The matrices W' and W'' are diagonal consisting of the weights w'_i and w''_i respectively. The Levenberg-Marquardt version of equation (11.34) is given by:

$$\begin{aligned} &\left[J'^T W' J' + J''^T W'' J'' + \lambda \mathrm{diag}\left(J'^T W' J' + J''^T W'' J''\right)\right]\Delta\beta \\ &\qquad + \left(J'^T W' \Delta J' + J''^T W'' \Delta J''\right) = 0. \end{aligned} \tag{11.44}$$

The Jacobian matrix J of the residual is of size n × p. The solution of the normal equations provides the estimates of the Cole-Cole parameters. The asymptotic parameter statistics are calculated as explained before for NLS.

As mentioned earlier, the major advantage of this fitting procedure is that the error bounds of the parameters are readily available from the fitting statistics. The disadvantage of this approach is that it is very difficult to make an initial guess of the parameter set, especially of α and τ, since no visual guideline is available. A poor guess may cause divergence. Also, the fitting does not provide the circle parameters; they need to be calculated from equation (11.32) before the semicircle can be drawn on the observed data. In the geometrical fitting procedure, the initial guess is easily made because the shape of the circle is clearly visible in the complex plane plot, and the convergence is almost always assured. The fitted circle is readily drawn on top of the measured data since the circle parameters are readily available from the fitting. This is the reason the geometrical fitting procedure is used most often. This problem can be avoided by adopting a two-step approach, first fit a circle geometrically to the data and then use the simultaneous fitting procedure using the parameters calculated from the circle fitting as initial guesses.

11.5 Graphical User Interface Implementation of the Nonlinear Least Square Procedures: Implementation of CNLS using MATLAB

The nonlinear regression analysis can be implemented using any commercial computing software, such as Matlab®. A set of functions (subroutines) can be developed to perform the intended steps. A flowchart of the overall program structure is shown in Figure 11.3. Matlab® has been chosen as the platform of data handling and CNLS analysis because of its scientific and technical computing capability with advanced data visualization. It is commonly used by scientists and engineers around the world and is readily available on campus of all academic institutions.

The major parts of the program are described here. Software codes written to perform Gauss-Newton and Levenberg-Marquardt procedures are presented here as example for the user. Important features of the program are described. It is to be noted that the procedures described herein may be implemented with any other standard computing software with visualization capabilities and modifications per user needs, such as Visual C++. The R-C parallel circuit shown in Figure 11.4 has been used as an example to describe the procedure. The procedure is demonstrated by fitting a semicircle to the impedance data (Z^*-plane). Examples of fitting in other complex planes are provided in the next section.

11.5.1 Input Data Generation

Input data file has been generated from the immittance response measurements using the impedance analyzer HP4192, but any other commercial

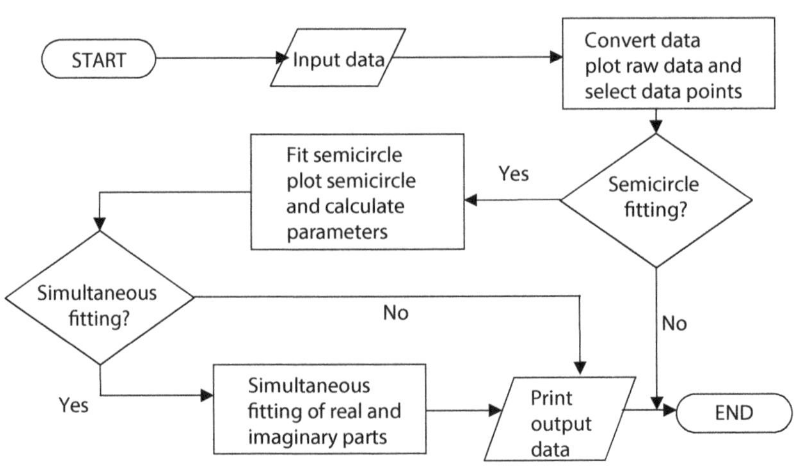

Figure 11.3 Flowchart of the program.

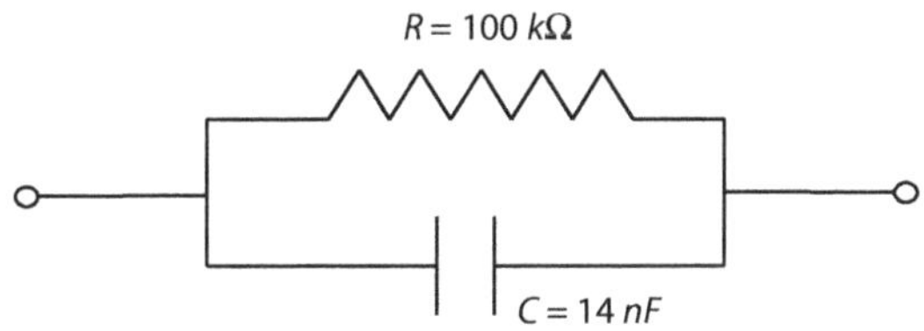

Figure 11.4 R-C parallel circuit with known values of the parameters used in data acquisition.

instrument can be used. Either parallel or series mode of measurements are possible with the impedance analyzer. In the examples parallel measurements have been used. Frequencies are varied from the lowest to the highest possible ranging from 5 Hz to 13 MHz.

Data are stored in an Excel spreadsheet in three columns: column 1 contains the frequency (f in Hz), column 2 contains the parallel capacitance (C_p in Farad or F) and column 3 contains the admittance (G_p in Siemens, S). The program requires that the data are entered in these standard units. If any other unit is displayed by the analyzer, it must be converted into these units while preparing the input file.

The response data collected for the R-C parallel circuit were for the conductance (G_p) and capacitance (C_p). Appendix - A shows Microsoft Excel® spreadsheet of the data collected from the impedance analyzer and the data supplied to the program. The data shown in columns 2, 3, 4 and 5 are as-displayed data by the analyzer and the columns 1, 6 and 7 contain the data supplied to the program. While converting the units into F and S, enough significant digits must be retained to avoid truncation of the measured data. Note that no truncation has occurred for the input data in the present case with five significant digits after the decimal.

Once the input data file is prepared, the program may be run from the Matlab® command prompt. A typical runtime display of Matlab® command window is shown in Figure 11.5.

11.5.2 Input Data Processing

The input file name and a choice of complex plane to be analyzed are required for processing the input data. The input measured data are converted into the real and imaginary parts for the selected complex plane and a matrix of size n × 4 is generated with columns containing the observation number, frequency, real part and imaginary part, n is the number of data points. This matrix is used in the subsequent analysis.

```
>> immispec
++++COMPLEX PLANE ANALYSIS++++
Want to convert data into selected complex plane (y/n)? y
Input file must have **THREE COLUMNS**in this order: Freq (Hz), Cp (F), Gp (S).
Enter input file name with **FULL PATH**:
  C:\...\...\...\...\...\inpdata.txt
Proceed (y/n)? y
Enter 1 for Z*-plane, 2 for Y*-plane, 3 for C*-plane, 4 for M*-plane: 1
Scale data (y/n)? y
Enter a factor to scale data (e.g. 1e-04): 1.E-04
Scale data (y/n)? n
Identify data points (y/n)?
Enter interval of data points for marking : 1
Continue for circle fitting (y/n)?
Enter individual data points or range of data points **TO BE IGNORED**; e.g.
[1,5,7:10,34,55:60]
Enter 0 (zero) if no data point to be ignored: 0
Enter initial estimates of circle parameters [x-coord; y-coord; radius]: [5; 0; 5]
Converged within maximum iteration limit.
See output table for details.
Continue to calculate Cole-Cole parameters (y/n)?
Continue to simultaneous fitting (y/n) ?
Converged within maximum iteration limit.
See output table for details.
Print Output file (y/n)?
```

Figure 11.5 On-screen display of Matlab® command window.

11.5.2.1 Visualization of the Measured (Raw) Data

The data obtained from the input file (in appropriate units) are plotted on four complex planes as shown in Figure 11.6. The imaginary part is plotted along the y-axis and the real part along the x-axis. The complex plane plots are made in equal scale axes. The equal sale is necessary to discern any semicircular behavior. As seen in Figure 11.6, semicircular behavior is observed in two complex planes, Z^* and M^*-plane. After visualizing the data in four complex planes, the user selects a complex plane for analysis. The data are then shown on that particular complex plane. The plotted data are scaled if needed. The data points are identified by their corresponding observation numbers. Figure 11.7 shows the measured data points for the R-C parallel circuit in Z^*-plane. A visual inspection of the plot is performed and a decision is made determining whether the data is suitable for further analysis. This completes the visualization part of the program and the user has the choice to proceed to NLS fitting if desired, or terminating the program at this level if the data are not suitable for analysis.

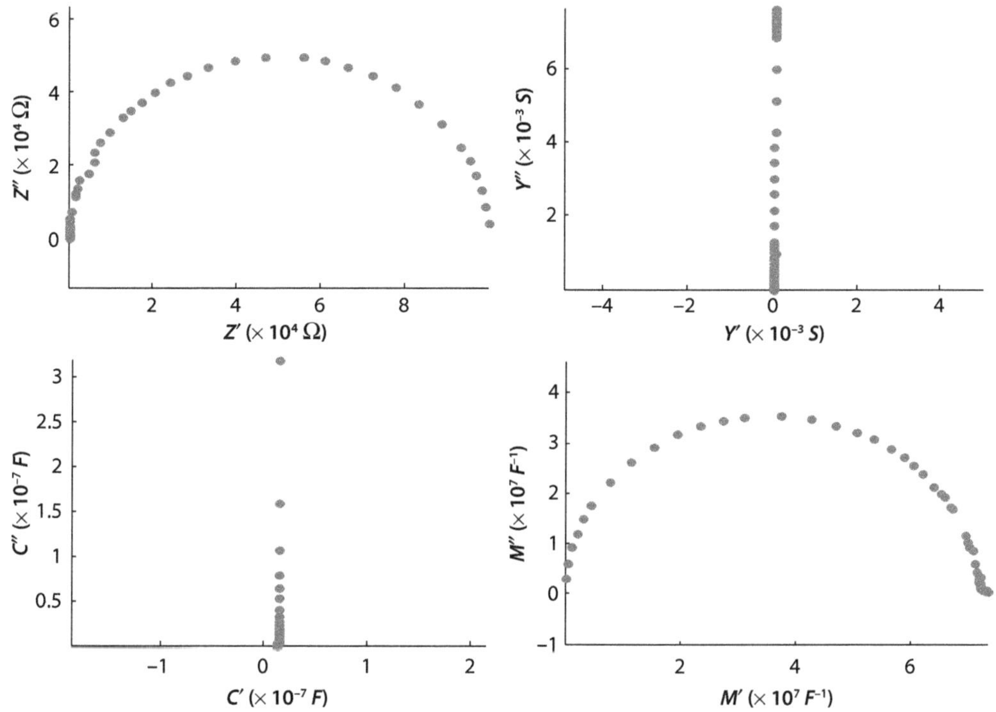

Figure 11.6 Four complex plane plots of the measured data for the R-C circuit shown in Figure 11.4.

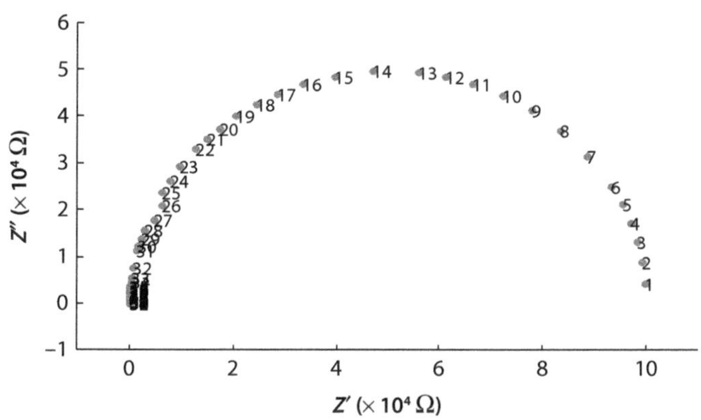

Figure 11.7 Measured data of the R-C circuit (Figure 11.4) displayed in the Z*-plane showing semicircular behavior.

11.5.2.2 Selection of Data Points for Fitting

A set of data points is selected by the user for which the semicircle will be fitted. This range is selected based on visual inspection of the data for which semicircular behavior is observed. The user has the option to

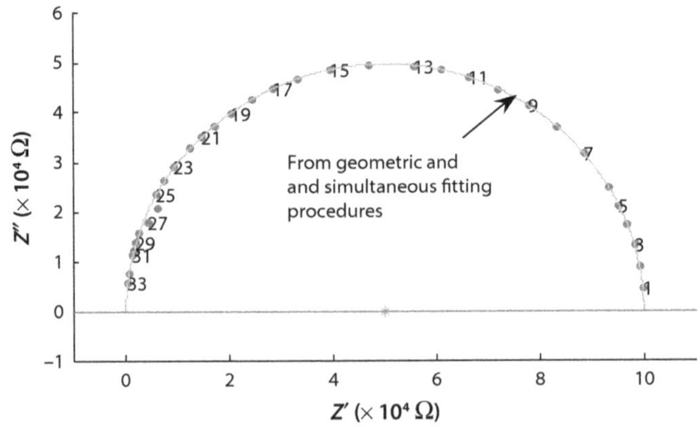

Figure 11.8 Fitted semicircles to the R-C circuit (Figure 11.4) data in Z*-plane using both geometric and simultaneous fitting procedures.

provide a range of data points or individual data points to be ignored. The ignored data points are marked on the plot and the selected data are stored separately. The selected data are plotted on a different figure window and marked; in the following step, the semicircle is plotted over this figure window. In the present example, data points marked 1 through 33 are selected as shown in Figure 11.8.

The range may need to be changed if fitting is unsatisfactory. If skewed behavior is observed, consider fitting Cole-Davidson or Havriliak-Negami relaxation function (which is beyond the scope of this chapter). If no familiar behavior is observed, check data acquisition procedure for the source of noise or interpret the data according to its own merit (e.g. a new material behavior).

11.5.2.3 Fitting of the Semicircle: Geometric Fitting

The semicircle is fitted on the selected data points. The main program at this stage calls a function which executes either the Gauss-Newton or the Levenberg-Marquardt algorithm. Initial estimates of the circle parameters are made by observing the shape of the curve generated by the selected measured data points. For the present problem, the initial estimates are entered as [5; 0; 5] for the x-coordinate, the y-coordinate and the radius. These values are chosen based on the visual inspection of the data in Figure 11.7. If convergence is achieved, the function returns the estimated parameters of the circle, analysis of variance table and statistics such as the error measures, variance-covariance matrix of the parameters, etc. On the selected data figure window, the semicircle is plotted on top of the selected measured data (Figure 11.8). If convergence is not achieved the program terminates with an error message. The user has the choice to terminate the program if the fitting is unsatisfactory

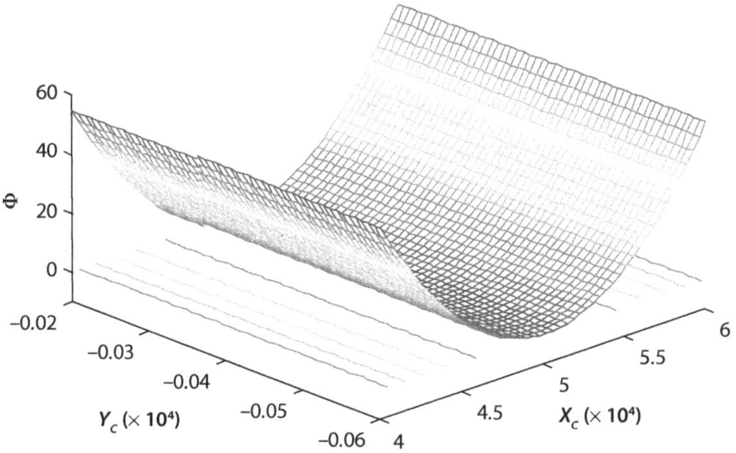

Figure 11.9 Surface of Φ obtained from the semicircle fitting of the R-C circuit of Figure 11.4 data in Z*-plane.

and different data points are selected in the next run. If the fitting is satisfactory, the next step of calculation of Cole-Cole parameters is entered.

The surface of Φ is plotted in Figure 11.9 as a function of the coordinates of the center. The illustration indicates a well defined minimum value of Φ which is obtained from the minimization procedure as described earlier in this chapter.

11.5.2.4 Calculation of the Parameters from the Semicircle Fitting

For the geometrical fitting procedure, the Cole-Cole parameters are calculated with equations (11.35), (11.36), (11.37), and (11.38). From the slope and intercept of the straight line the peak frequency is determined for which $\log(U/V) = 0$, which is then used to determine the relaxation parameter using equation (11.37). The depression angle is obtained from the circle fitting and is used in equation (11.36) to determine the parameter α. The intercepts are determined using the radius and coordinates of the center of the circle using equation (11.38). The circuit parameters are then determined using equation (11.39). Figure 11.10 shows the fitted straight line through the same set of data as in Figure 11.8 and the calculation of the frequency corresponding to the peak of the semicircle. Note the negative slope of the straight line. The slope is negative when plotting the data in Z*- and C*-planes (frequency increasing in the anti-clockwise direction) and is positive when plotting the data in Y*- and M*-planes (frequency increasing in the clockwise direction).

If both the semicircle and the straight line reasonably fit the data range selected, then the next step in section 5.2.5 is followed; otherwise the range

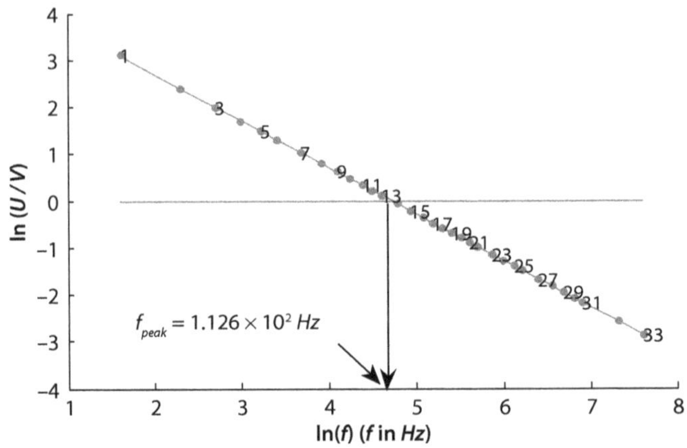

Figure 11.10 Calculation of peak frequency for the R-C circuit of Figure 11.4 data in the Z*-plane.

of the data is changed by either including more data or deleting some data points and refitting the semicircle. Repeat this procedure to determine the range of data for which addition or deletion of some data points on either end of the semicircle does not change the circle parameters significantly. Once this is achieved, the final circle parameters are obtained and the next step is performed. In the following step, the user decides whether to continue for simultaneous fitting of real and imaginary parts.

11.5.2.5 Calculation of the Parameters from the Simultaneous Fitting of Real and Imaginary Parts

The simultaneous fitting with the real and imaginary parts of the data may be performed if desired by the user. The estimated values of the parameters from the geometric fitting are used as initial guesses and the set of Cole-Cole parameters are obtained directly from the fitting. The circle parameters are then calculated based on equation (11.32) and the semicircle is plotted on the measured data. Figure 11.7 shows the semicircle obtained from the simultaneous fitting plotted on top of the semicircle obtained from the geometric fitting in Z*-plane. The two semicircles are indistinguishable. The outputs of the two procedures can then be compared to see the differences in results. This simultaneous fitting is an additional optional procedure which may be used to compare results with the geometric fitting procedure. As explained further, under certain circumstances the simultaneous fitting procedure may result in poor fitting in the complex plane and the parameters extracted from this procedure may not be useful.

The surface of the sum of squares, Φ, is plotted in Figure 11.11 as function of two parameters τ and α. A well-defined minimum value of Φ is visible and the corresponding values of the parameters are almost identical to the ones obtained from the semicircle fitting.

11.5.3 Output Generation: Output File

The parameters of Cole-Cole equation and equivalent circuit are calculated from the fitted parameters for both geometric and simultaneous fitting procedures. The output is printed in a text file. The output text file for the R-C parallel circuit, which is shown in Appendix - B, is for representation purpose only, showing the important output parameters to be obtained from the analysis. Various sections of the output file are described in sequence.

11.5.3.1 Parameters from the Semicircle Fitting

This section of the output file provides the parameters calculated using geometric fitting: coordinates of the center, radius, R^2 of the geometric fitting, depression angle θ, depression parameter α, relaxation time τ, peak frequency, slope and R^2 of line fitting, low frequency intercept, high frequency intercepts, relaxation resistance and relaxation capacitance. For the R-C parallel circuit, there is a small depression angle, 0.268143° corresponding to the depression parameter 0.997021 indicating an almost ideal Debye behavior.

11.5.3.2 Nonlinear Regression: Semicircle Fitting Output

The first three lines of the output file contain information regarding the number of parameters estimated and total number of data points used.

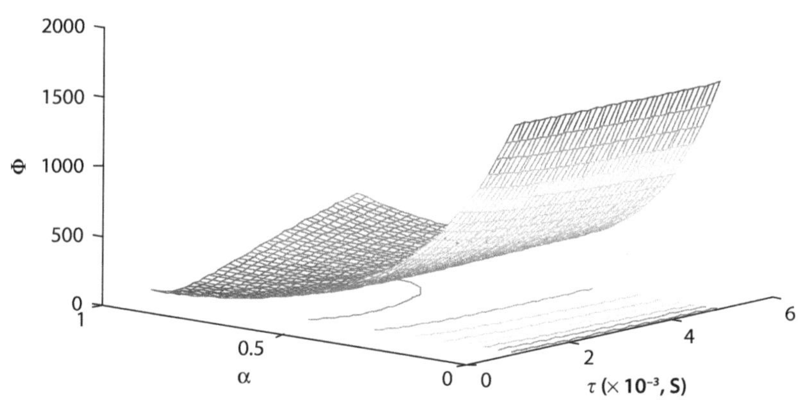

Figure 11.11 Surface of Φ obtained from the simultaneous fitting of the R-C circuit of Figure 11.4 data in Z*-plane.

It also indicates what type of weight matrix has been used; in the present case equal weights have been used. The three columns X, Y and WT contain real and imaginary parts of the data and the weights. Since equal weights have been used, $w_i = 1/\sigma_i^2 = 1/\sigma$. The nonlinear least square algorithm used in this example is Gauss-Newton and closed form functions of derivatives have been used to calculate the Jacobian matrix.

The iterations are shown next which displays the three parameters being estimated (parameter 1: x-coordinate of the center, parameter 2: y-coordinate of the center and parameter 3: the radius) and the sum of squares which is being minimized. The table of summary statistics shows that the fitted model is significant (large F-statistic of 38510.93633 and low p-value of <0.0001). The asymptotic parameter estimate table shows the final estimated values of the parameters with their standard error, confidence interval and significance (t-statistic and p-value). Note that the y-coordinate of the center has p-value greater than 0.05 indicating that the null hypothesis (y-coordinate of the center = 0) cannot be rejected, indicating ideal Debye behavior. Other parameters are highly significant (p-value < 0.0001 and small standard errors). The R^2 statistic of the semicircle fitting is 0.999611 indicating an excellent fitting.

11.5.3.3 Linear Regression: Line Fitting Output

This section of the output file provides the slope and intercept of the fitted line to ln(U/V) versus ln(f) data.

11.5.3.4 Parameters from Simultaneous Fitting of Real and Imaginary Data

This section of the output file contains results obtained from the simultaneous fitting of the data, if performed upon selection by the user. The same set of parameters displayed earlier in the section "Parameters from the Semicircle Fitting" is calculated. A direct comparison between the two procedures may be done at this level. It is seen that the two procedures provide output results which are extremely close within acceptable level of accuracy. The estimated depression angle is 0.413714° and the estimated depression parameter is 0.995403, indicating again a near perfect Debye behavior.

11.5.3.5 Nonlinear Regression: Simultaneous Fitting of Real and Imaginary Data Output

This section of the output file follows the same format as of "Nonlinear Regression: Semicircle Fitting Output". The real and imaginary parts of the data used in the analysis with the respective weights are displayed in columns. For this example all weights (both for real and imaginary data) are assumed equal, i.e. $w'_i = w''_i$ $1/\sigma^2$ where σ is the total sum of squares, including the sums of squares involving real and imaginary parts of the data. Gauss-Newton algorithm has been used. Jacobian matrices for the real and imaginary parts have been calculated using the finite difference method since the closed form derivatives of them are too complicated. Iterations are shown in next lines, followed by summary statistics and the parameter estimations. The parameters 1, 2, 3 and 4 represent I_{LI}, I_{RI}, τ and α, respectively. Note that the p-value of I_{LI} is higher than 0.05 (0.14583 > 0.05), and the confidence interval of the parameter includes zero (0) indicating that the null hypothesis $I_{LI} = 0$ cannot be rejected, i.e. the estimated left-intercept (high frequency intercept) is not statistically significant at 0.05 level of confidence. This again confirms the ideal Debye behavior of the R-C parallel circuit without any series resistance. Such observation was not available with the geometrical fitting procedure which extracted the almost same value of the parameter without providing any statistical significance test.

The non-zero value of the parameter was obtained from the measurement noise in the data. Therefore, the simultaneous fitting procedure is a useful tool to test the statistical validity of the parameters in Cole-Cole expression. The R^2 statistic is 0.999943 in the case of simultaneous fitting. However, in some circumstances the simultaneous fitting results may not agree with the geometric fitting results and produce misleading information about the parameters as discussed in the next section.

11.5.3.6 Measured Data used in Analysis

This section of the output file displays the actual measured data points with real and imaginary components and frequency used in the analysis. The comparison of the parameter values obtained from the two fitting procedures with the actual values for the ideal Debye R-C parallel circuit in Figure 11.4 is presented in Figure 11.12. The actual and estimated values match very closely.

Parameter	Ideal Debye R–C parallel circuit	Semicircle fitting	Simultaneous fitting
α	1.0	0.999611	0.995403
τ (sec)	1.4×10^{-3}	1.412485×10^{-3}	1.413935×10^{-3}
R (Ω)	10^5	0.998117×10^5	0.999599×10^5
C (F)	14×10^{-9}	14.15149×10^{-9}	14.14501×10^{-9}

Figure 11.12 Comparison of results from CNLS fitting with parameters from the R-C parallel circuit.

11.6 Effect of Fitting Procedure, Measurement Noise, and Solution Algorithm on the Estimated Parameters

To compare the two fitting procedures, synthetic data obtained from simulations of Cole-Cole equation in Z*- and C*-plane (frequencies increasing anti-clockwise) are used as input. Following values of the parameter were assumed in the simulation: $I_{RI} = 1.1$, $I_{LI} = 0.10$, $\alpha = 0.6$ and 0.8, $\tau = 10^{-3}$ sec. The results are shown in Figure 11.13. The results indicate a close match between the two procedures.

The effect of measurement noise in the data on the values of the estimated parameters is shown in Figure 11.14. Synthetic Cole-Cole data were again used with the following parameters: $I_{RI} = 1.1$, $I_{LI} = 0.10$, $\alpha = 0.8$, $\tau = 10^{-3}$ sec and measurement noise was added to both real and imaginary parts. The noise was assumed to be Gaussian distributed with zero mean and standard deviation equal to a percentage of the measurement. Figure 11.14 shows the comparison between the two procedures with 1% and 2% Gaussian noise in the synthetic data. The two procedures produced similar results. Figure 11.15 shows the values of the estimated parameters obtained using Gauss-Newton and Levenberg-Marquardt algorithms. The results indicate that two algorithms produced similar results.

Estimated parameters	$\alpha = 0.6$		$\alpha = 0.8$	
	Semicircle fitting	Simultaneous fitting	Semicircle fitting	Simultaneous fitting
I_{RI}	1.099983	1.100000	1.100000	1.100000
I_{LI}	1.000173×10^{-1}	0.999999×10^{-1}	0.100000	0.100000
α	0.600000	0.600000	0.800000	0.800000
τ	9.994370×10^{-4}	9.999998×10^{-4}	0.999977×10^{-3}	1.000000×10^{-3}

Figure 11.13 Effect of the parameter α and fitting procedure on the estimated values of the parameters for synthetic Cole-Cole data.

Estimated parameters	1% Gaussian noise		2% Gaussian noise	
	Semicircle fitting	Simultaneous fitting	Semicircle fitting	Simultaneous fitting
I_{RI}	1.099029	1.099276	1.089798	1.092162
I_{LI}	9.998211×10^{-2}	9.971883×10^{-2}	1.008924×10^{-1}	0.993121×10^{-1}
α	0.798561	0.797600	0.805485	0.800925
τ	1.005190×10^{-3}	1.003603×10^{-3}	9.755233×10^{-4}	9.714031×10^{-4}

Figure 11.14 Effect of measurement noise and fitting procedure on the estimated values of the parameters for synthetic Cole-Cole data.

Estimated parameters	Semicircle fitting		Simultaneous fitting	
	Gauss-Newton	Levenberg-Marquardt	Gauss-Newton	Levenberg-Marquardt
I_{RI}	1.099029	1.099032E+000	1.099276	1.099263E+000
I_{LI}	9.998211×10^{-2}	9.997407×10^{-2}	9.971883×10^{-2}	9.973214×10^{-2}
α	0.798561	0.798574	0.797600	0.797630
τ	1.005190×10^{-3}	1.005171×10^{-3}	1.003603×10^{-3}	1.003618×10^{-3}

Figure 11.15 Comparison between Gauss-Newton and Levenberg-Marquardt algorithms for the two fitting procedures.

Figure 11.16(a) and (b) show the contour plots of Φ obtained from geometric and simultaneous fitting of the synthetic data simulating Y^*- or M^*-plane (frequencies increasing clockwise) with 2% Gaussian noise. The plots show a well-defined Φ_{min} for both cases and the optimum values of the parameters obtained are same within acceptable level of accuracy. The difference between their values becomes vanishingly small when noise level is set to zero.

11.7 Case Studies: CNLS Fitting of the Measured Data in the Complex Planes

The fitting procedures described in the previous section are applied to several cases of measured data of different materials. The extracted Cole-Cole parameters from the geometric fitting procedure are listed in Figure 11.17 and those from the CNLS fitting are shown in Figure 11.18.

11.7.1 M^*-Plane Fitting: R-C Parallel Circuit

The measured data of the R-C parallel circuit of Figure 11.4 are plotted in four complex planes. Semicircular behavior is observed in the Z^*- and the M^*-planes as seen in Figure 11.6. In the M^*-plane the frequencies increase clockwise along the arc of the semicircle. The same range of data selected for fitting for both Z^*- and M^*-planes. Figure 11.19 shows the fitted semicircle from both procedures. A truncated version of the output file with relevant sections is displayed in Appendix - C.

It can be seen from the results in Figure 11.17 that there is a finite non-zero value of the left-intercept from the geometric fitting in both complex planes. Ideally this should be zero since there is no series resistance or capacitance with the R-C parallel circuit shown in Figure 11.4. This small value of the left-intercept (compared to the right-intercept) is obtained due to the measurement noise in the data. The statistical significance of this parameter can be seen from the asymptotic parameter estimation obtained from the CNLS fitting; the p-value of the parameter is 0.15 > 0.05 for Z*-plane and 0.28 > 0.05 for M*-plane shown in Figure 11.18. It indicates that the null hypothesis that the left-intercept is zero cannot be rejected (with critical p-value of 0.05) and the Debye circuit shown in Figure 11.4 is therefore obtained. The depression parameter, relaxation capacitance, relaxation resistance and relaxation time obtained from

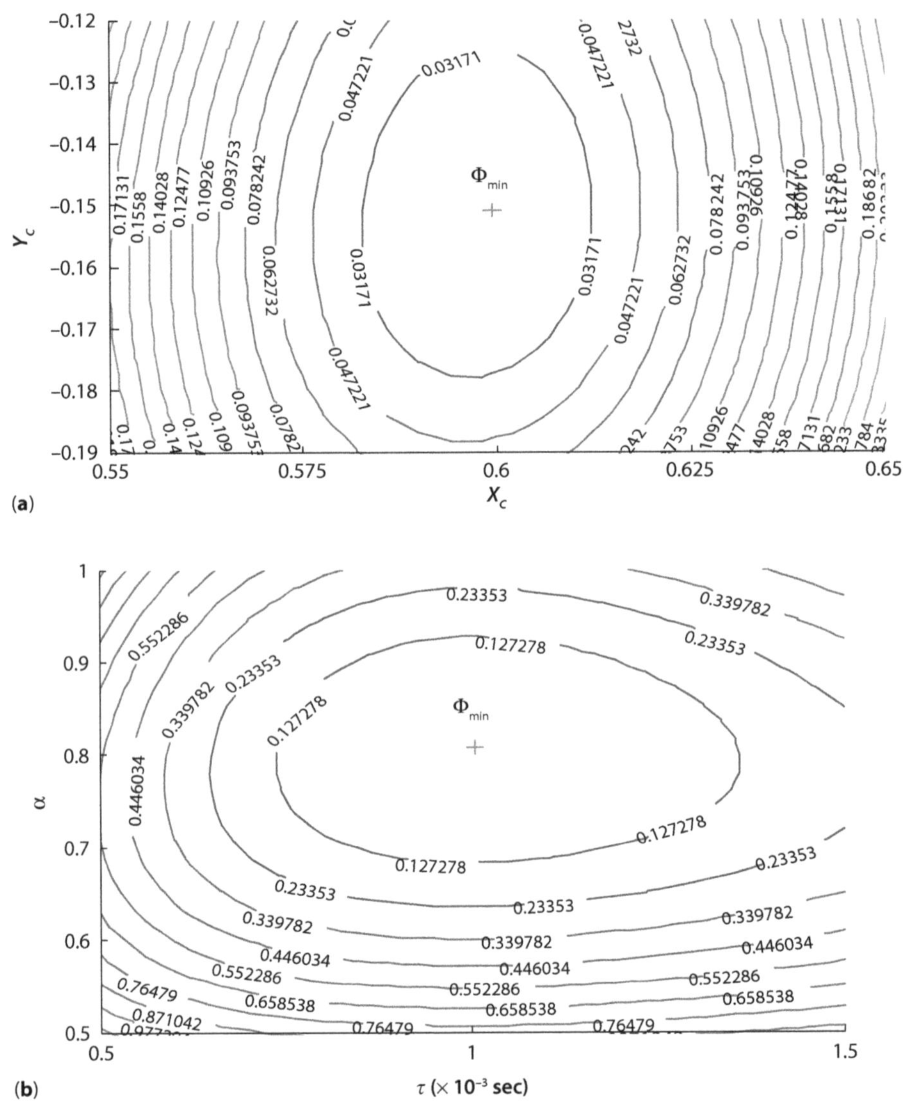

Figure 11.16 Contour of Φ for Cole-Cole synthetic data with 2% Gaussian noise simulating Y*- or M*-plane with $I_{LI} = 0.10$, $I_{RI} = 1.1$, $\alpha = 0.8$ and $\tau = 10^{-3}$; (a) as function of the coordinates of center obtained from semicircle fitting resulting $\alpha = 0.812188$ and $\tau = 1.002915 \times 10^{-3}$; (b) as function of α and τ obtained from simultaneous fitting resulting $\alpha = 0.807653$ and $\tau = 1.006462 \times 10^{-3}$.

Case	Z*-plane	Y*-plane	C*-plane	M*-plane
R-C Parallel circuit	I_{LI} = 232 Ω I_{RI} = 100 kΩ α = 0.99 τ = 1.4 ms R^2 (circle) = 0.999	No semicircle	No semicircle	I_{LI} = −0.21×10⁶ F⁻¹ I_{RI} = 72×10⁶ F⁻¹ α = 0.988 τ = 1.35 ms R^2 (circle) = 0.999
LN Crystal	No semicircle	No semicircle	I_{LI} = 38.2 pF I_{RI} = 371.7 pF α = 0.921 τ = 0.20 ms R^2 (circle) = 0.999	I_{LI} = 2.8×10⁹ F⁻¹ I_{RI} = 23.9×10⁹ F⁻¹ α = 0.962 τ = 11.2 μs R^2 (circle) = 0.999
Multilayered junction device	I_{LI} = 46.8 Ω I_{RI} = 323.2 Ω α = 0.748 τ = 16.8 μs R^2 (circle) = 0.998	I_{LI} = 3.3 mS I_{RI} = 17.8 mS α = 0.826 τ = 1.97 μs R^2 (circle) = 0.999	No semicircle	No semicircle

Figure 11.17 Estimated parameters in the complex planes from the geometric fitting.

Case	Z*-plane	Y*-plane	C*-plane	M*-plane
R-C Parallel circuit	I_{LI} = 16.3 Ω (0.15) I_{RI} = 100 kΩ α = 0.995 τ = 1.4 ms	No semicircle	No semicircle	I_{LI} = −0.51×10⁶ F⁻¹ (0.28) I_{RI} = 72.2×10⁶ F⁻¹ α = 0.979 τ = 1.37 ms
LN Crystal	No semicircle	No semicircle	I_{LI} = 27.6 pF I_{RI} = 385.9 pF α = 0.801 τ = 0.23 ms	I_{LI} = 2.4×10⁹ F⁻¹ I_{RI} = 25.3×10⁹ F⁻¹ α = 0.818 τ = 8.5 μs
Multilayered junction device	I_{LI} = 46.4 Ω I_{RI} = 322.5 Ω α = 0.748 τ = 16.6 μs	I_{LI} = 3.3 mS I_{RI} = 17.8 mS α = 0.827 τ = 1.97 μs	No semicircle	No semicircle

Note: p-values > 0.05 are indicated in the parentheses; p-values for other parameters (not shown) are < 0.0001.

Figure 11.18 Estimated parameters in the complex planes from the CNLS fitting.

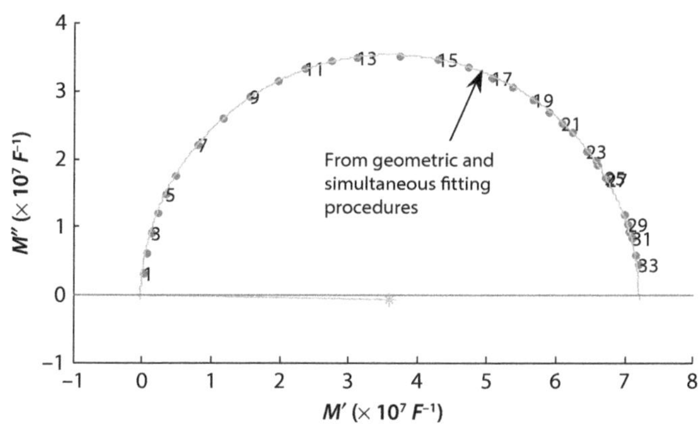

Figure 11.19 Both geometric and simultaneous fitting procedures for the data of the R-C parallel circuit of Figure 11.4 in the M*-plane.

both fitting procedures are identical. The relaxation resistance, relaxation capacitance and the relaxation time obtained from the fitting are exactly the same or extremely close to the actual circuit values.

11.7.2 C*- and M*-Plane Representations of the Lithium Niobate (LN) Crystal

Electrical characterization of the ferroelectric Lithium Niobate (LN) crystal [12] was performed with QuadTech 1920 precision LCR meter with frequency range 10 Hz $\leq$ f $\leq$ 10^6 Hz and temperature range 35 °C $\leq$ T $\leq$ 150 °C. The measured data were frequency (f), capacitance (C_p) and tangent of the phase angle (tan δ = ratio of imaginary and real parts of capacitance). The measured data were converted to all four complex planes and the resulting plots are shown in Figure 11.20 for the measurement temperature of 59 °C. The plots indicate semicircular behavior in C*- and M*-planes. Figure 11.21 shows the selected measured data points in C*-plane and the fitted semicircle using the geometric and simultaneous fitting procedures. The geometric fitting produced the following results: I_{RI} = 371.7 pF, I_{LI} = 38.2 pF, $\alpha = 0.92$ and $\tau = 0.2$ ms (Figure 11.17). When the same measured data are used with CNLS fitting procedure, the

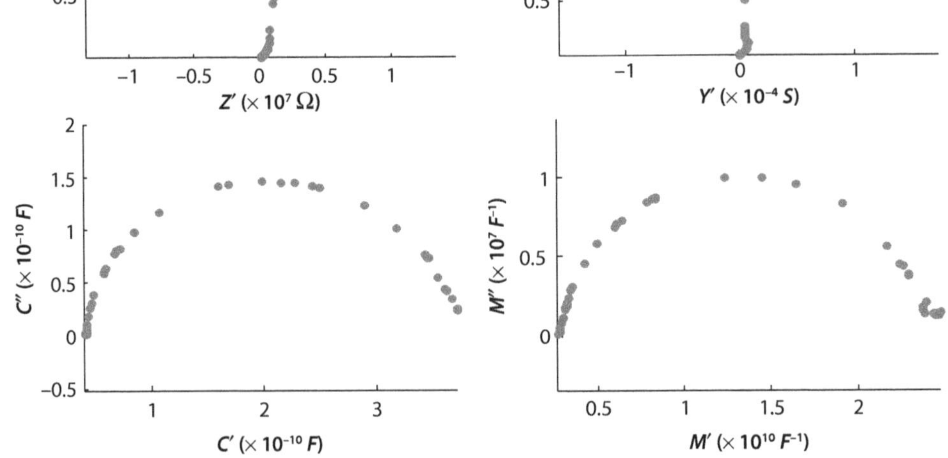

Figure 11.20 Complex plane representations of Lithium Niobate crystal data at 59 °C.

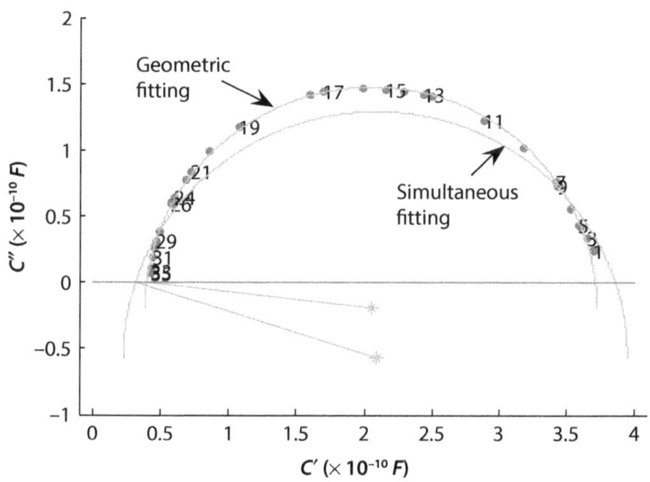

Figure 11.21 Fitted semicircles to Lithium Niobate data in C*-plane at 59 °C.

following results are obtained: I_{RI} = 385.9 pF, I_{LI} = 27.6 pF, α = 0.80 and τ = 0.23 ms (Figure 11.18).

While the values for I_{RI}, I_{LI}, and τ are close, the values of α from the two procedures do not match. As seen in Figure 11.21, the CNLS fitting procedure resulted in poor fitting in the complex plane while it minimized the total sum of squares obtained separately for the real and imaginary parts. It is to be noted that while both approaches minimize the sum of squares, the geometric fitting procedure works with the semicircular relationship between the real and imaginary parts while the CNLS fitting procedure fits the real and imaginary parts over the range of frequencies. A clear semicircular behavior is visible from the display of the data in the complex plane at all test temperatures that are provided earlier [34]. The CNLS fitting procedure produced results which were not meaningful. In this case, the geometric fitting (direct semicircle fitting in the complex plane) is the viable approach which resulted in meaningful extraction of Cole-Cole parameters for the present case.

For the R-C parallel circuit, the two procedures produced similar results within an acceptable level of accuracy because of the small level of measurement noise present in the R-C parallel circuit measurement data. In case of the simulated data the ideal random Gaussian noise was used. In the case of LN crystal, various sources may have contributed to the overall randomness of the data, such as: measurement noise, bias (both human and machine), temperature change during measurement, etc; the noise effect could not be classified as ideal Gaussian random. An accurate measurement of noise structure may be needed to correctly model the behavior [57].

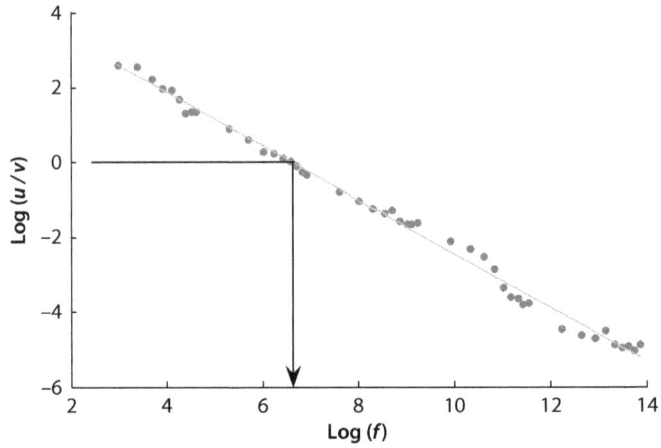

Figure 11.22 Determination of the peak frequency from the geometric fitting procedure in C*-plane for the Lithium Niobate crystal data at 59 °C.

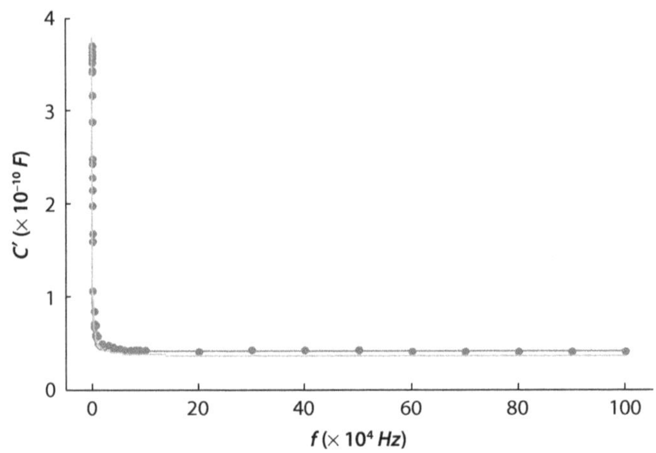

Figure 11.23 Real part of complex capacitance versus frequency for Lithium Niobate crystal data at 59 °C and the comparison between the geometric (blue) and the CNLS (red) fitting procedure.

The error in measurement is evident in Figure 11.22 which shows the plot of log(u/v) versus log(f) and the determination of peak frequency. As seen in the figure, there is considerable noise in the measurement. The Bode plots are shown in Figures 11.23, 11.24 and 11.25 where both methods of fitting are compared. While there is a good match between the geometric and CNLS fitting in the real part of the Bode plot of Figure 11.23, there is a lack in any match in the imaginary part of the Bode plot per Figure 11.24. In Figure 11.24, the peak of the fitted curve from CNLS fitting is situated below the peak of the measured data causing a depressed

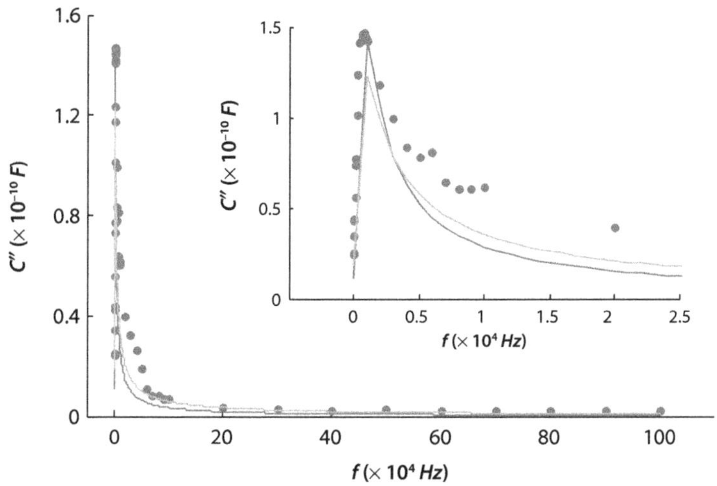

Figure 11.24 Imaginary part of complex capacitance versus frequency for Lithium Niobate crystal data at 59 °C and the comparison between the geometric (blue) and the CNLS (red) fitting procedure.

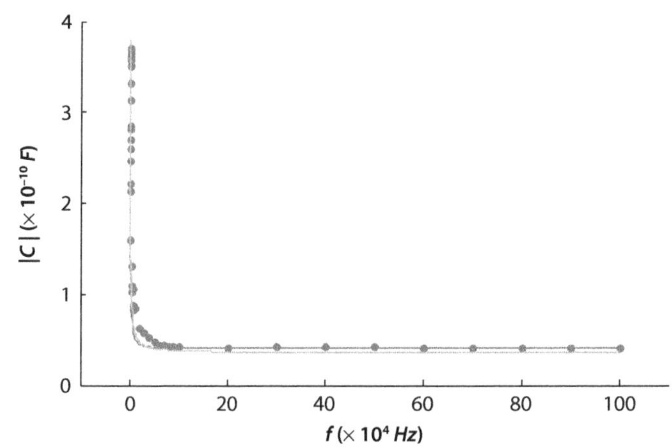

Figure 11.25 Magnitude of complex capacitance versus frequency for Lithium Niobate crystal data at 59 °C and the comparison between the geometric (blue) and the CNLS (red) fitting procedure.

semicircle (in red) shown in Figure 11.21. While the fitted curve from the geometric fitting matches the peak of the measured data, the error is larger in Bode plot compared to the CNLS fitted curve.

The fitting of the LN data in the M*-plane is shown in Figures 11.26, 11.27, 11.28, and 11.29. Similar Conclusions on the measurement noise can be drawn for the M*-plane as well. In Figure 11.29, the frequency

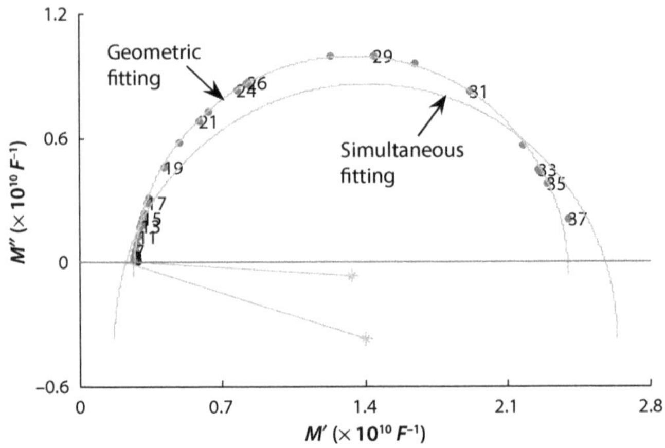

Figure 11.26 Fitted semicircles to Lithium Niobate data in M*-plane at 59 °C.

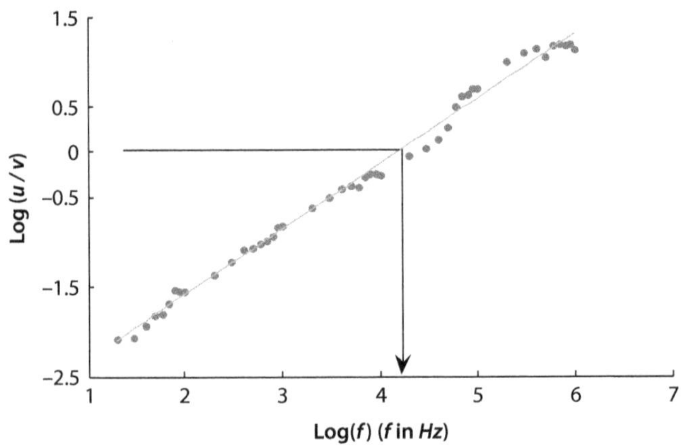

Figure 11.27 Determination of the peak frequency from the geometric fitting procedure in complex modulus plane for Lithium Niobate crystal data at 59 °C.

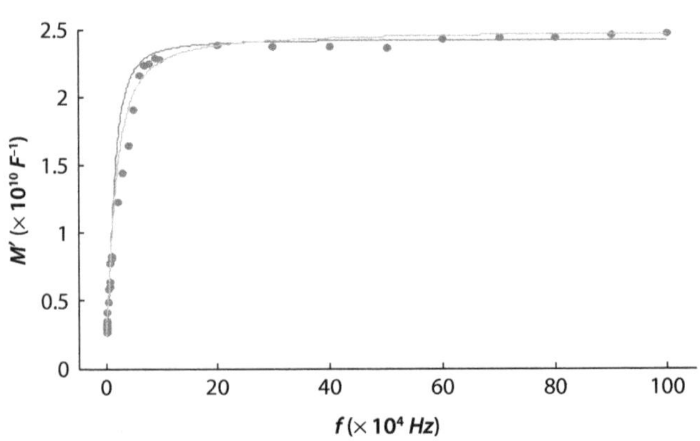

Figure 11.28 Real part of complex modulus versus frequency for Lithium Niobate crystal data at 59 °C and the comparison between the geometric (blue) and the CNLS (red) fitting procedure.

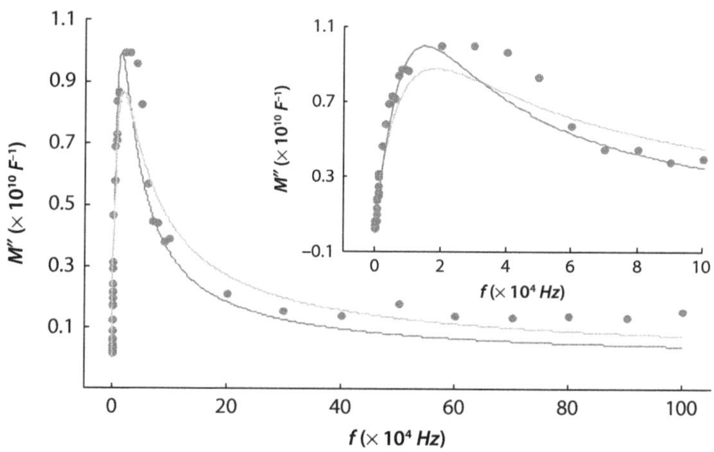

Figure 11.29 Imaginary part of complex modulus versus frequency for Lithium Niobate crystal data at 59 °C and the comparison between the geometric (blue) and the CNLS (red) fitting procedure.

corresponding to the peak of the fitted curves does not match the frequency of the peak of the measured data shown in the inset of Figure 11.29. The output file for C*-plane analysis is shown in Appendix – D.

More closely spaced frequencies were used in collecting the data for the multilayered junction device compared to the interval in the LN crystal measurement frequencies. From the previous discussion it can be concluded that two fitting procedures do not produce exactly identical results when: (1) there is measurement noise present in the data, and (2) enough closely spaced frequencies are not used during the data collection. In such a situation, the geometric procedure is expected to be the realistic method of determination of the Cole-Cole equation to obtain the equivalent circuit parameters. The output file for the Y*-plane analysis is shown in Appendix – E.

11.7.3 Z*- and Y*-Plane Representations of Multi-Layered Junction Device

Figure 11.30 shows the complex plane representation of the multi-layered semiconducting junction (SiO_2/SiO_2-Ge multi-layered) device. The data were obtained in the parallel mode at room temperature. A portion of the data in the Z*-plane shows semicircular behavior per Figure 11.31. Figure 11.32 represents the log(u/v) versus log(f) plot for this plane indicating very low level of the measurement noise. A portion of the data in the Y*-plane show semicircular behavior per Figure 11.33.

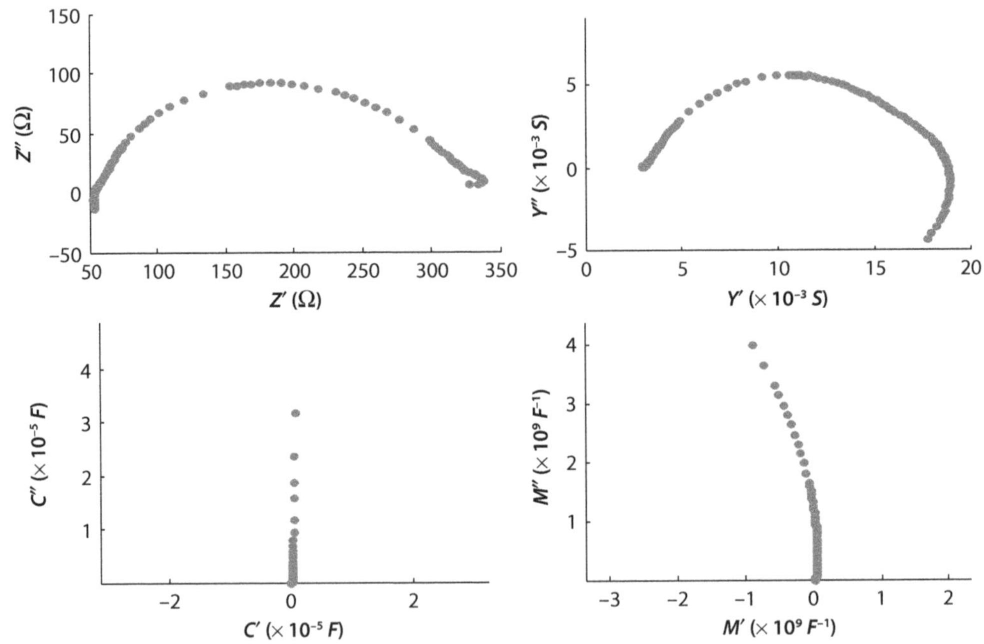

Figure 11.30 Four complex plane representations of the multi-layered junction device (SiO_2/SiO_2-Ge multi-layered un-bombarded thermoelectric device).

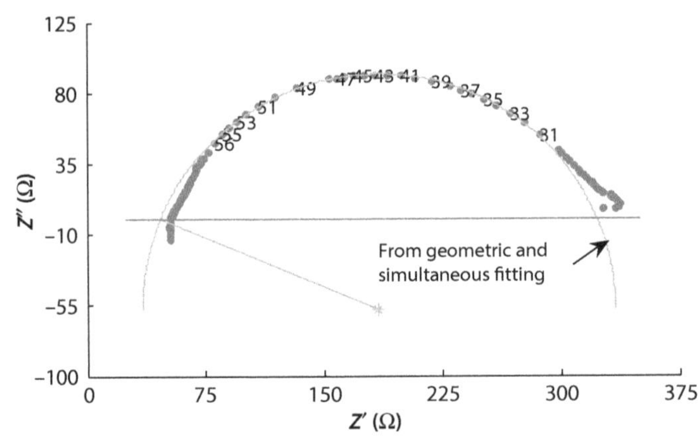

Figure 11.31 Fitted overlapping semicircles from geometric and CNLS fitting of the multi-layered junction device data in impedance plane. The data points used in fitting are denoted by their observation numbers.

Figure 11.34 represents the log(u/v) versus log(f) plot for this plane indicating very low level of the measurement noise. Overall, Figure 11.24 and Figure 11.26 display fitted semicircles in Z*- and Y*-planes with both geometric and CNLS fitting procedures. Both fitting procedures produced similar results and statistically significant parameter estimations

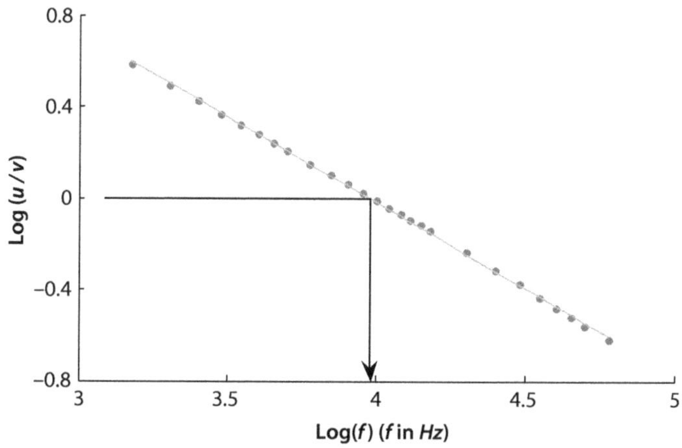

Figure 11.32 Determination of the peak-frequency from the geometric fitting in the impedance plane for the multi-layered junction device data.

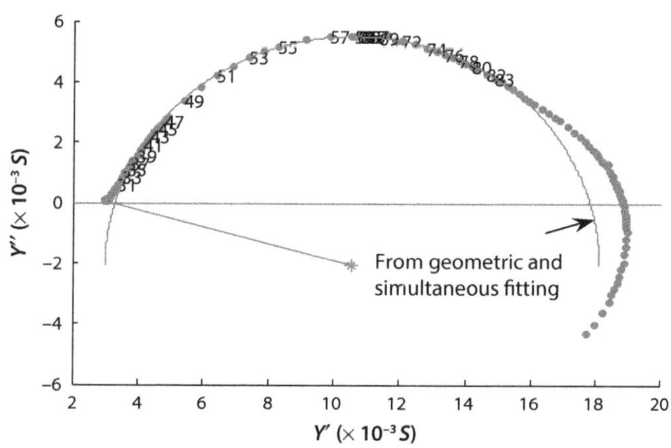

Figure 11.33 Fitted overlapping semicircles from geometric and CNLS fitting of the multi-layered junction device data in admittance plane. The data points used in fitting are denoted by their observation numbers.

are provided in Figure 11.17 and Figure 11.18. Figure 11.28 and Figure 11.29 are called the chord fitting [log(u/v) versus log(f)] plots for the two complex planes.

11.7.4 Y*-plane Representation of the *E. Coli* Bacteria in Brain Heart Infusion Medium

The electrical characterization of the suspended E-Coli bacteria in BHI (brain heart infusion) medium is shown in Figure 11.35 at different time of growth. The curves depict as-measured ac electrical data in the

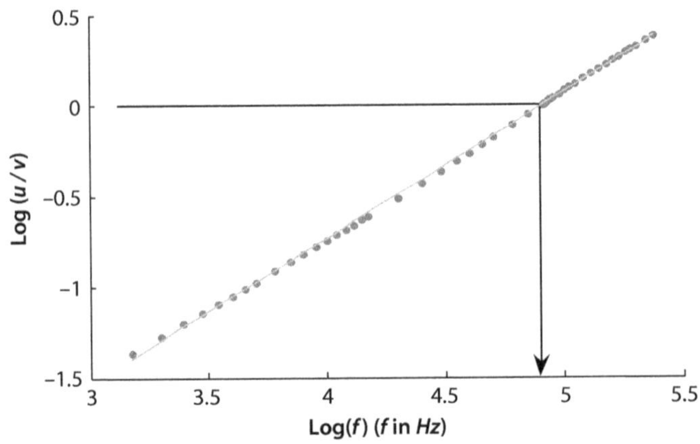

Figure 11.34 Determination of the peak-frequency from the geometric fitting in the admittance plane for the multi-layered junction device data.

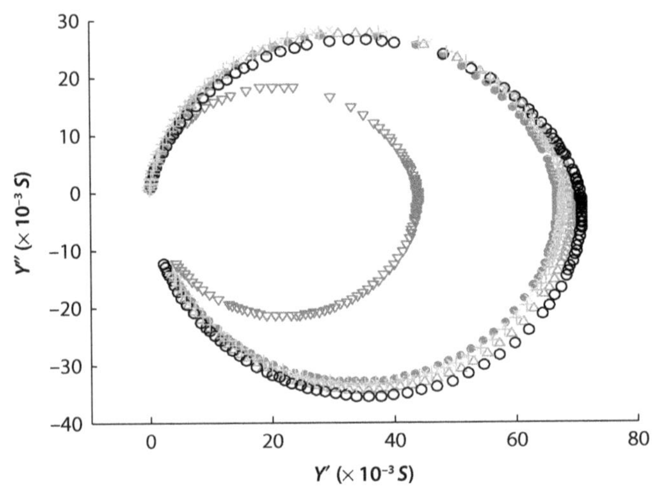

Figure 11.35 Measured data of the sterile BHI medium and *E. Coli* bacteria suspended in BHI medium in the Y*-plane. Legend: ∇- sterile BHI medium, *E. Coli* bacteria suspended in BHI medium after: 3 hours (•), 6.5 hours (+), 10 hours (×), 16.5 hours (Δ), and 24 hours (o) of growth.

admittance (Y*) plane of the sterile BHI medium as well as the growing bacteria suspended in the BHI medium. The observed relaxation is semicircular in both positive and negative domains of the Y*-plane.

The semicircular relaxation in the negative domain is attributed to the inductive response of the material system. The semicircular relaxation in the positive domain exhibits center below the x-axis indicating depressed semicircle for the non-Debye response. Thus, the semicircle in the negative domain is larger than the semicircle in the positive domain.

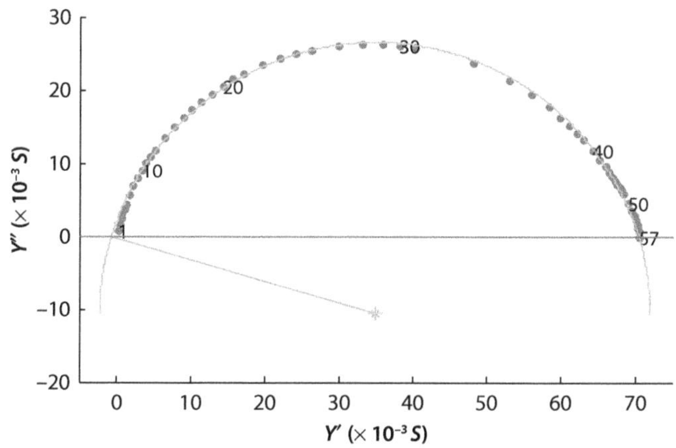

Figure 11.36 Fitted semicircles (overlapped semicircles from both procedures of fitting) in the Y*-plane for the data of the *E. Coli* suspended BHI medium taken after 24 hours of growth.

Overall, each curve portrays non-Debye response of the material system as a whole. The conductance is increasing with time reflecting gradual multiplication of the *E. Coli* bacteria. Two procedures of fitting produced identical overlapped semicircles in Y*-plane which are shown in Figure 11.36. Again, the effect of lack of noise and closely spaced data in frequency domain is clearly visible in identically produced semicircles. It is essential that the investigation requires closely spaced data acquisition in order to establish firm characterization of the material system.

11.8 Summary

A massive portion of the foregoing content is available [71] for the learners and investigators emphasizing both geometric method and CNLS method of data analysis. Thus, the analysis of the real world data of the solid and liquid systems for the semicircular relaxation fitting for each of the four complex plane formalisms is demonstrated using both methods. The distinction between the two methods is given as:

1. geometrical procedure by fitting semicircle for the measured data, and
2. complex nonlinear least square (CNLS) procedure by fitting real and imaginary parts of the immittance response.

Both methods provide identical information when the data collection is large in quantity having small interval in the successive frequencies.

Essentially for fewer data points there is a lag in matching implying that both methods do not provide accuracy for the analysis due to wide interval in frequencies. The geometric procedure seems to be better suited compared to the CNLS procedure for the fitting. Two procedures were accomplished using Gauss-Newton and Lavenberg-Marquardt nonlinear least square methods (algorithms).

In the case of material investigation, considering two procedures leads to somewhat dubious situation, and obviously inaccuracy of the interpretation becomes obvious. Nevertheless, a large collection of data as a function of measurement frequency certainly assures better accuracy and increases confidence in extracting equivalent circuit parameters. Fine tuning of the extracted parameters is likely to increase the performance and functional capabilities of the smart electronic materials and devices. Three case studies are presented with various material systems:

a) R-C parallel circuit,
b) Lithium Niobate (LN) crystal, and
c) multilayered semiconducting junction device (for thermoelectric purposes).

The R-C parallel circuit showed near perfect Debye behavior in both impedance and modulus planes, and both procedures provided identical fitting. The $LiNbO_3$ crystal showed depressed semicircle indicating non-Debye behavior in the capacitance and modulus planes. Two procedures provided different fitting parameters because of the lack of sufficient measurement data points that caused noise in the procedure. The multilayered device produced depressed semicircle indicating non-Debye behavior for a part of the data set in both impedance and admittance planes. Two procedures produced identical fitting.

It is visualized that both methods are valid depending on the confidence level of the investigators and researchers. Both methods lead to the accuracy to evaluate novel electronic materials and devices depending on the quality of the data set. The important aspect is that the geometric method yields the parameters upon fitting whereas the CNLS method assumes the parameters for exact fitting. In this way two methods are considered to be drastically different in the approaches. Therefore, investigators require a thorough understanding about the material system for which the investigation is undertaken.

Regarding the quality of the data it must be noted that the learners and investigators must check efficiently the nature of the data when the signal-voltage is varied. This means that the immittance data are always the same for each applied frequency at various small-signal voltages then the noise in the data can be assumed to be either minimum or zero. Until a saturated individual immittance value for each frequency is achieved it is more than unlikely that the noise in the data can be eliminated. Therefore, it is essential that the saturated individual immittance value for each frequency must be observed before finalization of the data acquisition process.

As a response to the available content, a computer software program has been developed for analyzing electrochemical impedance based on an expanded measurement model [72]. An equivalent circuit model consisting of a series of impedance elements such as resistors and capacitors are considered for both Debye and non-Debye type relaxations. The total response spectrum of the equivalent circuit was then determined from the sum of the individual response spectrum of the impedance elements.

The total impedance spectrum was determined as the superposition of individual element-based impedance spectrum. The primary goal of the development of the software is to provide a mechanism of easy estimation of the initial values of the equivalent circuit parameters through a graphical user interface (GUI). The GUI consisted of several displays indicating the Bode plots [ln(f) versus Z_{real} or ln(f) versus $Z_{imaginary}$], complex plane plot of the measured data and fitted curve, and residual plot of the fitting and the fitted parameter values.

The user can select a particular element by using a pull-down menu. Initial guess for the impedance elements are obtained either from a typical data point and the corresponding element or from the data point and the characteristic impedance of the element. The experimental data are fitted with a series of R-C parallel impedance element blocks in series with each other. The overall features of the procedure presented are:

a) spectral information of the individual elements and the measured data was available on screen,
b) the measured spectrum was modeled as a summation of impedance spectra using a series of impedance elements, and
c) the initial guesses of the parameters were obtained by visual observation of the GUI display with the individual impedance element and data points.

Problems

1. Consider the 1982 paper of Tsai and Whitmore. How can two semicircles be simultaneously fitted? How does simultaneous

fitting differ from the fitting of a single semicircle by iteration and selection of data points so that a common merging point is achieved on the x-axis? Can necessary software be developed for complex plane plotting?

2. Consider the 1941 Cole-Cole, 1950 Davidson-Cole, and 1966 Havriliak-Negami papers. Take the relaxation function from each empirical equation. Make a plot of the imaginary part of the relaxation function versus logarithmic of frequency for each. Take frequency between 10 Hz through 10 MHz so that a minimum of 10 data points per decade is created. Make the next plot using the imaginary part of the relaxation function versus the product of angular frequency and the time constant for each relaxation function.
3. Consider the Cole-Cole equation. Take own values systematically and rationally for each parameter so that the generated data set ultimately make three columns: frequency, parallel capacitance, and parallel resistance. Let the frequency run between 10 Hz through 10 MHz so that a minimum of 15 data points per decade is created. Use these data for making complex plane plots using all four of them. Also determine the Bode plots. Try to extract the equivalent circuit model and verify whether it matches with the values taken in the first place.
4. Consider the Davidson-Cole equation. Take own values systematically and rationally for each parameter so that the generated data set ultimately make three columns: frequency, parallel capacitance, and parallel resistance. Let the frequency run between 10 Hz through 10 MHz so that a minimum of 20 data points per decade is created. Use these data for making complex plane plots using all four of them. Also determine the Bode plots. Try to extract the equivalent circuit model and verify whether it matches with the values taken in the first place.
5. Consider the Havriliak-Negami equation. Take own values systematically and rationally for each parameter so that the generated data set ultimately make three columns: frequency, parallel capacitance, and parallel resistance. Let the frequency run between 10 Hz through 10 MHz so that a minimum of 25 data points per decade is created. Use these data for making complex plane plots using all four of them. Also determine

the Bode plots. Try to extract the equivalent circuit model and verify whether it matches with the values taken in the first place.

References

1. M. Andres-Verges and A. R. West; "Impedance and Modulus Spectroscopy of ZnO Varistors," *Journal of Electroceramics*, 1(2), 125–132, 1997.
2. D. C. Sinclair and A. R. West; "Impedance and Modulus Spectroscopy of Semiconducting $BaTiO_3$ Showing Positive Temperature Coefficient of Resistance," *Journal of Applied Physics*, 66, 3850–3856, 1989.
3. I. M. Hodge, M. D. Ingram, and A. R. West; "Impedance and Modulus Spectroscopy of Polycrystalline solid Electrolytes," *Journal of Electroanalytical Chemistry*, 74(2), 125–143, 1976.
4. I. M. Hodge, M. D. Ingram, and A. R. West; "New Method for Analyzing the ac Behavior of Polycrystalline Solid Electrolytes," *Journal of Electroanalytical Chemistry and Interfacial Electrochemistry*, 58(2), 429–432, 1975.
5. M. A. Seitz; "Study of Heterogeneous Composite Materials via Lumped Parameter/Complex Plane Analysis," *International Journal of Hybrid Microelectronics* (ISHM), 3, 1–7, 1980.
6. K. A. Abdullah, A. Bui, and A. Loubiere; "Low Frequency and Low Temperature Behavior of ZnO-Based Varistor by ac Impedance Measurements," *Journal of Applied Physics*, 69, 4046–4052, 1991.
7. A. Bui, K. A. Abdallah, A. Loubiere, M. Tao, and Q. C. Nguyen; "Impulse-Degradation Analysis of ZnO-Based Varistors by ac Impedance Measurements," *Journal of Physics D: Applied Physics*, 24, 757–762, 1991.
8. M. A. Alim; "Admittance-Frequency Response in Zinc Oxide Varistor Ceramics," *Journal of the American Ceramic Society*, 72(1), 28–32, 1989.
9. M. A. Alim, M. A. Seitz, and R. W. Hirthe; "Complex Plane Analysis of Trapping Phenomena in Zinc Oxide Varistors," *Journal of Applied Physics*, 63(7), 2337–2345, 1988.
10. M. A. Alim; "Electrical Characterization of Engineering Materials," *Active and Passive Electronic Components*, 19, 139–169, 1996.
11. M. A. Alim, S. Khanam, and M. A. Seitz; "Immittance Spectroscopy of Smart Components and Novel Devices," *Active and Passive Electronic Components*, 16, 153–170, 1994.
12. M. A. Alim, A. K. Batra, S. Bhattacharjee, and M. D. Aggarwal; "Complex Capacitance in the Representation of Modulus of the Lithium Niobate Crystals," *Physica* B, 406, 1088–1095, 2011.
13. M. A. Alim, S. R. Bissell, A. A. Mobasher; "Analysis of the AC Electrical Data in the Davidson–Cole Dielectric Representation," *Physica* B, 403, 3040–3053, 2008.

14. C. C. Chen, M. M. Nasrallah, H. U. Anderson, and M. A. Alim; "Immittance Response of $La_{0.6}Sr_{0.4}Co_{0.2}Fe_{0.8}O_3$ Based Electrochemical Cells," *Journal of the Electrochemical Society*, 142, 491–496, 1995.
15. A-. M. Azad, L. L. W. Shyan, and M. A. Alim; "Immittance Response of $CaSnO_3$ Prepared by Self-Heat-Sustained Reaction," *Journal of Materials Science,* 34, 1175–1187, 1999.
16. A-. M. Azad, L. L. W. Shyan, and M. A. Alim; "The AC Electrical Characterization of the Solid-State Reaction Derived $CaSnO_3$," *Journal of Materials Science,* 34, 3375–3396, 1999.
17. C. C. Wang, W. H. Chen, S. A. Akbar, and M. A. Alim; "High-Temperature A.C. Electrical Behaviour of Polycrystalline Calcium Zirconate," *Journal of Materials Science*, 32, 2305-2312, 1997.
18. C. C. Wang, V. D. Patton, S. A. Akbar, and M. A. Alim; "Effect of Zirconia Doping on the Electrical Behavior of Yttria," *Journal of Materials Research,* 11(2), 422–429, 1996.
19. W. Zhu, C. C. Wang, S. A. Akbar, R. Asiaie, P. K. Dutta, and M. A. Alim; "The AC Electrical Behavior of Hydrothermally Synthesized Barium Titanate Ceramics," *Japanese Journal of Applied Physics,* 35, 6145–6152, 1996.
20. C. J. Ping, C. J. Peng, H. Y. Lu, and S. T. Wu; "Lanthanum-Magnesium and Lanthanum-Manganese Donor-Acceptor Co-Doped Semiconducting Barium Titanates," *Journal of the American Ceramic Society,* 73, 329–334, 1990.
21. Y. T. Tsai and W. H. Whitmore, "Nonlinear Least-Squares Analyses of Complex Impedance and Admittance Data for Solid Electrolytes," *Solid. State Ionics*, 7, 129, 1982.
22. W. Jakubowski and D. H. Whitmore; "Analysis of Electrolyte Polarization in Polycrystalline Sodium β-Alumina by a Complex Admittance Method," *Journal of the American Ceramic Society,* 62, 381–385, 1979.
23. J. E. Bauerle; "Studies of Solid Electrolyte Polarization by a Complex Admittance Method," *Journal of Physics and Chemistry of Solids*, 30, 2657–2670, 1969.
24. F. A. Grant; "Use of Complex Conductivity in the Representation of Dielectric Phenomena," *Journal of Applied Physics*, 29(1), 76–80, 1958.
25. S. Havriliak, and S. Negami; "A Complex Plane Analysis of Dispersions in some Polymer Systems," *Journal of Polymer Science: Part C,* 14, 99–117, 1966.
26. T. M. Gur, I. D. Raistrick, and R. A. Huggins; "AC Admittance on Stabilized Zirconia with Porous Platinum Electrodes," *Solid State Ionics*, 1, 251–271, 1980.
27. E. A. Cooper, T. O. Mason, M. E. Biznek, and U. Balachandran; "Impedance Spectra during Solid-State Powder Reactions of Oxide Superconductors," *Journal of the American Ceramic Society*, 73, 154–157, 1990.
28. S. Trolier-McKinstry and R. E. Newnham; "Sensors, Actuators, and Smart Materials," *MRS Bulletin,* 18, 27–33, 1993.

29. R. E. Newnham and G. R. Ruschau; "Smart Electroceramics," *Journal of the American Ceramic Society*, 74, 463–480, 1991.
30. R. E. Newnham; "Composite Electroceramics," *Annual Review of Materials Science,* 16, 47–68, 1986.
31. K. S. Cole, and R. H. Cole; "Dispersion and Absorption in Dielectrics I: Alternating Current," *Journal of Chemical Physics,* 9, 341–351, 1941.
32. D. W. Davidson, and R. H. Cole; "Dielectric Relaxation in Glycerine," *Journal of Chemical Physics,* 18(10), 1417–1417, 1950.
33. J. R. Macdonald; "*Impedance Spectroscopy: Emphasizing Solid Materials and Systems*," edited, Wiley-Interscience, John Wiley & Sons, New York, 1987.
34. E. Barsoukov and J. R. Macdonald; "Impedance Spectroscopy: Theory, Experiment, and Applications," edited, Wiley-Interscience, John Wiley & Sons, New York, 2005.
35. M. E. Orazem and B. Tribollet; "Electrochemical Impedance Spectroscopy," Wiley-Interscience, John Wiley & Sons, New York, 2008.
36. V. F. Lvovich; *"Impedance Spectroscopy: Applications to Electrochemical and Dielectric Phenomena,"* Wiley-Interscience, John Wiley & Sons, New York, 2012.
37. J. R. Macdonald; "Impedance Spectroscopy: Old Problems and New Developments," *Elecrrochimica Acta,* 35(10), 1483–1492, 1990.
38. C. P. Smyth; "*Dielectric Behavior and Structure*," McGraw-Hill Book Company, New York, 1955.
39. J. Gimsa and D. Wachner; "A Unified Resistor-Capacitor Model for Impedance, Dielectrophoresis, Electrorotation, and Induced Transmembrane Potential," *Biophysical Journal*, 75, 1107–1116, 1998.
40. T. G. Paunescu and S. I. Helman; "cAMP Activation of Apical Membrane Cl2 Channels: Theoretical Considerations for Impedance Analysis," *Biophysical Journal,* 81, 838–851, 2001.
41. H. G. L. Coster; "The Double Fixed Charge Membrane: Low Frequency Dielectric Dispersion," *Biophysical Journal,* 13, 118–132, 1973.
42. Y.- Z. We and S. Sridhar; "A New Graphical Representation for Dielectric Data," *Journal of Chemical Physics,* 99(4), 3119–3124, 1993.
43. R. Thangavel, V. Sabarinathan, S. Ramasamy, and J. Kumar; "Investigations on the growth of Zinc Oxide Crystals from Molten Hydrous KOH Solution and on the Impedance Analysis of Zinc Oxide Crystals," *Materials Letters,* 61, 4090–4093, 2007.
44. N. Priyantha, P. Jayaweera, D. D. Macdonald, and A. Sun; "An Electrochemical Impedance Study of Alloy 22 in NaCl Brine at Elevated Temperature: I. Corrosion Behavior," *Journal of Electroanalytical Chemistry,* 572, 409–419, 2004.
45. D. D. Macdonald, A. Sun, N. Priyantha, and P. Jayaweera; "An Electrochemical Impedance Study of Alloy-22 in NaCl Brine at Elevated Temperature: II. Reaction Mechanism Analysis," *Journal of Electroanalytical Chemistry*, 572, 421–431, 2004.

46. L. Yang, C. Ruan, and Y. Li; "Detection of Viable Salmonella typhimurium by Impedance Measurement of Electrode Capacitance and Medium Resistance," *Biosensors and Bioelectronics*, 19, 495–502, 2003.
47. G. W. Walter; "A Review of Impedance Plot Methods used for Corrosion Performance Analysis of Painted Metals," *Corrosion Science*, 26(9), 681–703, 1986.
48. M. E. Orazem, P. Shukla, and M. A. Membrino; "Extension of the Measurement Model Approach for Deconvolution of Underlying Distributions for Impedance Measurements," *Electrochimica Acta*, 47, 2027–2034, 2002.
49. C. Andrade, V. M. Blanco, A. Collazo, M. Keddam, X. R. Novoa, and H. Takenouti; "Cement Paste Hardening Process Studied by Impedance Spectroscopy," *Electrochimica Acta*, 44, 4313–4318, 1999.
50. E. Siebert, A. Hammouche, and M. Kleitz; "Impedance Spectroscopy Analysis of $La_{1-x}SrMnO_3$-Yttira-Stabilized Zirconia Electrode Kinetics," *Elecrrochimica Acta*, 40(11), 1741 1753 (1995).
51. H. Nyquist, "Regeneration Theory," Bell System Technical Journal, 11, 126–147, 1932.
52. M. A. Alim, S. Bhattacharjee, S. Khanam, N. Muna, and S. Runa; "Ni-Cd Batteries for Automotive Applications: The ac Electrical Data of the Ni-Cd Battery in the Havriliak-Negami Dielectric Representation," in *Autonomous Vehicles: Intelligent Transport Systems and Smart Technologies*, N. Bizon, L. Dascalescu, and N. M. Tabatabaei edited, ISBN 978-1-63321-324-1, Nova Science Publishers, Inc., New York, Chapter 17, 453–483, 2014.
53. J. R. Macdonald; "Comparison and Application of Two Methods for the Least Squares Analysis of Immittance Data," *Solid State Ionics*, 58, 97–107, 1992.
54. P. B. Macedo, C. T. Moynihan, and R. Bose, "The Role of Ionic Diffusion in Polarization in Vitreous Ionic Conductors," *Physics and Chemistry Glasses*, 13, 171–179, 1972.
55. J. H. Sluyters; "*On the Impedance of Galvanic Cells, I*," Recueil des Travaux Chimiques des Pays-Bas, 79, 1092–1100, 1960.
56. J. R. Macdonald and J. A. Garber; "Analysis of Impedance and Admittance Data for Solids and Liquids," *Journal of the Electrochemical Society*, 124(7), 1022–1030, 1977.
57. Donald F. Sterba; "AC Electrical Impedance Monitoring and Modeling of Crop Growth and Development: Tomato Fruit Ripening" M.S. Thesis, Marquette University, Milwaukee, 1981.
58. K. Levenberg; "A Method for the Solution of Certain Non-Linear Problems in Least Squares," *The Quarterly of Applied Mathematics*, 2, 164–168, 1944.
59. D. W. Marquardt; "An algorithm for Least-Squares Estimation of Nonlinear Parameters," *Journal of the Society for Industrial and Applied Mathematics*, 11(2), 431–441, 1963.

60. N. S. Hari, P. Padmini, T. R. N. Kutty; "Complex Impedance Analyses of n-$BaTiO_3$ Ceramics Showing Positive Temperature Coefficient of Resistance," *Journal of Materials Science: Materials in Science*, 8, 15–22, 1997.
61. A. H. Eid, and S. Mahmoud; "The A.C. Impedance of Polycrystalline $CuInS_2$ Thin Films," *Journal of Material Science: Materials in Electronics*, 8, 259–264, 1997.
62. D. K. Pradhan, B. K. Samantaray, R. N. P. Choudhary, and A. K. Thakur; "Complex Impedance Analysis of $NaLaTiO_4$ Electroceramics," *Journal of Material Science: Materials in Electronics*, 17, 157–164, 2006.
63. M. A. Ponce, P. R. Bueno, J. Varela, M. S. Castro, and C. M. Aldao; "Impedance Spectroscopy Analysis of SnO_2 Thick-Films Gas Sensors," *Journal of Material Science: Materials in Electronics*, 19, 1169–1175, 2008.
64. P. Dhak, D. Dhak, M. Das, and P. Pramanik; "Dielectric and Impedance Spectroscopy Study of $Ba_{0.8}Bi_{2.133}Nb_{1.6}Ta_{0.4}O_9$ Ferroelectric Ceramics, Prepared by Chemical Route," *Journal of Material Science: Materials in Electronics*, 22, 1750–1760, 2011.
65. B. Pati, B. C. Sutar, B. N. Parida, P. R. Das, and R. N. P. Choudhury; "Dielectric and Impedance Spectroscopy of Barium Orthovanadate Ceramics," *Journal of Material Science: Materials in Electronics*, 24, 1608–1616, 2013.
66. D. K. Mahato, and T. P. Sinha; "Electrical Impedance and Electric Modulus Approach of Double Perovskite Pr_2ZnZrO_6 Ceramics," *Journal of Material Science: Materials in Electronics*, 24, 4399–4405, 2013.
67. M. Prabu, and A. Chandrabose; "Complex Impedance Spectroscopy Studies of PLZT (5/52/48) Ceramics Synthesized by Sol–Gel Route," *Journal of Material Science: Materials in Electronics*, 24, 4560–4565, 2013.
68. B. C. Sutar, R. N. P. Choudhary, and P. R. Das; "Dielectric and Impedance Spectroscopy of $Sr(Bi_{0.5}Ta_{0.5})O_3$ Electroceramics," *Journal of Material Science: Materials in Electronics*, 25, 4278–4285, 2014.
69. N. R. Draper, and H. Smith; "*Nonlinear Regression Analysis*," Wiley-Interscience, John Wiley & Sons, New York, 1998.
70. M. E. Orazem; "A Systematic Approach toward Error Structure Identification for Impedance Spectroscopy," *Journal of Electroanalytical Chemistry*, 572, 317–327, 2004.
71. S. Bhattacharjee and M. A. Alim; "Frequency Dependent Impedance and Admittance (Immittance) Data-Handling and Interpretation using Complex Plane Formalisms via Nonlinear Regression Analysis for Smart Electronic Materials and Devices: Overview and Case Studies," *Journal of Materials Science: Materials in Electronics*, 26, 4521–4540, 2015.
72. K. Kobayashi, Y. Sakka, and T. S. Suzuki; "Development of an Electrochemical Impedance Analysis Program Based on the Expanded Measurement Model," *Journal of the Ceramic Society of Japan*, 124(9), 943-949 (2016).

12

Appendices

12.1 Appendix – A: Sample Input Data for the R-C Parallel Circuit

Spreadsheet Data of the R-C Parallel Circuit Measurement with Impedance Analyzer in the Parallel Mode

Frequency (Hz) (1)	**As Displayed by HP-4192**				**Used in Analysis**	
	C_p (2)	**Unit (3)**	**G_p (4)**	**Unit (5)**	**C_p (F) (6)**	**G_p (S) (7)**
5	0.014	μ F	10.01	μ S	1.40000E-08	1.00100E-05
10	14.3	n F	10	μ S	1.43000E-08	1.00000E-05
15	14.3	n F	10	μ S	1.43000E-08	1.00000E-05
20	14.3	n F	10	μ S	1.43000E-08	1.00000E-05
25	14.2	n F	10	μ S	1.42000E-08	1.00000E-05
30	14.2	n F	10.01	μ S	1.42000E-08	1.00100E-05
Continued	Continued	Continued	Continued	Continued	Continued	Continued
84000	13.634	n F	0.059	μ S	1.36340E-08	5.90000E-05
85000	13.622	n F	0.06	μ S	1.36220E-08	6.00000E-05
86000	13.64	n F	0.06	μ S	1.36400E-08	6.00000E-05
87000	13.635	n F	0.061	μ S	1.36350E-08	6.10000E-05
88000	13.611	n F	0.061	μ S	1.36110E-08	6.10000E-05
89000	13.596	n F	0.061	μ S	1.35960E-08	6.10000E-05

12.2 Appendix – B: R-C Parallel Circuit Data Analysis Output in Z*-Plane

```
++++++++++PARAMETERS FROM SEMICIRCLE FITTING++++++++++
CIRCLE: CENTER                                       Xc =   5.013821E+004
                                                     Yc =  -2.335598E+002

CIRCLE: RADIUS                                          =   4.990641E+004
CIRCLE: R-SQUARE                                        =        0.999611

DEPRESSION ANGLE (THETA)                                =       -0.268143 DEG
DEPRESSION PARAMETER (ALPHA)                            =        0.997021

PEAK FREQUENCY                                     =  1.126773E+002 HZ,    7.079722E+002
RADIAN
TIME CONSTANT (TAU)                                     =   1.412485E-003 SEC

CHORD RATIO -VS - FREQ: INTERCEPT                       =   4.701543E+000
CHORD RATIO -VS - FREQ: SLOPE                           =  -9.951351E-001
CHORD RATIO -VS - FREQ: R-SQUARE                        =        0.999991

LOW FREQUENCY INTERCEPT                                 =   1.000441E+005 OHM
HIGH FREQUENCY INTERCEPT                                =   2.323382E+002 OHM

RELAXATION RESISTANCE (CHORD)                           =   9.981174E+004 OHM
RELAXATION CAPACITANCE                                  =   1.415149E-008 F

++++++++++NONLINEAR REGRESSION: SEMICIRCLE FITTING OUTPUT++++++++++
Number of observations = 33
Number of estimated parameters = 3
Equal weights used. W(i,i) = 1/Var[y(i)] = 1/Var[Err].
```

X	Y	WT
9.970761E+000	4.380989E-001	1.091535E+003
9.919917E+000	8.913001E-001	1.091535E+003
9.821599E+000	1.323699E+000	1.091535E+003
9.687184E+000	1.740778E+000	1.091535E+003
9.526053E+000	2.124816E+000	1.091535E+003
9.323381E+000	2.493038E+000	1.091535E+003
8.859938E+000	3.157242E+000	1.091535E+003
8.329075E+000	3.708235E+000	1.091535E+003
7.783869E+000	4.125198E+000	1.091535E+003
7.206699E+000	4.454589E+000	1.091535E+003
6.640166E+000	4.691736E+000	1.091535E+003
6.096532E+000	4.847093E+000	1.091535E+003
5.581531E+000	4.935232E+000	1.091535E+003
4.689237E+000	4.957566E+000	1.091535E+003
3.953659E+000	4.853093E+000	1.091535E+003
3.340309E+000	4.681314E+000	1.091535E+003
2.844449E+000	4.477065E+000	1.091535E+003
2.444838E+000	4.261150E+000	1.091535E+003
2.043429E+000	3.997068E+000	1.091535E+003
1.727248E+000	3.743940E+000	1.091535E+003
1.471549E+000	3.507714E+000	1.091535E+003
1.276547E+000	3.301193E+000	1.091535E+003
9.686879E-001	2.922328E+000	1.091535E+003
7.671919E-001	2.626065E+000	1.091535E+003
6.179237E-001	2.372573E+000	1.091535E+003
6.336404E-001	2.087854E+000	1.091535E+003
4.534711E-001	1.786379E+000	1.091535E+003
2.693952E-001	1.585962E+000	1.091535E+003

```
    2.085742E-001     1.396608E+000     1.091535E+003
    1.667560E-001     1.247928E+000     1.091535E+003
    1.365267E-001     1.128817E+000     1.091535E+003
    6.298514E-002     7.575475E-001     1.091535E+003
    3.639087E-002     5.714205E-001     1.091535E+003

Nonlinear Least Square Algorithm: Gauss-Newton.
Jacobian calculated using closed form derivatives.

  ITER            PARA1             PARA2             PARA3               SSE
     0    5.000000E+000    0.000000E+000     5.000000E+000    5.319424E-002
     1    5.013798E+000   -2.329950E-002     4.990570E+000    2.748429E-002
     2    5.013821E+000   -2.335598E-002     4.990641E+000    2.748423E-002
Converged within max iteration limit.

----SUMMARY STATISTICS
SOURCE          DF              SS               MS          F-RATIO          P-VALUE
Model            2       70.562899        35.281449     38510.936330          <0.0001
Residual        30        0.027484         0.000916
Total           32       70.590383

STDEV OF ERR =           0.030268,  R-SQ =    0.999611

----ASYMPTOTIC PARAMETER ESTIMATIONS
PARAMETER           ESTIMATE        STAND. ERR         LCL (0.95)       UCL (0.95)
T-VALUE      P-VALUE
           1     5.013821E+000    7.176428E-003    4.999166E+000    5.028475E+000
698.651258      <0.0001
           2    -2.335598E-002    1.808220E-002   -6.027984E-002    1.356787E-002
-1.291656    0.206339
           3     4.990641E+000    1.195688E-002    4.966226E+000    5.015057E+000
417.386493      <0.0001

++++++++++LINEAR REGRESSION: LINE FITTING OUTPUT++++++++++

----PARAMETER ESTIMATIONS
PARAMETER          ESTIMATE        STAND. ERR         LCL (0.95)        UCL (0.95)
T-VALUE  P-VALUE
         0    4.701543e+000    2.792732e-003     4.695847e+000    4.707240e+000
1683.492318  <0.0001
         1   -9.951351e-001    5.340334e-004    -9.962245e-001   -9.940457e-001
-1863.432222  <0.0001

++++++++++PARAMETERS FROM SIMULTANEOUS FITTING OF REAL AND IMAGINARY DATA++++++++++
CIRCLE: CENTER                                 Xc =    5.014318E+004
                                               Yc =   -3.608957E+002

CIRCLE: RADIUS                                      =   4.998128E+004
SIMULTANEOUS FITTING R-SQUARE                       =        0.999943

DEPRESSION ANGLE (THETA)                            =       -0.413714 DEG
DEPRESSION PARAMETER (ALPHA)                        =        0.995403

PEAK FREQUENCY                                 =  1.125617E+002 HZ,    7.072463E+002
RADIAN
TIME CONSTANT (TAU)                                 =   1.413935E-003 SEC

LOW FREQUENCY INTERCEPT                             =   1.001232E+005 OHM
HIGH FREQUENCY INTERCEPT                            =   1.632002E+002 OHM

RELAXATION RESISTANCE (CHORD)                       =   9.995996E+004 OHM
RELAXATION CAPACITANCE                              =   1.414501E-008 F
```

++++++++++NONLINEAR REGRESSION: SIMULTANEOUS FITTING OF REAL AND IMAGINARY DATA OUTPUT++++++++++

Number of observations = 33
Number of estimated parameters = 4
Equal weights used.

X	Y_REAL	Y_IMAG	WT_REAL	WT_IMAG
3.141593E+001	9.970761E+000	4.380989E-001	9.829999E+002	9.829999E+002
6.283185E+001	9.919917E+000	8.913001E-001	9.829999E+002	9.829999E+002
9.424778E+001	9.821599E+000	1.323699E+000	9.829999E+002	9.829999E+002
1.256637E+002	9.687184E+000	1.740778E+000	9.829999E+002	9.829999E+002
1.570796E+002	9.526053E+000	2.124816E+000	9.829999E+002	9.829999E+002
1.884956E+002	9.323381E+000	2.493038E+000	9.829999E+002	9.829999E+002
2.513274E+002	8.859938E+000	3.157242E+000	9.829999E+002	9.829999E+002
3.141593E+002	8.329075E+000	3.708235E+000	9.829999E+002	9.829999E+002
3.769911E+002	7.783869E+000	4.125198E+000	9.829999E+002	9.829999E+002
4.398230E+002	7.206699E+000	4.454589E+000	9.829999E+002	9.829999E+002
5.026548E+002	6.640166E+000	4.691736E+000	9.829999E+002	9.829999E+002
5.654867E+002	6.096532E+000	4.847093E+000	9.829999E+002	9.829999E+002
6.283185E+002	5.581531E+000	4.935232E+000	9.829999E+002	9.829999E+002
7.539822E+002	4.689237E+000	4.957566E+000	9.829999E+002	9.829999E+002
8.796459E+002	3.953659E+000	4.853093E+000	9.829999E+002	9.829999E+002
1.005310E+003	3.340309E+000	4.681314E+000	9.829999E+002	9.829999E+002
1.130973E+003	2.844449E+000	4.477065E+000	9.829999E+002	9.829999E+002
1.256637E+003	2.444838E+000	4.261150E+000	9.829999E+002	9.829999E+002
1.413717E+003	2.043429E+000	3.997068E+000	9.829999E+002	9.829999E+002
1.570796E+003	1.727248E+000	3.743940E+000	9.829999E+002	9.829999E+002
1.727876E+003	1.471549E+000	3.507714E+000	9.829999E+002	9.829999E+002
1.884956E+003	1.276547E+000	3.301193E+000	9.829999E+002	9.829999E+002
2.199115E+003	9.686879E-001	2.922328E+000	9.829999E+002	9.829999E+002
2.513274E+003	7.671919E-001	2.626065E+000	9.829999E+002	9.829999E+002
2.827433E+003	6.179237E-001	2.372573E+000	9.829999E+002	9.829999E+002
3.141593E+003	6.336404E-001	2.087854E+000	9.829999E+002	9.829999E+002
3.769911E+003	4.534711E-001	1.786379E+000	9.829999E+002	9.829999E+002
4.398230E+003	2.693952E-001	1.585962E+000	9.829999E+002	9.829999E+002
5.026548E+003	2.085742E-001	1.396608E+000	9.829999E+002	9.829999E+002
5.654867E+003	1.667560E-001	1.247928E+000	9.829999E+002	9.829999E+002
6.283185E+003	1.365267E-001	1.128817E+000	9.829999E+002	9.829999E+002
9.424778E+003	6.298514E-002	7.575475E-001	9.829999E+002	9.829999E+002
1.256637E+004	3.639087E-002	5.714205E-001	9.829999E+002	9.829999E+002

Nonlinear Least Square Algorithm: Gauss-Newton.
Jacobian of real part of I* calculated using finite difference method.
Jacobian of imaginary part of I* calculated using finite difference method.

ITER	PARA1	PARA2	PARA3	PARA4	SSE
0	2.323382E-002	1.000441E+001	1.412485E-003	9.970206E-001	3.065703E-002
1	1.634244E-002	1.001229E+001	1.413932E-003	9.954059E-001	2.950154E-002
2	1.632002E-002	1.001232E+001	1.413935E-003	9.954032E-001	2.950153E-002

Converged within max iteration limit.

----SUMMARY STATISTICS

SOURCE	DF	SS	MS	F-RATIO	P-VALUE
Model	3	518.060465	172.686822	169751.125042	<0.0001
Residual	29	0.029502	0.001017		
Total	32	518.089966			

```
STDEV OF ERR =          0.031895,  R-SQ =   0.999943

----ASYMPTOTIC PARAMETER ESTIMATIONS
PARAMETER        ESTIMATE      STAND. ERR         LCL (0.95)          UCL (0.95)
T-VALUE     P-VALUE
           1  1.632002E-002  1.091955E-002   -6.010460E-003    3.865049E-002
1.494569   0.145830
           2  1.001232E+001  1.320575E-002    9.985310E+000    1.003932E+001
758.178509    <0.0001
           3  1.413935E-003  3.295259E-006    1.407196E-003    1.420673E-003
429.081426    <0.0001
           4  9.954032E-001  1.639056E-003    9.920513E-001    9.987550E-001
607.302824    <0.0001

++++++++++MEASURED DATA USED IN ANALYSIS++++++++++
DATA POINT      FREQ (Hz)           REAL        IMGINARY
         1  5.000000E+000  9.970761E+004  4.380989E+003
         2  1.000000E+001  9.919917E+004  8.913001E+003
         3  1.500000E+001  9.821599E+004  1.323699E+004
         4  2.000000E+001  9.687184E+004  1.740778E+004
         5  2.500000E+001  9.526053E+004  2.124816E+004
         6  3.000000E+001  9.323381E+004  2.493038E+004
         7  4.000000E+001  8.859938E+004  3.157242E+004
         8  5.000000E+001  8.329075E+004  3.708235E+004
         9  6.000000E+001  7.783869E+004  4.125198E+004
        10  7.000000E+001  7.206699E+004  4.454589E+004
        11  8.000000E+001  6.640166E+004  4.691736E+004
        12  9.000000E+001  6.096532E+004  4.847093E+004
        13  1.000000E+002  5.581531E+004  4.935232E+004
        14  1.200000E+002  4.689237E+004  4.957566E+004
        15  1.400000E+002  3.953659E+004  4.853093E+004
        16  1.600000E+002  3.340309E+004  4.681314E+004
        17  1.800000E+002  2.844449E+004  4.477065E+004
        18  2.000000E+002  2.444838E+004  4.261150E+004
        19  2.250000E+002  2.043429E+004  3.997068E+004
        20  2.500000E+002  1.727248E+004  3.743940E+004
        21  2.750000E+002  1.471549E+004  3.507714E+004
        22  3.000000E+002  1.276547E+004  3.301193E+004
        23  3.500000E+002  9.686879E+003  2.922328E+004
        24  4.000000E+002  7.671919E+003  2.626065E+004
        25  4.500000E+002  6.179237E+003  2.372573E+004
        26  5.000000E+002  6.336404E+003  2.087854E+004
        27  6.000000E+002  4.534711E+003  1.786379E+004
        28  7.000000E+002  2.693952E+003  1.585962E+004
        29  8.000000E+002  2.085742E+003  1.396608E+004
        30  9.000000E+002  1.667560E+003  1.247928E+004
        31  1.000000E+003  1.365267E+003  1.128817E+004
        32  1.500000E+003  6.298514E+002  7.575475E+003
        33  2.000000E+003  3.639087E+002  5.714205E+003
```

12.3 Appendix – C: R-C Parallel Circuit Data Analysis Output in M*-Plane

```
++++++++++PARAMETERS FROM SEMICIRCLE FITTING++++++++++
CIRCLE: CENTER                                    Xc =   3.588919E+007
                                                  Yc =  -6.790303E+005

CIRCLE: RADIUS                                       =   3.610219E+007
CIRCLE: R-SQUARE                                     =        0.999923

DEPRESSION ANGLE (THETA)                             =       -1.077715 DEG
DEPRESSION PARAMETER (ALPHA)                         =        0.988025

PEAK FREQUENCY                                  =  1.176833E+002 HZ,    7.394258E+002
RADIAN
TIME CONSTANT (TAU)                                  =   1.352401E-003 SEC

CHORD RATIO -VS - FREQ: INTERCEPT                    =  -4.657808E+000
CHORD RATIO -VS - FREQ: SLOPE                        =   9.768900E-001
CHORD RATIO -VS - FREQ: R-SQUARE                     =        0.998274

LOW FREQUENCY INTERCEPT                              =  -2.066153E+005 1/F
RECIPROCAL OF LOW FREQUENCY INTERCEPT                =  -4.839912E-006 F

HIGH FREQUENCY INTERCEPT                             =   7.198500E+007 1/F
RECIPROCAL OF HIGH FREQUENCY INTERCEPT               =   1.389178E-008 F

CHORD                                                =   7.219161E+007 1/F
RELAXATION CAPACITANCE (RECIPROCAL OF CHORD)         =   1.385202E-008 F
RELAXATION RESISTANCE                                =   9.763199E+004 OHM

----ASYMPTOTIC PARAMETER ESTIMATIONS
PARAMETER          ESTIMATE         STAND. ERR        LCL (0.95)        UCL (0.95)
T-VALUE     P-VALUE
          1    3.588919E+000    2.252299E-003    3.584320E+000    3.593518E+000
1593.446809    <0.0001
          2   -6.790303E-002    5.798526E-003   -7.974362E-002   -5.606244E-002
-11.710395    <0.0001
          3    3.610219E+000    3.887716E-003    3.602280E+000    3.618158E+000
928.622175    <0.0001

++++++++++PARAMETERS FROM SIMULTANEOUS FITTING OF REAL AND IMAGINARY DATA++++++++++
CIRCLE: CENTER                                    Xc =   3.585222E+007
                                                  Yc =  -1.168260E+006

CIRCLE: RADIUS                                       =   3.638469E+007
SIMULTANEOUS FITTING R-SQUARE                        =        0.998651

DEPRESSION ANGLE (THETA)                             =       -1.840001 DEG
DEPRESSION PARAMETER (ALPHA)                         =        0.979556

PEAK FREQUENCY                                  =  1.158701E+002 HZ,    7.280330E+002
RADIAN
TIME CONSTANT (TAU)                                  =   1.373564E-003 SEC

LOW FREQUENCY INTERCEPT                              =  -5.137080E+005 1/F
RECIPROCAL OF LOW FREQUENCY INTERCEPT                =  -1.946631E-006 F

HIGH FREQUENCY INTERCEPT                             =   7.221816E+007 1/F
RECIPROCAL OF HIGH FREQUENCY INTERCEPT               =   1.384693E-008 F

CHORD                                                =   7.273186E+007 1/F
```

```
RELAXATION CAPACITANCE (RECIPROCAL OF CHORD)   =    1.374913E-008 F
RELAXATION RESISTANCE                          =    9.990188E+004 OHM

----ASYMPTOTIC PARAMETER ESTIMATIONS
PARAMETER          ESTIMATE        STAND. ERR        LCL (0.95)          UCL (0.95)
T-VALUE    P-VALUE
          1   -5.137080E-002    4.651309E-002   -1.464901E-001    4.374846E-002
-1.104438  0.278483
          2    7.221816E+000    3.927463E-002    7.141499E+000    7.302132E+000
183.879931   <0.0001
          3    1.373564E-003    1.592382E-005    1.341000E-003    1.406128E-003
86.258481   <0.0001
          4   -9.795555E-001    8.013781E-003   -9.959437E-001   -9.631674E-001
-122.233879   <0.0001
```

12.4 Appendix – D: Lithium Niobate Crystal Data Analysis Output in C^*-Plane

```
++++++++++PARAMETERS FROM SEMICIRCLE FITTING++++++++++
CIRCLE: CENTER                                    Xc =   2.053563E-010
                                                  Yc =  -1.930463E-011

CIRCLE: RADIUS                                       =   1.667964E-010
CIRCLE: R-SQUARE                                     =        0.998967

DEPRESSION ANGLE (THETA)                             =       -6.646176 DEG
DEPRESSION PARAMETER (ALPHA)                         =        0.926154

PEAK FREQUENCY                                  =  7.487859E+002 HZ,    4.704760E+003
RADIAN
TIME CONSTANT (TAU)                                  =   2.125507E-004 SEC

CHORD RATIO -VS - FREQ: INTERCEPT                    =   4.761237E+000
CHORD RATIO -VS - FREQ: SLOPE                        =  -7.193882E-001
CHORD RATIO -VS - FREQ: R-SQUARE                     =        0.991586

LOW FREQUENCY INTERCEPT                              =   3.710318E-010 F
HIGH FREQUENCY INTERCEPT                             =   3.968080E-011 F

RELAXATION CAPACITANCE (CHORD)                       =   3.313510E-010 F
RELAXATION RESISTANCE                                =   6.414668E+005 OHM

----ASYMPTOTIC PARAMETER ESTIMATIONS
PARAMETER       ESTIMATE     STAND. ERR     LCL (0.95)      UCL (0.95)       T-VALUE
P-VALUE
           1     2.053563E+000   3.674606E-003   2.046059E+000    2.061067E+000
558.852627     <0.0001
           2    -1.930463E-001   9.986039E-003  -2.134378E-001   -1.726548E-001
-19.331621     <0.0001
           3     1.667964E+000   6.267709E-003   1.655165E+000    1.680763E+000
266.120203     <0.0001

++++++++++PARAMETERS FROM SIMULTANEOUS FITTING OF REAL AND IMAGINARY DATA++++++++++
CIRCLE: CENTER                                    Xc =   2.086076E-010
                                                  Yc =  -5.670622E-011

CIRCLE: RADIUS                                       =   1.858977E-010
SIMULTANEOUS FITTING R-SQUARE                        =        0.985415

DEPRESSION ANGLE (THETA)                             =      -17.760570 DEG
DEPRESSION PARAMETER (ALPHA)                         =        0.802660

PEAK FREQUENCY                                  =  6.806049E+002 HZ,    4.276367E+003
RADIAN
TIME CONSTANT (TAU)                                  =   2.338434E-004 SEC

LOW FREQUENCY INTERCEPT                              =   3.856453E-010 F
HIGH FREQUENCY INTERCEPT                             =   3.156994E-011 F

RELAXATION CAPACITANCE (CHORD)                       =   3.540754E-010 F
RELAXATION RESISTANCE                                =   6.604338E+005 OHM

----ASYMPTOTIC PARAMETER ESTIMATIONS
PARAMETER       ESTIMATE     STAND. ERR     LCL (0.95)      UCL (0.95)       T-VALUE
P-VALUE
```

1	3.156994E-001	5.817367E-002	1.967343E-001	4.346646E-001	5.426844	<0.0001
2	3.856453E+000	7.761560E-002	3.697730E+000	4.015177E+000	49.686577	<0.0001
3	2.338434E-004	1.541548E-005	2.023187E-004	2.653680E-004	15.169388	<0.0001
4	8.026603E-001	3.119692E-002	7.388626E-001	8.664580E-001	25.728830	<0.0001

12.5 Appendix – E: Multilayer Junction Thermoelectric Device Data Analysis Output in Y*-Plane

```
++++++++++PARAMETERS FROM SEMICIRCLE FITTING++++++++++
CIRCLE: CENTER                                   XC =   1.054861E-002
                                                 YC =  -2.036936E-003

CIRCLE: RADIUS                                      =   7.545683E-003
CIRCLE: R-SQUARE                                    =        0.999429

DEPRESSION ANGLE (THETA)                            =      -15.661126 DEG
DEPRESSION PARAMETER (ALPHA)                        =        0.825987

PEAK FREQUENCY                                 =  8.080343E+004 HZ,    5.077029E+005
RADIAN
TIME CONSTANT (TAU)                                 =   1.969656E-006 SEC

CHORD RATIO -VS - FREQ: INTERCEPT                   =  -9.111637E+000
CHORD RATIO -VS - FREQ: SLOPE                       =   8.063556E-001
CHORD RATIO -VS - FREQ: R-SQUARE                    =        0.999451

LOW FREQUENCY INTERCEPT                             =   3.283058E-003 S
RECIPROCAL OF LOW FREQUENCY INTERCEPT               =   3.045941E+002 OHM

HIGH FREQUENCY INTERCEPT                            =   1.781416E-002 S
RECIPROCAL OF HIGH FREQUENCY INTERCEPT              =   5.613512E+001 OHM

CHORD                                               =   1.453110E-002 S
RELAXATION RESISTANCE (RECIPROCAL OF CHORD)         =   6.881791E+001 OHM
RELAXATION CAPACITANCE                              =   2.862127E-008 F

----ASYMPTOTIC PARAMETER ESTIMATIONS
PARAMETER          ESTIMATE         STAND. ERR        LCL (0.95)        UCL (0.95)
T-VALUE   P-VALUE
          1     1.054861E+001    2.012135E-002    1.050816E+001    1.058906E+001
524.249628   <0.0001
          2    -2.036936E+000    4.620618E-002   -2.129820E+000   -1.944053E+000
-44.083630   <0.0001
          3     7.545683E+000    4.114990E-002    7.462964E+000    7.628403E+000
183.370638   <0.0001

++++++++++PARAMETERS FROM SIMULTANEOUS FITTING OF REAL AND IMAGINARY DATA++++++++++
CIRCLE: CENTER                                   XC =   1.055106E-002
                                                 YC =  -2.021466E-003

CIRCLE: RADIUS                                      =   7.531200E-003
SIMULTANEOUS FITTING R-SQUARE                       =        0.999721

DEPRESSION ANGLE (THETA)                            =      -15.569806 DEG
DEPRESSION PARAMETER (ALPHA)                        =        0.827002

PEAK FREQUENCY                                 =  8.068278E+004 HZ,    5.069448E+005
RADIAN
TIME CONSTANT (TAU)                                 =   1.972601E-006 SEC

LOW FREQUENCY INTERCEPT                             =   3.296220E-003 S
RECIPROCAL OF LOW FREQUENCY INTERCEPT               =   3.033778E+002 OHM

HIGH FREQUENCY INTERCEPT                            =   1.780589E-002 S
RECIPROCAL OF HIGH FREQUENCY INTERCEPT              =   5.616118E+001 OHM
```

```
CHORD                                            =   1.450967E-002 S
RELAXATION RESISTANCE (RECIPROCAL OF CHORD)      =   6.891954E+001 OHM
RELAXATION CAPACITANCE                           =   2.862180E-008 F

----ASYMPTOTIC PARAMETER ESTIMATIONS
PARAMETER       ESTIMATE     STAND. ERR     LCL (0.95)     UCL (0.95)       T-VALUE
P-VALUE
           1      3.296220E+000   2.360123E-002   3.248749E+000   3.343692E+000
139.663092     <0.0001
           2      1.780589E+001   5.981393E-002   1.768558E+001   1.792620E+001
297.688045     <0.0001
           3      1.972601E-006   1.306826E-008   1.946316E-006   1.998887E-006
150.945926     <0.0001
            4     -8.270022E-001   3.441348E-003  -8.339241E-001  -8.200802E-001
-240.313449     <0.0001
```

Index

Others

Lightning Source UK Ltd.
Milton Keynes UK
UKHW032235160419
341131UK00004B/9/P

9 781119 18485